KU-007-380

Willis's
Practice and Procedure
for the Quantity Surveyor

Twelfth Edition

Allan Ashworth
Keith Hogg

UNIVERSITY OF WOLVERHAMPTON
LEARNING & INFORMATION
SERVICES

ACC NO
2467275

CLASS 062

CONTROL NO.
1405145781

692.
5
ASH

DATE
-3 JUL 2009

SITE
WV

Blackwell
Publishing

Twelfth Edition © 2007 Blackwell Publishing Ltd & C.J. Willis, J.A. Willis, A. Ashworth, K.I. Hogg

Blackwell Publishing editorial offices:
Blackwell Publishing Ltd, 9600 Garsington Road, Oxford OX4 2DQ, UK
Tel: +44 (0)1865 776868
Blackwell Publishing Inc., 350 Main Street, Malden, MA 02148-5020, USA
Tel: +1 781 388 8250
Blackwell Publishing Asia Pty Ltd, 550 Swanston Street, Carlton, Victoria 3053, Australia
Tel: +61 (0)3 8359 1011

The right of the Author to be identified as the Author of this Work has been asserted in accordance with the Copyright, Designs and Patents Act 1988.

All rights reserved. No part of this publication may be reproduced, stored in a retrieval system, or transmitted, in any form or by any means, electronic, mechanical, photocopying, recording or otherwise, except as permitted by the UK Copyright, Designs and Patents Act 1988, without the prior permission of the publisher.

Designations used by companies to distinguish their products are often claimed as trademarks. All brand names and product names used in this book are trade names, service marks, trademarks or registered trademarks of their respective owners. The Publisher is not associated with any product or vendor mentioned in this book.

This publication is designed to provide accurate and authoritative information in regard to the subject matter covered. It is sold on the understanding that the Publisher is not engaged in rendering professional services. If professional advice or other expert assistance is required, the services of a competent professional should be sought.

First published by Crosby Lockwood & Son Ltd 1951, Second Edition 1957, Third Edition 1963, Fourth Edition 1966, Fifth Edition (metric) 1969, Sixth Edition 1972, Seventh Edition by Crosby Lockwood Staples 1975, Eighth Edition by Granada Publishing 1980, Reprinted by Collins Professional and Technical Books 1985, Ninth Edition 1987, Reprinted by BSP Professional Books 1990, 1992, Tenth Edition by Blackwell Scientific Publications 1994, Eleventh Edition by Blackwell Science Ltd 2002, Twelfth Edition by Blackwell Publishing Ltd 2007

2 2008

ISBN: 978-1-4051-4578-7

Library of Congress Cataloging-in-Publication Data

Ashworth, A. (Allan)
   Willis's practice and procedure for the quantity surveyor / Allan Ashworth, Keith Hogg. — 12th ed.
       p. cm.
   Includes bibliographical references and index.
   ISBN: 978-1-4051-4578-7 (pbk. : alk. paper)
   1. Building—Estimates—Great Britain. I. Hogg, Keith. II. Title. III. Title: Practice and procedure for the quantity surveyor.
   TH435.W6853 2007
   692'.50941—dc22

                                                                    2007002030
A catalogue record for this title is available from the British Library

Set in 10/12.5pt Palatino by SNP Best-set Typesetter Ltd, Hong Kong
Printed and bound in Great Britain by TJ International Ltd, Padstow, Cornwall

The publisher's policy is to use permanent paper from mills that operate a sustainable forestry policy, and which has been manufactured from pulp processed using acid-free and elementary chlorine-free practices. Furthermore, the publisher ensures that the text paper and cover board used have met acceptable environmental accreditation standards.

For further information on Blackwell Publishing, visit our website:
www.blackwellpublishing.com/construction

Contents

Preface   xi

1 The Work of the Quantity Surveyor   1
   Introduction   1
   A changing industry   3
   Construction sectors   6
   The role of the quantity surveyor   8
   Skills, knowledge and understanding   15
   What's in a name?   18
   Discussion topic   20
   References and bibliography   23

2 Education, Training and Employment   24
   Introduction   24
   Chronology of quantity surveying education 1960–2006   24
   Quantity surveyors in education   27
   Partnership and accreditation   28
   Non-cognate disciplines   30
   National vocational qualifications (NVQs)   30
   Assessment of professional competence (APC)   31
   Continuing professional development (CPD)   32
   Construction Industry Council (CIC)   33
   Changing work patterns   34
   The professions   44
   Role of the RICS   48
   Discussion topic   50
   References and bibliography   52

3 Organisation and Management   53
   Introduction   53
   Staffing   53
   Office organisation   55
   Employer's responsibilities   57
   Public relations and marketing   61
   Quality management   63
   Time and cost management   65
   Education and training   66

Finance and accounts   68
Discussion topic   73
References and bibliography   77

4   **The Quantity Surveyor and the Law**   **78**
Introduction   78
The quantity surveyor and the client   78
Collateral warranties   85
Performance bonds   87
Professional indemnity insurance   88
Contracts of employment   88
The Disability Discrimination Act 2004   90
Discussion topic   91
References and bibliography   93

5   **Research and Innovation**   **95**
Introduction   95
RICS   96
Classification of research and development   97
Research and development in the construction and property
  industries   98
Rethinking construction innovation and research   99
Changing role of the quantity surveyor   102
Research and development in quantity surveying practice   103
Academic research   106
Research dissemination   107
Information and communication technologies (ICTs)   107
Quantity surveying practice   108
Major ICT issues   110
The future of ICT   112
The importance of change   112
Innovation   114
Conclusions   116
Discussion topic   117
References and bibliography   119

6   **Cost Control**   **121**
Introduction   121
Project cost control   121
Cost advice   122
Precontract methods   123
General considerations   127
Accuracy of approximate estimates   130
Preparing the approximate estimate   130
Whole life costing   131
Value management   132

Risk analysis 132
Best value 132
Taxation 133
Financial assistance for development 136
Post-contract methods 138
Discussion topic 141
References and bibliography 144

7   **Whole Life Costing** **146**
Introduction 146
Brief history 147
Government policy 147
Whole life value 148
Whole life costing applications 149
Whole life costs 151
Main factors to consider 152
Targeting the major elements of costs-in-use 154
Depreciation and obsolescence in buildings 154
Long life, loose fit and low energy 155
Calculations 155
Forecasting the future 157
Whole life cost forum (WLCF) 160
Conclusions 161
Discussion topic 161
References and bibliography 164

8   **Value Management** **166**
Introduction 166
Background 167
Terminology 167
When should surveyors use value management? 168
The application of value management 171
Functional analysis 178
Supporting the case for value management 183
Professional development and accreditation 185
Discussion topic 186
References and bibliography 189

9   **Risk Management** **190**
Introduction 190
When should surveyors use risk management? 191
The application of risk management 194
Risk analysis 195
Risk registers 197
Expected monetary value (EMV) 200
Simulation (quantitative risk analysis) 200

Risk management   203
Appraisal of risk management options   206
Considerations in risk allocation   207
Merging risk management and value management opportunity?   208
Discussion topic   209
References and bibliography   213

10   Procurement   **215**
Introduction   215
General matters   216
Standards forms of contract   218
Methods of price determination   219
Contractor selection and appointment   221
Procurement options   226
Contract strategy   232
Client procurement needs   233
Partnering   234
The Private Finance Initiative (PFI)   237
The role of the quantity surveyor   237
Discussion topic   238
References and bibliography   241

11   Contract Documentation   **243**
Contract documents   243
Coordinated project information   244
Form of contract   245
Contract drawings   247
Schedules   247
Contract bills   248
Methods of measurement   251
Contract specification   252
Schedules of work   253
Master programme   254
Information release schedule   254
Discrepancies in documents   254
Discussion topic   254
References and bibliography   257

12   Preparation of Contract Bills   **258**
Appointment of the quantity surveyor   258
Receipt of drawings   259
Taking-off   261
Contract bills   265
Invitation to tender   269
Receipt of tenders   274
E-tendering   280

Discussion topic   281
References and bibliography   283

13 Cost Management   **285**
Introduction   285
Valuations   286
Valuation on insolvency   302
Cost control and reporting   302
Discussion topic   306
References and bibliography   309

14 Final Accounts   **310**
Introduction   310
Variations   311
Procedure for measurement and evaluation   313
Pricing variations   316
Provisional sums   321
Fluctuations   323
Completing the account   327
Audit   328
Timing and resources   330
Discussion topic   331
References and bibliography   334

15 Insolvency   **335**
Introduction   335
The role of the quantity surveyor   337
Scenario   338
The role of the liquidator   339
The law   340
Determination of contract (contractor insolvency)   340
Provision in the forms of contract   341
Factors to consider at insolvency   342
Completion of the contract   344
The employer's loss   345
Expenditure involved   345
Termination of contract (employer insolvency)   346
Insolvency of the quantity surveyor or architect   348
Performance bonds   348
Discussion topic   349
References and bibliography   352

16 Contractual Disputes   **353**
Introduction   353
Why disputes arise   353
Litigation   355

Arbitration   356
Adjudication   358
Alternative dispute resolution   358
Expert witness   361
Lay advocacy   362
Claims   363
Discussion topic   370
References and bibliography   373

17   Project Management   **375**
Introduction   375
Justifying project management by adding value   376
Terminology   377
Attributes of the project manager   378
Duties and responsibilities of the project manager   385
Quantity surveying skills and expertise   391
Fees   391
Education and training for the project manager   392
Discussion topic   393
References and bibliography   396

18   Facilities Management   **398**
Introduction   398
The work of the facilities manager   399
Sustainability   413
Facilities management opportunities for the quantity surveyor   415
Education and training for the facilities manager   417
Discussion topic   417
References and bibliography   420

Index   423

Preface

In writing the twelfth edition of this book we have recognised that the role of the quantity surveyor continues to evolve by building on the sound foundations laid down earlier by the profession. It is sometimes incorrectly assumed that change is only a modern day phenomenon, but change has always been present in our world and its nature means that the future will be a fascinating place in which to live and work.

The apparent demise of the quantity surveyor has been very much over-exaggerated. Articles in all sorts of publications have forecast the end of this profession over the past twenty years or more. But they have all been wrong! There are those who have attempted to keep quantity surveying entirely within the building industry, for example, but the authors' own experience thirty years ago saw the rapid and welcome growth of quantity surveyors in civil engineering. Some assumed that the introduction of computers and information technology would spell the end, but if anything, these have provided an added impetus for growth and there are now many more quantity surveyors in the world than at any time in history. The phrase was often quoted, 'no bills, no fees and no quantity surveyors'. But whilst the use of bills has diminished in the UK, and the fees for their production have declined considerably, the processes of using information technology have greatly enhanced the process and removed a fair amount of the drudgery involved.

The land of China has more recently, against these predictions, also shown a particular interest in the way that quantity surveyors analyse construction works, calculate costs and add value. Quantity surveyors are now working in more countries than ever before. Even just a few years ago they would have been restricted to those countries that were a part of the former Commonwealth. Not so today. Countries that were hitherto considered to be *foreign* have now adopted and adapted some of the quantity surveying practices under the banner of their construction management consultancy.

In the present economic world climate, the future for quantity surveying is excellent with these skills being in high demand amongst a diverse range of clients and for a wide range of activities. For some of these clients they require a rather different service to that practised even just a couple of decades ago. For others, there remains a high demand for more traditional services, where in some cases quantity surveyors are now being employed directly by contractors to assist them with pre-contract tendering activities.

What is clear is that the role and activities of quantity surveyors have now become extremely diversified, with a range of employers to match. Their skills have been enhanced to meet these needs. Within this book we have sought to reflect the work and role of quantity surveyors in these different roles. We have retained as a basis the skills on which the profession is founded and are still practised, whilst at the same time referring to those activities of changing practices and procedures.

We have included a discussion topic in each chapter that we feel will be of considerable interest, not only to students, but also to others as these enhance the text.

This textbook was first published under the authorship of Arthur J. Willis in October 1950. Much has changed in quantity surveying practice and procedure since then. Willis was a household word in the profession. Students and practitioners would often ask when faced with a problem, 'What does Willis say?' Both his son Christopher and his grandson Andrew have played an important part in the previous editions of this textbook.

We want to express our thanks in the preparation of this edition to John Pearson BSc, LLB (Honours), MEd, FRICS of Northumbria University for his kind assistance in updating the two chapters on law and to Andrew Willis BSc, FRICS for his valuable comments. We also would like to express our gratitude to students and colleagues for the encouraging comments that they have expressed about the earlier editions of this book.

Allan Ashworth, DUniv (Hon), MSc, MRICS – University of Salford
Keith Hogg, BSc, FRICS – Northumbria University

1 The Work of the Quantity Surveyor

Introduction

In 1971, the Royal Institution of Chartered Surveyors (RICS) published a report titled, *The Future Role of the Quantity Surveyor*, which defined the work of the quantity surveyor as:

> 'ensuring that the resources of the construction industry are utilised to the best advantage of society by providing, *inter alia*, the financial management for projects and a cost consultancy service to the client and designer during the whole construction process.'

The report sought to identify the distinctive competencies or skills of the quantity surveyor associated with measurement and valuation, in the wider aspects of the construction industry. This provides the basis for the proper cost management of the construction project in the context of forecasting, analysing, planning, controlling and accounting. Many reading this will reflect that this is no longer an adequate description of the work of the quantity surveyor.

From the 1970s onwards the profession began to evolve rapidly and, in 1983 the RICS prepared another report that would explore further the work of the quantity surveyor and at the same time attempt to assess its future potential and directions. This report, *The Future Role of the Chartered Quantity Surveyor*, identified a range of skills, knowledge and expertise provided by the quantity surveyor and indicated a greater expansion of possible services that could be provided both inside and outside of the construction industry. This report began to examine the changing and shifting scene, the requirements of clients and their dissatisfaction with the services provided by construction professionals generally and their frequent disappointment with the products that they received.

Almost ten years later in 1991 the Davis, Langdon and Everest consultancy group produced *QS2000* on behalf of the RICS. This report began to describe the threats and opportunities that were facing the profession at the end of the twentieth century. Again its key message related to change and in ensuring the services provided recognised that the status quo no longer applied. Clients were demanding more for their fees. Fee scales for their services had been abandoned many years earlier and were continuing to fall. This in itself was sufficient for quantity surveyors to examine their role and work in the construction industry. The changes identified in this report included:

- *Changes in markets:* outlining the previous performance and trends in workloads across the different sectors and the importance of the changing international scene, particularly the challenges arising from the deepening European Union.
- *Changes in the construction industry:* through the changing nature of contracting, an emphasis upon management of construction, the comparison with other countries abroad and the competition being offered from non-construction professionals.
- *Changes in client needs:* with an emphasis in terms of the value added to the client's business; they want purchaseable design, procurement and management of construction. Many now want the long-term view beyond the initial design and construction phase.
- *Changes in the profession:* noting employment patterns, the growth in graduate members, the impact of fee competition, the ways in which the quantity surveyor is now appointed, and changes in their role and practice with changing attitudes and horizons.

Towards the end of that decade, the former Quantity Surveyor's Division of the RICS produced a report titled, *The Challenge of Change* (Powell 1998). This report provided stark warnings to the profession, almost as a final warning that if the profession did not adapt to change then it would not exist in the future. This would also be the last report from the Quantity Surveyor's Division, since this marked the end of the divisional structure within the RICS. The report focused on five important areas, as follows.

Business world

This suggested that there would be economic stability in many countries around the world, thereby implying and encouraging construction activity to take place in the early part of the twenty-first century. Economic analysis suggested an end to boom and bust philosophies, and the greater stability that would allow the industry to prosper. By 2006, the industry had enjoyed over 10 years of stability. The distinction between the contracting and professional services functions will be less well defined.

Customers

The clients of the industry will remain diverse but the larger ones will increase their international perspective. They will require all of their business partners to add value, take risk and in some cases probably share equity as well.

Projects

Information and communication technology will raise the expectations amongst all client groupings. They will expect improved value, higher quality and earlier completion. This will apply to all types of project and will not be restricted to capital cost projects alone.

Skills

The requirement for skills is both changing and increasing. Some of the technical functions will be transferred to computer applications, other will be downgraded to a technical role and the distinction between technician and professional roles will become more emphasised. A broader knowledge of construction and property, increased communication abilities and a greater commercial acumen will become necessary along with some specialisation. The importance of life long learning will be fundamental to career prospects.

Information and communication technology (ICT)

The ICT revolution will continue to develop at an exponential rate. This will help to increase productivity and expand the range of information available and the services provided by quantity surveyors. The integration of computer applications will save time and increase speed through electronic data interchange. The application of consistency within the overall process and at every stage will generally increase the overall efficiency and effectiveness.

A changing industry

The prospects of the construction industry are intrinsically linked to those of a country's economy. In times of recession the industry's major employers are reluctant to invest and this has an immediate knock-on effect on the fortunes or otherwise of the construction industry. As a proportion of GDP, the output of the construction industry in the United Kingdom is comparatively stable at about 8%. The construction industry, for example, has not suffered the considerable and terminal decline of engineering, especially ship building and coal mining.

However, the industry is changing shape. As a result of privatisations over the past 20 years the share of the public sector's construction portfolio has been considerably reduced. At its peak in the 1970s this represented almost 50%. It is now less than half of this figure. Coupled with this have been strategic changes in the procurement of public sector building and civil engineering projects, through for example the introduction of the Private Finance Initiative (PFI). This has assisted the industry to refocus on longer term measures, such as the consideration of whole life costing. There is also an increasing use being made of design and build, a trend that is likely to continue.

In 1994, the Latham Report, *Constructing the Team*, was published with far reaching consequences on the construction industry and those employed in it, including quantity surveyors. Its chief aim was to attempt to change the culture of the industry and thus increase the performance of construction activities and the final product. The Latham Report, for example drew comparisons with motor car manufacturing and how this had changed to improve the product for the customer. By comparison the construction industry had not changed significantly or fast enough and was being regarded by some as little more than a handicraft industry (Harvey

& Ashworth 1997). Other reports followed with similar and uncomfortable themes. These included a report by the Royal Academy of Engineering (Barlow 1996) and *Rethinking Construction* (Egan 1998). In order to achieve the objectives set out in the above reports, the whole design and construction process, including the work of the quantity surveyor needs to be re-engineered. The large British contracting firms would also copy the car industry; fewer of them and foreign owned. More mergers and acquisitions are on the horizon.

The National Audit Office provided yet a further review of the construction industry in a report titled *Modernising Construction* (2001). This report focused on procurement and the delivery of construction projects in the United Kingdom and how it can be modernised. This has yet further implications for the way in which quantity surveyors carry out their work in advising and assisting clients of the industry. This report also highlights the need for major savings and repeats the 30% reduction in the costs of construction from the Latham Report. Some of the main recommendations from the Report are:

- Collaboration
- Competition
- Appropriate risk sharing
- Value for money
- Clear understanding of the project's requirements
- Transparency in respect of costs and profits
- Clearly understood rights and obligations
- Appropriate incentives
- Early involvement of the whole construction team
- Operational efficiency of completed buildings

The report places an important emphasis on making greater use of innovation, disseminating good practices more widely and actively measuring improvements in construction performance.

The construction industry is continuing to go through major change that is being driven by government, regular-procuring clients and initiatives outlined in the reports noted above.

Many of these industry reports have highlighted the inefficiencies of traditional methods of procuring and managing major projects. It is important to remember in this context that the majority of projects that the industry carries out do not fall into this category and the recommendations may therefore not be entirely appropriate in very many cases, however, the key principles behind them remain valid.

Characteristics of the construction industry

The total value of the construction output in the UK is currently £90bn, which is approximately 8% of GDP. The industry offers direct employment to around two million people and to others in supporting occupations. It is the fourth largest construction industry in Europe and represents about 9% of total output. The public sector includes a diminishing share of the work in the UK. In addition many UK firms and practices, including quantity surveyors have an international perspective often through offices overseas or through associations with firms abroad. There has, for example, been an increasing and expanding role of activities on mainland Europe. Approximately 80% of the UK workload is on building projects as distinct from engineering works. New construction projects account for about 50% of the workload of the industry. The repair and maintenance sector will remain an important component for the foreseeable future as clients place greater emphasis upon the improved long-term management of such major capital assets. A detailed analysis of the industry can be found in Harvey and Ashworth (1997).

The industry is characterised by the following:

- The physical nature of the product
- The product is normally manufactured on the client's premises, i.e. the construction site
- Many of its projects are one-off designs in the absence of a prototype model
- The traditional arrangement separates design from manufacture
- It produces investment rather than consumer goods
- It is subject to wider swings of activity than most other industries
- Its activities are affected by the vagaries of the weather
- Its processes include a complex mixture of different materials, skills and trades
- Typically, throughout the world, it includes a small number of relatively large construction firms and a very large number of small firms
- The smaller firms tend to concentrate on repair and maintenance.

There have been recent developments in procurement management, supply chain management and the management of projects. In the case of PFI (private finance initiative) projects, the management of the asset becomes the financial responsibility of the consortium involved in the initial design and construction. PFI has allowed public clients to build many projects including health, education and road and rail infrastructure projects that otherwise might never have been envisaged. Some commentators suggest that one of the reasons for the sustained growth in the construction industry is due at least in part to the PFI concepts. PFI allows for continuity and integrity of delivery and maintenance. This at least should provide for a greater consideration of long-term needs and benefits to clients. PFI is one of the reasons for current interest, development and application of whole life costing. PFI is also driving a longer-term view of capital investment, with a greater appreciation of value for money, a greater understanding of risk, finance and taxation.

PFI specifies a method, developed initially by the UK government, to provide financial support for public–private partnerships (PPPs) between the public and private sectors. This process has now been adopted in several countries overseas, for example, Canada, France, the Netherlands, Norway, Australia, Japan and Singapore, as part of a wider reform programme for the delivery of public services which is driven by the World Trade Organisation (WTO), International Monetary

Fund (IMF) and the World Bank as a part of their deregulation and privatisation drive.

These types of project aim to deliver all kinds of works for the public sector, together with the provision of associated operational services. In return, the private sector receives payment, linked to its performance in meeting agreed standards of initial provision and longer term operation and maintenance.

Construction sectors

Within the construction industry quantity surveyors are involved in the following four main areas of work.

Building work

The employment of the quantity surveyor on building projects today is well established. The introduction of new forms of contract and changes in procedures continue to alter the way in which quantity surveyors carry out their duties and responsibilities. They also occupy a much more influential position than in the past, particularly when they are involved at the outset of a project. In some cases, they may be the client's first point of contact.

Quantity surveyors are the cost and value experts of the construction industry. Their responsibilities include advising clients on the cost and value implication of design decisions and the controlling of construction costs. Great importance is now attached to the control of costs on the majority of projects. Clients and designers are prone to making changes after the contract has been signed, and to order additional works that were not envisaged. This sometimes gives the incorrect impression that the quantity surveyor has not done the work correctly.

Building engineering services

Whilst this work is very much a part of the building project, it has tended to become a specialist function for the quantity surveyor, especially on large complex projects. An ever-increasing amount is expended on the elements that constitute this work. Traditionally, much of this work was included in bills of quantities as prime cost sums. It was largely presented in this way for three main reasons: building services engineers often failed to provide the appropriate details in time for quantification purposes; traditionally it was not the custom to measure this type of work; and contractors often preferred to offer lump sum quotations on the basis of drawing and specifications only. More enlightened clients realised that this approach was not very satisfactory in determining where the actual costs for this work are being expended. Whilst there is sometimes resistance to detailed quantification from some building services consultants, there is now a clear preference for a systematic breakdown of costs that can be properly compared and evaluated.

It is also accepted that to provide a rigorous cost control function for only part of a building project is unsatisfactory. The building services engineering work is frequently more extensive and expensive and its costs, value and cost control must be as rigorous as the methods applied on the remainder of the construction project. Quantity surveyors employed in this discipline have had to become more conversant with engineering services in their science, technology and terminology, in order to interpret engineering drawings correctly.

Civil engineering

It is difficult to define the line of demarcation between building and civil engineering works. The nature of civil engineering works often requires a design solution to take into account physical and geological problems that can be very complex. The scope, size and extent of civil engineering works are also frequently considerable. The problems encountered can have a major impact on the cost of the solution, and the engineer must be able to provide an acceptable one within the limits of an agreed budget, in a similar way that buildings are cost planned within cost limits. However, because of the nature of civil engineering works, they can involve a large amount of uncertainty and temporary works can be considerable, representing a significant part of the budget.

Civil engineering projects use different methods of measurement. In the UK, this might be either the Civil Engineering Standard Method of Measurement or Method of Measurement for Roads and Bridgeworks, although, in addition, other methods are also available. Different forms and conditions of contract are also used. These to some extent represent the different perception of civil engineering works. The work is more method-related than building works, with a much more intensive use of mechanical plant and temporary works. Bills of quantities, for example, comprise large quantities of comparatively few items. Because much of the work involved is at or below ground level, the quantities are normally approximate, with a full remeasurement of the work that is actually carried out. Also as there is not the same direct relationship between quantity and costs, contractual claims are potentially a more likely event.

Quantity surveyors working in the civil engineering industry provide similar services to those of their counterparts working on building projects. In addition to the methods of measurement and conditions of contract, quantity surveyors must also be conversant with the different working rule agreements, daywork rates and other documents such as *Civil Engineering Procedure*, which is published by the Institution of Civil Engineers.

Quantity surveyors have been employed by civil engineering contractors and design consultants since the turn of the twentieth century. Engineers also value the advice that quantity surveyors are able to provide on costs, value, contractual and other relevant matters. Engineers recognise the benefits of the quantity surveyor's specialised skills and knowledge in respect of the cost and financial aspects of construction. Many promoters, the civil engineering client, rely on pre-contract and post-contract services provided by quantity surveyors.

Heavy and industrial engineering

This work includes such areas as onshore and offshore oil and gas, petro-chemicals, nuclear reprocessing and production facilities, process engineering, power stations, steel plants, and other similar industrial engineering complexes. Quantity surveyors have been involved in this type of work for a great number of years, and as a result of changing circumstances within these industries a greater emphasis is also being placed on value for money. In an industry that employs a large number of specialists, quantity surveyors, with their practical background, commercial sense, cost knowledge and legal understanding, have much to offer.

The work involved is generally classified as cost engineering. Modern-day cost engineers may come from a variety of different professions but a considerable number have their roots in quantity surveying. The professional cost engineer is widely employed in the USA and many countries of Europe, and this continues to be a growth profession. The RICS and the Association of Cost Engineers prepared a Standard Method of Measurement for Industrial Engineering Construction (SMMIEC) in 1984. Standard forms of contract are also published by the Institution of Chemical Engineers among others.

The basic methods employed are similar to those used in other quantity survey-ing work. They may be more numerically based and offer different forms of analy-sis which lend themselves to computerised measurement and cost administration systems. Bills of approximate quantities are often produced from sketches and drawings provided. Otherwise, performance specifications or schedules of rates or one of the variety of cost-reimbursable contracts may be employed. Alternatively take-off schedules may be prepared for the purchase of materials only.

Essentially, quantity surveyors who are employed on this type of work must be able to adapt to new methods of measurement, cost analysis, contract procedures and cost engineering practices. There is also a likelihood of being involved in a wider range of activities than those encountered on building projects.

The role of the quantity surveyor

Traditional role

The traditional role of the quantity surveyor has been described elsewhere and in previous editions of *Practice and Procedure for the Quantity Surveyor*. The history of the quantity surveyor from the middle of the seventeenth century is briefly described in Seeley and Winfield (1999) and in Ashworth and Hogg (2000). This traditional role, that is still practised by some and especially on small to medium sized projects, can be briefly described as a measure and value system. For a more nostalgic view of the quantity surveyor, readers should refer to Nisbet (1989). Approximate estimates of the initial costs of building are prepared using a single price method of estimating (see Chapter 6), and where this cost was acceptable to the client then the design was developed by the architect. Subsequently the quan-tity surveyor would produce bills of quantities for tendering purposes, the work would be measured for progress payments and a final account prepared on the

- Single rate approximate estimates
- Cost planning
- Procurement advice
- Measurement and quantification
- Document preparation, especially bills of quantities
- Cost control during construction
- Interim valuations and payments
- Financial statements
- Final account preparation and agreement
- Settlement of contractual claims

Fig. 1.1 Traditional quantity surveying activities (circa 1960).

basis of the tender documentation (see Fig. 1.1). The process was largely reactive, but necessary and important. During the 1960s, to avoid tenders being received that were over budget, cost planning services were added to the repertoire of the duties performed by the quantity surveyor employed in private practice (PQS). The contractor's surveyor was responsible for looking after the financial interests of the contractor and worked in conjunction with the PQS on the preparation of interim payments and final accounts. On occasions, contractors felt that they were not being adequately reimbursed under the terms of the contract and submitted claims for extra payments. This procedure was more prevalent on civil engineering projects than on building projects, although the adversarial nature of construction was increasing all the time.

Pragmatism and realism are some of the qualities most highly valued by clients in quantity surveyors (Davis, Langdon and Everest 1991). Some will argue that the distinctive competence found in quantity surveyors relies heavily on their analytical approach to buildings and that this stems directly from their ability to measure construction works. Furthermore, the detailed analysis of drawings leads to a deep understanding of the design and construction which enables them to contribute fully to the process. This intimate knowledge of projects is at the root of the contribution the quantity surveyor can make to the value of the client's business through the provision of the services shown in Fig. 1.1.

Evolved role

In response to the potential demise of bills of quantities, quantity surveyors began exploring new potential roles for their services. Procurement, a term not used until the 1980s, became an important area of activity, largely because of the increasing array of options that were available. Increased importance and emphasis were also being placed upon design cost planning as a tool that was effective in meeting the client's objectives. This coupled with whole life costing (Chapter 7), value management (Chapter 8) and risk analysis and management (Chapter 9) were other tools being used to add value for the client. As buildings became more engineering services orientated increased emphasis was being placed on the measurement,

- Investment appraisal
- Advice on cost limits and budgets
- Whole life costing
- Value management
- Risk analysis
- Insolvency services
- Cost engineering services
- Subcontract administration
- Environmental services measurement and costing
- Technical auditing
- Planning and supervision
- Valuation for insurance purposes
- Project management
- Facilities management
- Administering maintenance programmes
- Advice on contractual disputes
- Planning supervisor
- Employers' agent

Fig. 1.2   Evolved role (circa 2006).

costs and value of such services. Quantity surveyors had historically dealt with this work through prime cost and provisional sums, but in today's modern buildings to describe the work in this context is inadequate. Other evolved roles have included project and construction management and facilities management (see Fig. 1.2). Because of the inherent adversarial nature of the construction industry they are also involved in contractual disputes and litigation.

Some QS practices became very nervous about the apparent demise of bills of quantities. It remains an *apparent* demise since at the commencement of the twenty-first century they still represent a significant proportion of work and associated fees for some quantity surveying firms. The wheel may have turned in some respects since instead of preparing bills for clients, quantity surveyors are preparing bills for contractors. This was the practice used during the nineteenth century! Also countries that hitherto had not yet embraced quantity surveying, and the number of these countries is continuing to decline, were beginning to see the benefits of schedules of quantities for tendering and contractual administration purposes.

Significant interest in the techniques of whole life costing and value management were being actively pursued by practices and actively supported by academic research.

Developing role

The future development of quantity surveying services is likely to be influenced by the following important factors:

- Client focus
- Development and application of information and communication technologies
- Research and its dissemination

- Reduced time scales
- Practical completion must mean total completion, not 'nearly ready'
- Simplified process
- Complete understanding of the procurement process
- Comprehensive service including mechanical and electrical installations
- Excluding the exclusions
- Effective change management
- Solutions not projects

Fig. 1.3 Basic requirements of clients (*Source:* Powell 1998).

- Choice
- Co-investment and risk taking
- Commitment
- Credibility
- Competence
- Clarity and accountability
- Consistency

Fig. 1.4 Client needs – the seven C's (*Source:* Powell 1998).

- Graduate capability
- Practice size.

Client focus

The construction costs of capital works projects will always be an important component for clients in their decision to build. These costs will include whole life costs. There can be only a few clients with the motto of Cheops, the builder of the great pyramid, who said, 'I don't care how much it costs or how long it takes!' Consultants are sometimes seen as adding excessive costs to projects and contractors, offering services that are late, of poor quality and of indeterminate high costs. Clients will always be prepared to pay for services that are able to demonstrate a financial benefit. The basic requirements of clients are shown in Fig. 1.3.

The client needs of the early part of this century are broadly identified in Fig. 1.4. Experience suggests that whilst improvement amongst the leaders in the construction industry can be expected to match the best in the world, the improvement generally will take time and will involve radical changes in culture and probably its structure. Barlow (1996) has suggested some of the examples of best practices that can be learned from the manufacturing industry sector (Fig. 1.5). There is a lack of understanding of best practices or even an awareness in some situations that change is the order of the day and urgent. There is a determination to survive but a lack of realisation of international competition.

- **Focus on customer satisfaction.** Recognise that clients want buildings and support after completion, at the right price, to the appropriate quality and standards, on time and meeting their needs.
- **Attention to the process as well as the product.** Product design has now become a byword in manufacturing industry, where the process used has contributed towards increasing the appropriateness of the product. Research is necessary in the construction industry in process analysis.
- **Concept of total quality approach and attitudes.** The total quality approach should not be confused with total quality management or quality assurance, which are now widely accepted in the industry. Total quality programmes are often expensive to implement relying on extensive training to bring about a shift in culture. It represents a continuous improvement programme.
- **Benchmarking.** The practice of benchmarking all of a company's activities against the best competition or against organisations who are known to be industry leaders is now commonplace in some quarters (see Chapter 18). A characteristic of many of these companies is their willingness to share knowledge with others.
- **Team-working and partnering, including supply chains.** This aims to make every individual feel worthwhile and as such it leads to greater pride within a company. It aims to harness the intelligence and experience of the whole workforce. It also extends beyond the individual company to include consultants, contractors, subcontractors and suppliers. For large clients they are also often part of the team (see Chapter 10).
- **Information technology.** The construction industry must welcome the more widespread use of information technologies and embrace the current Technology Foresight Initiative.

Fig. 1.5   Learning from manufacturing industry (Barlow 1996).

A growing number of the large quantity surveying practices, that have grown to include a wide range of different types of professional staff, are members of the Management Consultancies Association (MCA). This organisation represents members who employ over 40 000 consultants and generate £6.5 billion in fees each year. MCA defines management consultancy as, the creation of value for organisations through the application of knowledge, techniques and assets to improve performance. This is achieved through the rendering of objective advice and the implementation of business solutions.

Development and application of information and communication technologies

The implications associated with information and communication technologies are considered in Chapter 5.

Research and its dissemination

The importance of research and its dissemination is fundamental to any profession, especially one that is facing a change in direction and its practices. These are also considered in Chapter 5.

Graduate capability

The number of graduates in quantity surveying is unlikely to change significantly in the short term from the reduced numbers experienced in the late 1990s. The relative shortage in supply has already had the effect of increasing graduate salaries. Those graduates who have a good technical understanding, a broader use of business skills and a commitment towards life long learning are likely to be in high demand. For other graduates they will need to make themselves either more valuable to practices and contractors or less expensive.

The issues facing the future of the construction industry and in attracting able graduates to its various professions including quantity surveying, are many and varied and include the following:

- Under capitalisation exacerbated by fragmentation and the large numbers of small firms
- Low technology, labour intensive and traditionally craft based
- Litigious basis for settling differences and disputes
- Lack of prototype development resulting in untested and ill-specified components and technologies
- Low level of computing and use of information technology
- The use by government of the construction industry as an economic regulator contributes towards a cyclical workload
- High number of company business failures
- Poor image, working practices and employment conditions
- Difficulties of recruiting, training and retaining a skilled and committed workforce

However, since the beginning of the twenty-first century the attractiveness of quantity surveying has increased markedly amongst school leavers and also amongst graduates from non-cognate subject areas. Some of the issues identified above, where possible, are being addressed and the economic stability of the industry together with some recent high profile and imaginative design solutions, such as the new football stadiums at Wembley and for Arsenal FC, the Swiss Re building, the Scottish Parliament, Terminal Five and the Millennium bridge over the river Tyne, are helping to attract individuals to the profession. The selection of London for the Olympic Games in 2012 has also helped to make the construction industry and the profession more attractive.

There is currently a worldwide demand for quantity surveying graduates and the opportunities that currently exist have not been available for a long time, if ever before, for them to make their mark in the industry. One cause of the shortage is the success of the quantity surveying profession in broadening the range of services it offers, such as project management and advice on PFI and taxation. The demand for quantity surveyors is also particularly evident amongst contracting firms. There has been a huge growth throughout the UK in the numbers of part-time students on quantity surveying programmes. However, in the short term employers are increasingly looking towards other professions to fill gaps in their workforce.

Practice size

The particular advantages of small firms are their in-depth understanding of the local market and their low cost base. The traditional small practice has historically relied on document preparation and traditional quantity surveying activities. In order for these practices to survive in the longer term they will need to diversify and develop stronger links with a larger number of clients. Some small practices have been able to develop specialist activities by directing their activities away from the traditional work towards niche areas, such as taxation, dispute resolution and information technology for the industry. There is a view (Powell 1998), that many of the smaller firms will increasingly work on a subcontract basis for the larger firms, although the evidence for this at the present time is not convincing. Strategies might include some of the following:

- Being responsive to change and having the capability to move into new markets
- Developing links with a wider client base
- Investing in modern information communication technologies
- Managing the needs of the local market
- Seizing opportunities afforded by on-line working.

In growing a practice from a small firm to a large firm, one of the biggest hurdles to overcome is that of being a medium sized firm. Many of the larger firms have already diversified into multiple discipline practices. The difficulty for the medium sized firms is the retaining of high quality staff, forging links with larger clients and competing with the smaller firms with their lower overheads. The larger clients, for example, are likely to want to work with those practices that are able to offer the more comprehensive services, in essence a single point of professional responsibility. Strategies to overcome this difficulty might include some of the following:

- Avoiding competition with larger firms
- Developing market strategies to differentiate them from the larger firms
- Specialising and developing in-depth knowledge
- Forming alliances with overseas partners
- Entering joint ventures with complementary firms.

The large practices are expected to grow in size through expansion and mergers. They are likely to continue to diversify as the role of the quantity surveyor develops. The particular education, training and expertise will be used to increase the range of services that are provided. They will begin to challenge other established firms and in some cases will become management consultants to the construction industry. They will retain the specialised knowledge to allow them to compete with other disciplines that have established themselves with large clients. This move of providing broad based business solutions will continue despite the increasing competition from contractors and other large consultancy organisations.

The role of quantity surveyors in the public sector will continue to be threatened by private consultants, through fee competition and best value solutions. The

- Automated measurement and quantification
- Environmental and sustainability analysis
- Advice on information and communications technology
- Taxation and investment advice relating to projects
- Supply chain management
- Facilities management
- Legal services
- Quality management
- Niche markets

Fig. 1.6   Developing role.

smaller public sector quantity surveying departments are likely to disappear entirely as capital works are either carried out by the larger county authorities or awarded to private consultancies. Within the larger public sector organisations, longer term business solutions will become increasingly important and comparative benchmarking practices used more widely in order to establish best value solutions.

The clear differences between contracting and consulting have already become blurred. This process is likely to continue until demarcation lines between roles, activities and practices and the best attributes become more distinctive in newly divided sectors. The smaller contracting firms are all but disappearing, as subcontracting has become the normal pattern. These firms however will continue to use in-house surveyors to perform a wide range of managerial functions. The number of larger firms is becoming fewer and more of these are becoming part of multinational organisations. The remainder of contracting firms will become more specialised focusing on a limited range of activities, product specifications or type of project that they carry out.

Historically private and public practice and contracting have frequently been seen as the two distinct parts of the profession. The work undertaken by each was also considered to be different in terms of the practices involved and the ethical considerations of doing the job. Increasingly clients have and are opting for single-point responsibility solutions through, for example, design and build and other contractor orientated practices such as PFI. In the past there might have been considered possible conflicts of interest by working for clients as well as for contractors. Such a concept has now all but disappeared.

The developing role of the quantity surveyor in these various and different organisations will be an expansion or further expansion in the areas of activities that are shown in Fig. 1.6.

Skills, knowledge and understanding

The Royal Institution of Chartered Surveyors published a report in 1992, titled *The Core Skills and Knowledge Base of the Quantity Surveyor*. This examined the needs of quantity surveyors in respect of their education, training and continuing profes-sional development. This reflected the requirements in the context of increasing changes and uncertainties in the construction industry and more importantly within the profession. Since the publication of this report, there have been a number of government and educational initiatives with regard to the implementation of life long learning. All subject disciplines have in recent years placed an increasing emphasis on the development of generic and specialist skills within their respec-tive curricula. A course audit of such skills was first initiated by the Business and Technology Education Council (now EDEXCEL) and has since been accepted as an important component on all undergraduate and postgraduate courses in every uni-versity. The RICS report identified a range of skills that the profession would need to continue to develop if it wished to maintain its role within the construction industry. The report identified a knowledge base that includes:

- Construction technology
- Measurement rules and conventions
- Construction economics
- Financial management
- Business administration
- Construction law

and a skill base that includes:

- Management
- Documentation
- Analysis
- Appraisal
- Quantification
- Synthesis
- Communication.

All of these remain valid requirements ten years later, although their relative impor-tance has changed to suit changing needs and aspirations. The report also sought to set out and to identify the present market requirements and to anticipate expec-tations in the future. It also examined constraints that might in some way inhibit quantity surveyors from achieving their full and expected potential.

The report developed earlier themes from reports published by the RICS and others. These included *The Future Role of the Quantity Surveyor* (RICS 1971), *The Future Role of the Chartered Quantity Surveyor* (RICS 1983), *Quantity Surveying 2000* (Davis, Langdon & Everest 1991) and *Quantity Surveying Techniques: New Directions* (Brandon 1992). The report examined key trends in the demand for construction activity and the needs of professional services. The report also made reference to the wider opportunities that may lie beyond the horizon of construction, where the skills and knowledge base could be applied.

In analysing the knowledge base and accepting that this would be incremental and on a need-to-know basis, the report identified four key areas:

- *Technology:* This relates to process used and the product achieved
- *Information:* The requirement for sources and information management
- *Cultural:* The organisational and legal framework context
- *Economic:* The increasing importance of business and finance.

The report discusses the differences between skills and techniques. Quantity surveying has developed its own repertoire of techniques. Skills occur in respect of the levels of ability required to apply these techniques in an expert way. The different array of skills is assimilated with the knowledge base through education, training and practice. Whilst there is a general agreement about the skills and knowledge base required, different surveyors will place different emphases upon the relative importance in practice. The report concludes with a forecast of the future importance of the different core skills and knowledge requirements in a changing environment.

Powell (1998) further emphasised the importance of skills required of the chartered quantity surveyor tomorrow. Tomorrow has now already arrived! This report emphasised the need to:

- Develop a greater understanding of business and business culture
- Develop strong communications and ICT skills
- Challenge authoritatively the contributions of other team members
- Understand that value can be added only by managing and improving the client's customers and employer's performance
- Develop skills to promote themselves effectively
- See qualifications only as the starting point
- Recognise the need to take action now
- Become champions of finance and good propriety.

The skills of the quantity surveyor that were very important 50 years ago still remain important but their relative importance has declined to be replaced by new skills. This is evident across a whole range of industries and professions. The skills of today will also need to be enhanced as demands continue to evolve and change in the future.

For many undergraduates leaving university at the start of the twenty-first century the notion of skills has been extended beyond those developed in the RICS report of twenty years earlier. It has long been recognised that the development of skills has a much longer life span than knowledge since the latter are changing and frequently being updated. The Skills Plus Project (William 2003) identified a comprehensive listing of 39 skills organised under three headings of personal qualities, core skills and process skills. The personal qualities included, for example:

- Independence
- Adaptability
- Initiative taking
- A willingness to learn
- An ability to reflect on what has and what has not been achieved.

Core skills included the obvious three 'Rs' but also:

- The ability to present clear information when in a group
- Self-management
- Critical analysis
- The ability to listen to others.

The last of these, whilst obvious, is a skill that is frequently much under-developed in people generally. Common listening styles include the juggler (distracted listening), the pretender (pretend listening), the hurry-up-er (impatient listening), the rehearser (switched off listening) and the fixer (fix-it listening). Listening, by allowing others to speak, is a very powerful tool.

The process skills include:

- Computer literacy
- Commercial awareness
- Prioritising
- Acting morally and ethically
- Coping with ambiguity and complexity
- Negotiating.

Some of these are very difficult to develop in an academic setting. There is thus a growing awareness of recognising that links between universities and quantity surveying practice need to be strengthened for the benefit of all parties involved. Commercial awareness is often not fully understood and applied even by many working in practice.

What's in a name?

When we started to write the eleventh edition of *Practice and Procedure for the Quantity Surveyor* in 2002 there was much consideration given to a possible change in the title of the book since the name quantity surveyor was thought to be a little old fashioned. It was, and is, but so also are the names of many of the professions in that they do not now adequately describe what they do. We decided to retain the book's title and have hardly discussed the name in this edition.

Since that time of course the RICS, after a lapse period of six years, have renamed their Construction Faculty, 'Quantity Surveying and Construction'. The RICS about-turn stemmed from a sense that quantity surveyors should be judged on the value of their services, not their job title. The name change represents a positive and forward looking step for the profession, for the Faculty and for the wider RICS. The quantity surveyor has a varied and unique skill set to offer clients, whether providing feasibility advice, a procurement programme strategy or a whole life cost analysis. Many RICS members feel that they have been given their identity back.

The origins of quantity surveying as a distinct activity are hard to trace back in time much further than the Great Fire of London. However in the New Testament there is a story there about counting the cost before you build (Luke's Gospel chapter 14). Perhaps quantity surveyors can trace their routes back to more than 2 000 years ago! The building boom that followed the Great Fire of London encouraged the emergence of the architect and the growth of the single trades, contracting for their own part of the building work.

The measurers had to be invented, if they did not already exist. There was a real need for someone to ensure impartiality between the proprietor and the workmen. The rest is history. In 1834, the fire that destroyed the Palace of Westminster was partially responsible for the use of the quantity surveyor on a major scale. Charles Barry won the competition to replace it and was asked to prepare an estimate of cost. Although detailed drawings were not yet prepared, a quantity surveyor, Henry Hunt, came up with an estimated cost of £724,984. Whilst this figure was basically accurate, changes made by Parliament resulted in a final cost closer to £1.5 m. Nothing new then with the Scottish Parliament building! Incidentally it was not until 1834 that the RIBA was formed.

The name 'quantity surveyor' conjures up a variety of different images in people's imaginations. For some, the term quantity surveyor is an outmoded title from the past. It certainly no longer *accurately* describes the duties that are performed. When the term was first applied to the profession, the work of the quantity surveyor was vastly different to that now being carried out and anticipated in the twenty-first century. New titles have been suggested which include 'construction cost consultant', as being the most preferred. 'Building economist', 'construction accountant', 'contractual and procurement specialist' are others that have been disregarded.

Some practices have also deleted cost entirely from the title, describing themselves simply as 'construction consultant', or 'construction management consultant'. Others have added a multiplicity of titles aimed at reflecting the practice's work and capturing new work from clients. Some practices have also opted for the now somewhat vague title of 'chartered surveyor', but this title has little to recommend it, unless the practice is solely practising in the United Kingdom or a commonwealth country. Even then it is confusing, since the title 'chartered surveyor' conjures up different meanings. To the man on the Clapham omnibus it is invariably confused with an estate agent, since these are the surveyors with whom the general public have the greatest contact! Project management is also a desired title by some practices, but this title too is imprecise, since it might reflect a wider scope than the construction industry alone. 'Management consultant' is an all-embracing term and is used by at least one of the large quantity surveying practices. Quantity surveyors have perhaps been slow to work in this area of commerce and industry.

Adopting a title is important, but it must also seek to reflect the work carried out by the particular practice. In an age of diversification it may be desirable to use two or more titles to describe the work that is undertaken. Interestingly, the major reorganisation of the RICS in 2000 has moved away from divisions into faculties. The former Quantity Surveyor's Division, which was the second largest, initially re-branded itself primarily as the Construction Faculty. This shift in title was not supported by everyone in the profession and the recent renaming of the Quantity Surveying and Construction Faculty reflects practitioners' concerns. There are increasingly some regrets about the dissolution of the former Institute of Quantity Surveyors in the mid-1980s. The Quantity Surveying and Construction Faculty is the largest faculty in the RICS. Quantity surveyors are, of course, involved in a number of the other faculties within the RICS.

Contractors' surveyors have found a new title perhaps easier to acquire. Commercial management appears now to be their preferred description and title. This is a good title, reflecting well the work that they carry out for a construction company.

It should also be accepted that this dilemma of name is not unique to the quantity surveying profession. Many other professions have also diversified from their initially intended roles to offer commerce and industry a broader and sometimes all-inclusive service. The large accountancy practices have grasped the opportunities of management consultancy on a big scale. In fact most of the professions, including accountants and solicitors, with whom surveyors are sometimes grouped, now undertake work that is far removed from the simple dictionary definition of their description. It is also becoming increasingly common to find quantity surveyors at work in these multidisciplinary, multi-industry practices.

So the debate on the name for the quantity surveyor will continue. There are no prizes for the best title, but the authors of this book would be interested to hear the views of its readers.

Discussion topic

The role of the quantity surveyor is, like many other professional disciplines, continuing to evolve to meet changing situations. What are the influences that are causing such changes to take place and how will these affect the work of the quantity surveyor in the future?

In 1992, the RICS published a report, titled, *The Core Skills and Knowledge Base of the Quantity Surveyor*. This provided a sound base for future developments within the profession, and along with a number of different reports and articles produced by the profession, focused on an insatiable appetite for change. The quantity surveying profession has also shown that it has been resilient to change and the demands made upon it. Some commentators have predicted its long-term demise and whilst some of its work has changed in practice it has adapted and applied its skills and knowledge base to new situations and demands by clients both large and small, both government and private. To misquote Mark Twain, the death of the quantity surveyor has been an over-exaggeration.

Information technology (IT) is clearly one of those areas that has impacted on all of our lives in different ways. It has clearly affected the way in which quantity surveyors have carried out their work. In some cases this has removed some of the tedium, and speeded up the processes that are used whilst in other spheres it has enabled quantity surveyors to both enhance and offer a wider range of services that their clients consider to be important. Information technology applies not only to using computers for their work but also in terms of communicating with the outside world through the internet and the electronic exchange of information amongst the team and the different professions involved. There is a continued need to integrate much technology involved to allow different systems to be able to communicate effectively with each other. Using IT for improved knowledge management in firms to create an environment in which projects can thrive has been a shift in the emphasis of project management (Love, Fong & Irani 2005). Somerville and Craig (2003) have clearly shown that cost savings are achieved when moving from a paper-based system to an electronic document management system.

A second area that has influenced change has been discovering how other countries around the world deal with costs and values of construction projects and how they deal with forecasting and the payment for the work that is carried out. In the shrinking world in which we live we now have an understanding of professional experiences from different parts of the world. The commonwealth has traditionally adopted UK patterns of practice and although the quantity surveyor may sometimes be seen in a less important role than in the UK, the practices are comparable. This is not so in other countries such as mainland Europe, the USA and Japan for example. Here practices differ vastly to those of traditional quantity surveying but it has been observed that these practices and the range of services offered have and are continuing to change. Even a casual observation, however, indicates that several quantity surveying firms are flourishing in some of these locations which have hitherto been foreign to the profession. Davis and Baccarini (2004) have carried out a study on the use of bills of quantities in Australia that indicates a similar evolution and situation to that occurring in the UK.

The needs of clients have also changed markedly over the last fifty years. The large regular-procuring clients of the construction industry are increasingly pursuing innovative approaches to the way in which their projects are planned, designed and delivered to facilitate their business strategies. They tend to work more closely with a smaller number of organisations and more closely with their supply chains to maximise value and achieve continuous improvement in performance both of their construction processes and buildings when in use. A number of clients have welcomed the initiatives promoted by the Latham (1994) and Egan (1998) reports, the Construction Best Practice Programme and the Rethinking Construction movement. These are not a panacea for all problems but they have focused on key aspects of quality, delivery, cost and value. The industry too has responded to current changes in practice through for example the principles of lean construction (Ashworth 2006).

Quantity surveyors have worked in multi-disciplinary practices for many years. However, at the beginning of the twenty-first century, even more than in the past, some of the large quantity surveying firms are now amalgamating or being taken over by other consulting groups. Some of these arrangements have worked very well to the benefit of all parties; in other cases they have not be so successful, with conflicting interests and differing levels of responsibilities making such multi-disciplinary organisations too difficult to operate effectively. In some notable cases there have even been de-mergers to allow the different disciplines to focus more clearly on their own client issues.

The impact of changes in education practices that have taken place over the last thirty years in the case of undergraduate education and as little as ten years in respect of postgraduate education cannot be underestimated. Recent years have seen a considerable growth in research interest in industry and commerce. Much of the interest has been stimulated through advances in technology, but also through the wider opportunities fostered by education. Brandon (1992) commented almost fifteen years ago that the attitude of the construction industry and its allied professions prevented major shifts, although there were signs of change in the future. Considerable changes have in recent years been evident in many of the practices of the quantity surveying professions and some of these are due to changes in education, both in respect of the curriculum and teaching and learning styles. Also tasks that were at one time within the professional remit are now more clearly within the description of the role of a technician.

It is always difficult to predict how the role of any profession will change. Even those who are described as visionaries often get it wrong! Davies, writing in RICS Business (2006), refers to quantity surveyors now being the key advisers on construction and development strategies, but that they must continue to reinvent themselves to remain in this position. Ed Badke, the Director for Construction and the Built Environment at the RICS, commented that the QS was once only involved with measuring and valuing of construction work carried out under a building contract. However, two important changes took place in the 1980s. The QS firms found themselves facing severe fee competition and their clients were looking at new ways of managing contracts. The QS practices that survived then looked overseas, especially to the boom in the Middle East and at the same time began extending their services. The firms that failed to respond and to adapt to these new situations probably either reduced in size or disappeared entirely.

A key feature for the future is the need to continually add value and enhance professional services. All firms, whether in the public or private sectors are being challenged to do more for less. The QS practice that is able to provide clients with this has a secure future. Also many QS firms are now retained to help clients develop their strategy for managing all of their construction projects.

Other professions represent a greater threat to QS practices in the future than they do today. Several accountancy practices and management consultants already employ quantity surveyors. These practices often have access to firms and organisations through their auditing function. In the past they may have been asked to recommend a quantity surveying firm. In the future they may elect to endeavour to do the work themselves because of the professional relationships that have been established. There are also other firms who would be only too happy to move into the areas of advice that are currently the domain of the quantity surveyor.

The other threat facing the quantity surveying profession is manpower. There are insufficient numbers of graduates entering the profession to replace those coming to the end of their careers and for expansion into a currently booming construction industry worldwide. Whilst recruitment to courses is now vastly improved it will take four years for the increased numbers of graduates to enter the profession. This partially explains the huge increase in student numbers on degrees by part-time study.

References and bibliography

Ashworth A. *Contractual Procedures in the Construction Industry.* Pearson. 2006.

Ashworth A. and Hogg K. *Added Value in Design and Construction.* Pearson. 2000.

Barlow J. *A Statement on the Construction Industry.* The Royal Academy of Engineering. 1996.

Brandon P.S. (ed.) *Quantity Surveying Techniques: New Directions.* Blackwell Science. 1992.

Davies R. *The QS transformation.* RICS Business. March 2006.

Davis P. and Baccarini D. *The use of bills of quantities in construction projects – an Australian survey.* Proceedings of the RICS Foundation, Construction and Building Research Conference, Leeds. 2004.

Davis, Langdon and Everest *Quantity Surveying 2000.* Royal Institution of Chartered Surveyors. 1991.

Egan J. *Rethinking Construction.* Department of Industry and the Regions. 1998.

Harvey R.C. and Ashworth A. *The Construction Industry of Great Britain.* Butterworth Heinemann. 1977.

ICE Civil Engineering Procedure. Institution of Civil Engineers. 1996.

Latham M. *Constructing the Team.* HMSO. 1994.

Love P., Fong P.S.W. and Irani Z. *Management of knowledge in project environments.* Elsevier. 2005.

National Audit Office. *Modernising Construction.* HMSO. 2001.

Nisbet J. *Called to Account: Quantity Surveying 1936–1986.* Stoke Publications. 1989.

Pountney N. *The Right Formula.* Royal Institution of Chartered Surveyors. 1999.

Powell C. *The Challenge of Change.* Royal Institution of Chartered Surveyors. 1998.

QS Strategies. The Builder Group. 1999.

RICS. *The Future Role of the Quantity Surveyor.* Royal Institution of Chartered Surveyors. 1971.

RICS. *The Future Role of the Chartered Quantity Surveyor.* Royal Institution of Chartered Surveyors. 1983.

RICS. *The Core Skills and Knowledge Base of the Quantity Surveyor.* Royal Institution of Chartered Surveyors. 1992.

Seeley I. and Winfield R. *Building Quantities Explained.* Macmillan. 1999.

Somerville J. and Craig N. *Cost savings from electronic document management systems: The hard facts.* Proceedings of the RICS Foundation, Construction and Building Research Conference, Wolverhampton. 2003.

William A. *Skills plus: Tuning the undergraduate construction curriculum.* CIB W89 International Conference on Building Education and Research (BEAR). 2003.

2 Education, Training and Employment

Introduction

Courses in quantity surveying are offered at various levels and modes in universities and colleges in the UK. These courses are also replicated in many of the former commonwealth countries around the world. They are typically undergraduate courses of three (full-time), four (sandwich) or five (part-time) years' duration. Two-year technician courses are likely to be replaced with foundation degrees and these may allow for advanced entry to the undergraduate programmes.

Historically, quantity surveying courses were offered on a part-time basis through day release studies or through correspondence courses or distance learning. Students from these courses then sat the professional examinations of the Royal Institution of Chartered Surveyors or those of the former Institute of Quantity Surveyors until these professional bodies merged in 1983. The examinations were in three parts and these to some extent replicated the three years of undergraduate degrees offered by universities. However, since a majority of the study was on a part-time basis it usually took at least five years to complete the final qualifying examinations.

The undergraduate degree is the typical course that is studied by the majority of those who wish to become quantity surveyors. However it must be one that is recognised in partnership with or accredited by the Royal Institution of Chartered Surveyors. These courses allow students who obtain such a degree to become eligible for corporate membership, upon passing the Assessment of Professional Competence (APC). Programmes accredited by the RICS can be found in universities throughout the UK and in many parts of the world (RICS 1999).

Chronology of quantity surveying education 1960–2006

The historical development of quantity surveying has been well documented by Ashworth (1994) and this is summarised in Fig. 2.1. Surveyors traditionally entered the profession through an articled pupil scheme, whereby a partner or senior surveyor would personally supervise the embryonic surveyor. This system was later augmented through correspondence courses and day-release at a local college for preparation for professional examinations for institution membership.

The 1960s

- Quantity surveyors study on a day release basis from employment or through correspondence courses for Institution's examinations
- National Joint Committee is formed for technician qualifications
- Designation of the Council for National Academic Awards (CNAA)
- Introduction of full-time exempting quantity surveying diploma courses
- Initial debate on two-tiered profession

The 1970s

- 30 polytechnics formed in 1970
- RICS introduces its centres of excellence policy
- Development of full-time degree courses
- Part-time exempting diploma courses mainly for IQS qualifications
- Control of course expansion through the Department of Education and Science (DES), Her Majesty's Inspectorate (HMI), CNAA, and RICS
- Limited new course developments, although high demand by students

The 1980s

- Merger of the RICS and IQS
- Part-time degrees replace diplomas and external examinations
- Quantity surveying courses are mainly sandwich and part-time
- Recruitment and development of courses continues to expand

The 1990s

- Polytechnics become universities
- Popularity of quantity surveying courses among students
- Continuing professional development for all surveyors
- Development of higher degrees in construction management
- Continued role of accreditation in the UK and overseas

The 2000s

- Popularity of quantity surveying courses from an earlier low demand
- University partnerships established with the RICS
- Continued expansion of courses overseas to meet local demands
- The technician body, MSST, merges with the RICS and a new class of membership is formed
- Further debate on the need for a two-tiered profession
- Consideration of higher degree for membership of the RICS
- Increased growth of postgraduate programmes for non-cognate students
- High employability for quantity surveyors throughout the world

Fig. 2.1   Chronology of quantity surveying education.

The introduction of full-time and sandwich degree courses in quantity survey-ing did not occur until the late 1960s, and even then in only a small way. The wide-spread introduction of surveying undergraduate programmes took place in the 1970s. In parallel with these developments an expansion in higher education was also taking place in other subjects. Student participation in higher education was at that time about 7%. This compares with in excess of 40% in 2006.

The development of undergraduate courses in quantity surveying should not be underestimated. While at the time of their introduction only 10% of the population had degrees, today that figure is now approaching 40%. To have attempted to develop such programmes today would have been much more difficult. It should also be recognised that the introduction of undergraduate courses in replacing the Institution's own external examinations was not simply a change in name. The focus of study moved from one that was concerned with relatively narrow train-ing to that which provided a broader and more rounded education. As with change this did not suit everyone at the time. In hindsight it was a good policy. Figure 2.2 illustrates the expectations to be received from a university education.

Education continues to evolve rapidly in a changing world. The subject content, for example, is considerably different to that of 50 years ago. Changes have occurred in the relative importance of subjects and knowledge, new skills have been intro-duced and research has increased considerably. There has been a discernible shift towards problem solving and away from a reliance on memory recall. There is an emphasis towards understanding and the application of knowledge and use of a wider range of key and transferable skills. The delivery of programmes is also now more clearly focused on student centred learning and the development of study skills. The assessment of students places some importance, not just on an end of course examination, but on a combination of this and continuous assessment

---

The development of the trained mind, which includes:

- critical thinking and reasoning skills
- ability to think conceptually
- intellectual perspective and independence of thought

The acquisition of knowledge through the:

- exposure to different domains of knowledge
- respect for cultural traditions
- use of knowledge to form sound judgements

Personal development which includes:

- moral, social, aesthetic and creative dimensions of personality
- development of attributes and skills

Development of a base for life long learning:

- learning how to learn

Fig. 2.2 Expectations from a university education.

through a more holistic approach. Workplace learning needs to see a step change in practices.

Quantity surveyors in education

There are currently about 100 quantity surveying academics working full-time in the UK and a further 40 in the rest of the world. A number of practitioners also contribute to programmes on a part-time basis. A proportion of these are actively involved in research. This relatively small body of dedicated academics has a large and increasing clientele. The number of quantity surveying students, including full-time, part-time and those on masters programmes is probably at the highest level ever. Whilst in the early 1980s one former polytechnic department had in excess of twenty quantity surveying lecturers, today the maximum number hardly anywhere gets into double figures and typically quantity surveying teaching teams are no more than four or five strong. This, to some extent, reflects the changing nature of the curriculum and the emphasis that is given towards studying a wide range of subjects. However, it may also suggest a difficulty, experienced by some university departments, in trying to recruit quantity surveying practitioners as lecturers. The recognised advantages of moving from practice to education are no longer quite so obvious as they used to be. The difficulties in the recruitment and retention of quantity surveying lecturers is acknowledged as one of the most important issues now facing built environment departments.

Should any of this be of concern to those who practise quantity surveying and who, to a large extent, rely on universities to educate and to contribute towards the training of the future members of the profession? Since quantity surveying practice is only as good and as profitable as its staffing resource allows, the answer to this must clearly be an emphatic Yes.

Thirty years ago the typical quantity surveying lecturer would have professional qualifications that had been acquired through passing the Institution's external examinations and also a substantial amount of workplace experience. This provided the street credibility that is impossible to replicate within education alone. Such individuals knew not only the theoretical methods described in the text books and journals, but also how practice worked and how quantity surveyors actually carried out their work. From this position they were able to present to students aspects of both theory and application.

During that time it was rare for any quantity surveyors to have a degree qualification and debates raged about whether quantity surveying could ever become an academic discipline. But all that is now history. The arguments of whether a degree or professional apprenticeship produced the best surveyors is also history and we doubt whether any forward looking quantity surveyor would want to turn the clock back to those now outdated practices. Whatever we have lost with the demise of the professional articles approach is far outweighed by the quality of teaching, learning and support offered to students in 2006 by modern universities and the broader range of knowledge, skills and understanding that graduates now acquire.

Almost all quantity surveying lecturers now have a degree and a large number also have a recognised teaching qualification. Many, for example, have acquired higher degrees in law, or business or IT. It would be very odd if it were any different especially since within a few years time up to 50% of the population will have obtained some kind of higher education qualification. Professional bodies will soon need to consider how its future members should be differentiated from the population at large.

There are about forty universities in the UK that offer quantity surveying and many of these now offer postgraduate programmes and research degrees. Whilst recruitment to these programmes is currently good, the numbers of students from the indigenous population who choose to do research after their first degree are relatively small. This may pose further questions for the would-be educators of the future. Universities, when faced with appointing well-qualified staff or well-experienced staff are likely to give the most careful consideration to the former. This is for a number of good reasons, in particular, evidence of academic stature, but also in research capacity building that is now so important to a university's standing and success. The lack of such higher degrees may make it difficult in the future for universities to be able to appoint those from practice alone.

There is an urgent need for practice and education to work closer together. This must be seen as a two-way relationship between both parties. Whilst research and student placement help to bridge the gap, there needs to be a closer alignment between the subjects in the curriculum and the needs of professional practice. There is always the danger of skimping over the fundamental principles and skills, which are sometimes seen as a chore. For those who have never experienced practice, these may be dismissed as unimportant and something that practice can deal with at a later date anyway. But these aspects must be seen as valuable and adding value if we are to ensure that professional practice and education work more closely together in the future in helping to achieve the same goal and outcomes.

Bringing new talent into higher education is a constant concern around the world. Uncompetitive salaries, compared to those in the private sector, have meant that universities have had to struggle to recruit and retain highly qualified staff in some subject areas, including surveying (RICS 2006).

Partnership and accreditation

The RICS has, since the introduction of undergraduate degree courses in the 1960s, had a process of accrediting such courses. This allowed students from such courses to be exempt from the former RICS external examinations and to become corporate members upon satisfying the requirements of the Test of Professional Competence which subsequently became the Assessment of Professional Competence (APC).

Under its Agenda for Change initiative (Kolesar 1999), the RICS has attempted to review every aspect of the surveying profession. During this review it found that the status of the profession was in decline and that measures were needed to attract more able entrants to the profession. Whilst the accreditation of its courses was set to maintain standards in education suitable for the profession, the RICS believes that this existing process is unlikely to achieve continued improvement in quality without a radical change.

In response to this, the RICS has revised its accreditation policies and replaced these with partnership arrangements with selected universities. This new policy was introduced in 2001. In order for a university to be eligible to form a partnership with the RICS, four criteria are considered, as shown in Fig. 2.3. Each of these criteria must be met in full, regardless of how well other criteria may be achieved. The scale or reputation of a department or school must be of a sufficiently high standard in respect of research and teaching quality in order to attract the right calibre of staff and students. In order to remain a significant partner they must be capable of keeping pace with the RICS's drive towards higher standards of achievement.

The aims of each individual partnership arrangement will be to:

- Maintain standards
- Attract bright entrants to the profession
- Promote research in surveying related areas
- Respond to the needs of the profession with course development
- Improve profession and educational links.

In addition, individual partnership agreements will seek to develop niche areas of activity of an education provider.

The RICS partnership and accreditation board, in respect of the UK partnerships, will be responsible for:

- Analysing the total provision across the UK in respect of needs and demands
- Spreading best practices
- Approving and briefing practitioner external examiners
- Monitoring the threshold agreements.

Student entry	Average GCE A-Level points scores for entrants for the top 75% of entrants on to each relevant course.
Research	This is based on the latest research estimate scores or their equivalent which is currently defined in the Research Assessment Exercise (RAE).
Quality	The latest Teaching Quality Assessment (TQA) or equivalent. This generally refers to that which was carried out in 1996–98.
Employment	The Higher Education Statistical Agency (HESA) scores for the destination of graduates.

Whilst there are current targets for these four criteria, they are kept under regular review in order to meet the changing needs and demands of the profession.

Fig. 2.3 RICS partnership criteria.

Courses in quantity surveying have been accredited by the RICS in Hong Kong, Australia, New Zealand, Sri Lanka, Singapore, South Africa, Malaysia, USA and China. Professional reciprocal agreements also exist with the respective national professional bodies, whereby members of a national professional body who practise in the UK can also achieve an RICS qualification. Courses in quantity surveying exist in all former commonwealth countries, although some may have only a limited relationship with the RICS.

Non-cognate disciplines

The RICS (Venning 1989) has long recognised the importance of recruiting members from a wide range of different disciplines. Whilst the majority of its recruitment will come from applicants who have studied on surveying courses, a number will be recruited from graduates in a non-cognate subject. Postgraduate (diploma and MSc) courses have been developed to meet this need. These are essentially intensive fast track courses that enable such students to become chartered surveyors. Upon the completion of these courses, the students need to pass the APC. There has been a considerable growth in the number of students on these programmes due to the opportunities in practice and the increased attractiveness of the profession. More innovative approaches incorporate significant online learning.

National vocational qualifications (NVQs)

The NVQ framework is aimed at creating a coherent classification for qualifications across five levels, from level 1 to level 5. The highest level corresponds to graduate or professional level. The curriculum is a modular structure and is designed to facilitate transfer and progression. Core skills are included in all qualifications to enable recognition of aspects of achievement that may, in the past, have been unacknowledged. Outcomes are based on skills, knowledge, understanding and ability in application. The skills, or competence tests, are as far as possible conducted in the workplace although this has sometimes been found to be unrealistic in logistical terms.

The design and structure of the NVQs on a unit basis has meant that a variety of routes to the qualification are available. This, in turn, means that the curriculum can be offered as flexible learning programmes unlike the more rigid traditional programmes. Students can negotiate the units that they wish to achieve, the order in which they achieve them, and their pace of study. Thus, learning programmes are tailored to individual needs. The development of the NVQs has resource implications although co-operation between institutions may reduce the overall increase in resource demands. The unit structure of the NVQs lends itself to credit accumulation and transfer systems and, in particular, the effective exploitation of the accreditation of prior experience and learning (APEL). Advantages of the NVQs are:

- A flexible programme of study
- A nationally recognised qualification
- Prior experience and learning can be accredited
- Study may be undertaken at a pace chosen by the student
- Transfer and progression are facilitated
- Core skills are integrated into programmes of study.

Disadvantages of NVQs are:

- The emphasis on competence skills has been at the expense of knowledge
- Competences are too narrow and simplistic
- An increased level of resources is required.

Assessment of professional competence (APC)

The different professional bodies continue to test the professional capabilities through the form of a single examination which is designed to assess an individual's competence to practise. In the RICS, this is the Assessment of Professional Competence (APC) and is more concerned with training than education. It is now widely accepted that training is only effective *on the job*. Whilst some colleges have provided simulated work experience of good quality, this could not replace the experience gained in a surveyors' office or on site with a contractor. Simulation is never the real thing!

Good training programmes offer the trainee a variety of different experiences. The RICS training log expects this to be the case, and trainees whose experiences have been too narrow may be requested to resubmit their log books after a broader experience has been achieved. The training programmes provided by contractors are often more varied and are frequently beyond those activities that can properly be described as quantity surveying.

Ideally, the experience offered should give to the trainee an opportunity to develop professional practical skills, put theory into practice, make decisions, solve problems and evaluate the functioning of the industry. The RICS Recorded Experience separately identifies building from engineering and subdivides the latter between civil engineering and mechanical engineering. There are five areas of approved experience to be gained and these include:

- Cost advice and cost planning
- Contract documentation
- Tendering and contractual arrangements
- Contract services
- Specialisations.

The latter covers a wide range of activities including taxation, insurance, litigation, technical audits, etc.

In deciding whether an individual meets the requirements for corporate membership the following are assessed:

- Application of theory into practice
- Awareness of the RICS Rules of Conduct and the possession of integrity
- Importance of accuracy to safeguard employers and clients
- Able to communicate orally and in writing.

In addition, the young surveyor is expected to develop to be a good ambassador for the profession, professionally and commercially aware, clear on clients' and employers' objectives, up to date with knowledge and skills, a team member, and able to work confidently without supervision.

The RICS Practice Qualification Group (formerly the Inter-Divisional APC Committee) introduced in October 2000 the need for all APC candidates to be employed in an organisation having a structured training framework or, at the very least a competency achievement plan. Competencies are designated as levels 1, 2 and 3. A good spread of level 1 competencies with some at level 2 can be achieved through sandwich work placements. Competencies at level 3 are probably achievable for students having considerable relevant experience. Before registering on the APC it is strongly recommended that employers and students are familiar with the latest requirements from the RICS. These requirements are described in the following publications:

- *APC Candidates and Employers Guide* (fourth edition)
- *APC Requirements and Competencies* (fourth edition)

In addition an employer is also recommended to become familiar with the following:

- *APC Guidance to Developing a Structured Training Framework*

The following guides are also available:

- *A Guide to the Recruitment and Selection of Graduate Trainees*
- *A Guide to the Induction of Graduate Trainees*
- *A Guide to the Coaching of Graduate Trainees*

The RICS has a number of regional training advisers who can offer help and assistance to employers. The training of staff is also recognised as a qualifying continuing professional development (CPD) activity and they can thus obtain personal benefits in developing a structured training framework.

Continuing professional development (CPD)

It used to be the pattern of practice that once the final examinations were passed then the formal education of the quantity surveyor was complete. It was recognised that through practice there would be a continuous development and that this would be stimulated by the demands and changes in practice and project experience. The general updating would be left to the responsibility of the individual surveyor. Those who were dedicated sought their own personal development through further study, or perhaps the acquisition of additional qualifications. Training programmes or formal staff development were not generally provided through employment.

With the pace of change in the industry accelerating, many of the professional bodies in the construction sector have recognised the importance of CPD and life long learning. It became mandatory for corporate membership of the RICS in 1984. It was accepted that if it were to become effective for the profession and the individual, then some form of registration was necessary. This requirement ensured that the benefits of institutional membership could only be retained if an adequate amount of CPD was undertaken over a period of time. The benefits claimed for CPD include higher productivity and profitability, lower staff turnover and absenteeism, innovation, improved client services, higher quality and improved job satisfaction.

Construction Industry Council (CIC)

The Construction Industry Council (CIC) was established in 1988 with five founder members. Since then it has grown in size and influence to become the largest pan-industry body concerned with all aspects of the built environment.

CIC occupies a unique role within the UK construction industry. Its members represent over 500 000 professionals working for, and in association with, the construction industry and more than 25 000 construction firms. It is the representative forum for the industry's professional bodies, research organisations and specialist trade associations.

The breadth and depth of its membership means that CIC is the only single body able to speak with authority on the diverse issues connected with construction without being constrained by the self-interest of any particular sector of the industry. The mission statement of CIC is:

- To serve society by promoting quality and sustainability in the built environment
- To give leadership to the construction industry, encouraging unity of purpose, collaboration, continuous improvement and career development
- To add value and emphasis to the work of members.

CIC, like other organisations, produces a range of publications, some of which can be downloaded freely from its website at www.cic.org.uk.

In 1997, the Common Learning Outcomes covering built environment higher education programmes, were signed up to by the major professional bodies. Through consultation with these and universities and colleges, a revised set of outcomes was developed to reflect the changing needs of the industry. The construction and built environment sector recognises that improvements in construction practice can only be fully effective when all of the stakeholders, clients, fund-holders, regulators, planners, designers, procurers, suppliers, constructors, operators and maintainers work together and recognise the distinctiveness of all professionals within the industry.

The new set of graduate Common Learning Outcomes provides a valuable benchmark for construction and built environment graduates. The criteria within these outcomes set out the personal skills and levels of technical and professional awareness that new graduates should have achieved as they embark on their professional careers.

Changing work patterns

The pattern of quantity surveying employment and work patterns are being affected by many factors in the early part of the twenty-first century. For example, the implications of information and communication technologies and the application of added value services will become of increasing importance. Powell (1998) has suggested that those who are committed to life long learning to ensure that their knowledge and skills are up to date and relevant will have the greatest opportunity of meeting the needs of clients and employers. Changing employment patterns was also a major theme in *QS2000* (Davis, Langdon & Everest 1991).

The vast majority of quantity surveyors are employed in private or public practice or in a contractor's organisation. In addition, quantity surveyors have been appointed to a variety of executive positions throughout the construction and other industries. In many instances, although their education and training as a quantity surveyor has been an asset in attaining a particular position, the role which they now perform may well have little or nothing to do with surveying practice.

Private and public practice

Traditionally, the principal difference between private and public practice was that whereas private practices are businesses, the main function of public practice was to ensure the accountability of public finances. This difference has narrowed considerably over recent years with the privatisation of many local and central government quantity surveying departments and those of the former statutory authorities. The aims and functions of the two sectors are therefore now similar to some extent. They are now able to compete against each other for work, and their livelihoods depend upon their performance measured against a range of indicators.

The status of the quantity surveyor is such that there is a need to provide clear, impartial and unequivocal professional cost and contractual advice. This means that a balance has to be drawn between maximising profits and maintaining a good service to clients. For those in the public sector this also includes accountability for public spending.

Private practice

Traditionally most quantity surveying firms have practised as partnerships. However, following changes in the RICS by-laws, some practices have decided to operate as limited-liability companies. The principal differences between the two are summarised as follows.

Partnership

Where two or more persons enter into partnership they are jointly and severally responsible for the acts of the partnership. Further, they are each liable to the full extent of their personal wealth for the debts of the business. There is no limit to their liability, as there is for directors of a limited company, to whom failure may only mean the loss of their shares in the company. All partners are bound by the individual acts carried out by each partner in the course of business. They are not, however, bound in respect of private transactions of individual partners. Partnerships come into being, expand and contract for a variety of reasons:

- As a business expands there is a need to divide the responsibility for management and the securing of work
- Through pooling of resources and accommodation economy in expenditure can be achieved
- The introduction of work or the need to raise additional capital may result in new partners being introduced.

The detailed consideration of partnerships falls outside the scope of this book. Suffice it to say that while it is not a legal necessity, a formal partnership agreement should be in place which legally sets down how the partnership will operate and covers such details as partners' capital and profit share.

Limited liability

Formerly the by-laws of the RICS prohibited members from parting with equity or shares to parties not actively involved in the practice. Increasingly it was felt that chartered quantity surveyors should be able to structure their practices so as to allow them to raise finance in ways that would enable them to improve their efficiency and effectiveness. It was also considered that both large and small firms would benefit from such a change and be better able to compete more effectively with other organisations which represent a fast-growing competition in their markets. Consequently, in 1986 the RICS removed the restrictions on limited liability. The changes involved the following:

- Removing the requirement for issued and paid-up share capital of the company to be not less than £25000
- Bringing the professional indemnity insurance requirements for surveyors practising with limited liability into line with the requirements of other surveyor principals

- Removing the restrictions on the transfer of outside share capital in surveyors' limited-liability companies, and on accepting instructions from outside shareholders
- Removing the requirement for surveyors wishing to practise with limited liability to apply to the Institution for permission to do so
- Imposing on surveyors who are directors of either limited or unlimited companies a requirement to ensure that they have full responsibility for professional matters
- Imposing an obligation on such surveyor directors of companies to include clauses in their memorandum and articles of association which provide for matters to be conducted in accordance with the Institution's Rules of Conduct.

Several leading quantity surveying practices had already begun to practise as limited-liability companies prior to these changes being implemented. Deregulation allowed them a great deal more flexibility in the running of their companies, and led to more practices electing to operate in this way.

Limited liability partnerships

A relatively recent development has been the promotion by the UK Government of the concept of a limited liability partnership (LLP), which came into force in April 2001. The key features of LLPs are:

- They will be separate legal entities distinct from the owners (members)
- Members of LLPs will not be jointly and severally liable in the normal course of business
- They will be treated as partnerships for the purposes of UK income tax and capital gains tax
- A partnership which evolves into an LLP will not undergo a 'deemed cessation' for income tax purposes
- They will have to file audited public accounts similar to a limited company
- They will require two or more designated members who will carry out tasks similar to those of a company secretary, for example, signing the annual return.

Therefore, the new LLPs combine certain crucial structural features of both companies and partnerships, the general intention being that the LLP will have the internal flexibility of a partnership but external obligations equivalent to those of a limited company. In common with partnerships, the members of an LLP may adopt whatever form of internal organisation they choose. However, they are similar to limited companies in that the members' liability for the debts of the business will be limited to their stakes in it and, therefore, they will be required regularly to publish information about the business and its finances (including the disclosure of the amount of profit attributable to the member with the largest share of the profits). Also, they will be subject to insolvency requirements broadly equivalent in effect to those that apply to companies.

Public service practice

The same general principles of practice and procedure apply to both private and public practice, with the obvious exception of the financial responsibility of the principal and the differences in the character and requirements of the respective clients.

For many years the amount of building work for which public authorities were responsible grew continually, until by the mid-1970s it represented almost 50% of the nation's construction output. Since then the workload in this sector has declined considerably. There has been some reduction in the requirement for public buildings following privatisation, and public sector housing is now to a large extent managed by housing associations. The effect of all this has been a large reduction in the number of persons employed in public service. However, the service still exists, and what follows highlights the significant differences between public and private practice.

Organisation

Public practice tends to reflect the organisation of the large private firm, and staff are subdivided by their job function.

The difficulty of private practitioners spreading their work evenly is solved in the public service by engaging only a proportion of the staff required to deal with the total requirements of the particular office. The remainder of the work is then outsourced to the private sector.

The range of quantity surveying functions undertaken by public practice offices varies according to policy, volume of work and staff available. The function most commonly assigned to the private practitioner is the preparation of contract documentation, but any or all of the various functions may be so assigned whether for a proportion or for all projects. In some instances the estimating, cost planning, valuations and final accounts are carried out by the department surveyors, but only a small proportion of the contract documentation required is prepared by its own staff. At the other end of the scale the surveyors in some offices act as coordinators of the activities of private surveyors, to whom all work is let.

The introduction of fee competition and the requirement to comply with the EC Directives on Public Supplies, Works and Services has led to all publicly funded work being advertised and fee bids invited. The recent introduction of best value for public sector projects, although an excellent ideal, is difficult and bureaucratic to administer. This principle also extends to the public departments themselves. Whilst in theory it removes the need to award projects on the basis of initial costs alone, and this is admirable, its application in practice is complex.

Conditions of employment

Private practices pay market rate salaries in order to attract and retain their staff. In the public sector a surveyor's salary depends on the grade point on appointment and the limit within a published scale. The grade payable is subject to the provisions for frequent reassessment of salaries to take account of changes in responsibility, additional experience and qualifications. Alterations to or improvements in the pay and grading structure are the subject of negotiations between representatives of a staff organisation and the employer's side. In most local authorities the negotiating body is the National Association of Local Government Officers (NALGO) but in the Civil Service different trade unions act for staff on different grades. The trade union bodies are responsible for a whole range of staff in addition to surveyors. Appointments and promotions in the public service require a formal interview before an examining board. In 2006, salaries within the public sector are significantly below those of private practice.

Duties

The extent of the work carried out by surveyors in public practice is to a large extent the same as that of surveyors in private practice. They are responsible for projects from inception to completion. In some authorities, the close collaboration between the different government officers may mean that the quantity surveyor is involved in the project at the very beginning.

In 'controlling' government departments the surveyor's work is restricted to the examination of estimates prepared by surveyors for subordinate authorities, and of the tenders subsequently received with a view to the department's approval being given.

Forms of contract

Most local authorities and statutory bodies use the Standard Form of Contract (JCT) or one based on that form. Central government work, however, is carried out under the New Engineering Contract, which may be used for both building and civil engineering work. Under this form of contract, the quantity surveyor performs duties that are similar to those under the Standard Form of Contract.

No matter how elaborate or comprehensive a building contract is thought to be, there are frequently unforeseen circumstances which require individual judgement and the client's approval. When dealing with an individual private client it is easier for the surveyor to discuss the matter with a view to recommending a course of action. The client is not likely to be creating a precedent and will therefore generally make a quick and usually reasonable decision. With government and, to a lesser extent, local authority work, the quantity surveyor's recommendation (whether it be a private firm or staff surveyor) often has to be referred to a contracts directorate or a committee. This is important since it may have repercussions on other projects being carried out by the authority concerned. The spending of public funds and their subsequent auditing may result in the quantity surveyor's calculations and recommendations being subject to considerable scrutiny. Whereas an individual client may be ready to meet a payment that is justifiable on moral grounds, officers in public service, who exercise the client's functions, will be reluctant to depart from the strict interpretation and guidelines of the contract, irrespective of the surveyor's recommendation. When they do apply these judgements, the payment is invariably described as ex-gratia to avoid creating a precedent.

Public service as a client

Such differences as exist between practice for a private client and for a public body largely result from the safeguards concerned with public spending and the possibility of fraud or corruption. Because of this, administrative procedures have been established to consider, examine and approve building projects and their estimated costs. This creates a requirement for professional people, including surveyors, to submit proposals in a prescribed form.

The accepting of tenders, honouring of certificates and paying of final accounts, require methods of financial control to ensure that every process has been properly authorised, within the amount approved, and with little possibility of fraud or wilful negligence taking place.

The administrative and financial controls are operated by staff who have limited training and little knowledge of quantity surveying or building contracts, beyond what they gradually acquire through their experience. Yet they find themselves answerable to committees and district auditors in local government or, in central government, to the Treasury, Select Committee on Estimates, Exchequer and Audit, the Public Accounts Committee and even Parliament itself. It is often the case that what seems a straightforward case to the surveyor may take a considerable time to reach settlement. The system of controlling public expenditure, particularly in government service, is over 100 years old, and some administrators and economists think that fundamental changes are now due. Whatever the procedure, the surveyor must provide the information, advice and help considered necessary in the interests of the public purse.

Two requirements of the public service as a client particularly affect the surveyor, and no distinction is made between the surveyor in private practice or in public service. When a project is financed from public funds a specified sum is allocated for the purpose. The sum is set within cost limits, which generally should not be exceeded, and the surveyor is expected to assist the architect or engineer in keeping the cost of the project within the allotted sum. If additional money is required, a special case has to be fully substantiated and approval obtained before the authority is committed to the additional expenditure. This is particularly important when successive Chancellors of the Exchequer aim at reducing public expenditure. With private clients, although the aspect of cost control is of equal importance, the obtaining of additional funding is unlikely to be subject to such rigorous procedures. The second requirement that affects the surveyor is that of audit. Most public offices, other than government departments, have their accounts audited twice a year, once internally by members of the finance officer's staff and then externally by independent auditors. The officers who carry out the audit are mostly accountants, but others without financial or quantity surveying training are also employed. Their task is to ensure that the financial provisions of the contract have been strictly adhered to, that all payments have been properly made, and that any unexpended allowances have been recovered.

The purpose of auditing is to ensure that the correct procedures have been properly followed. In addition, the auditors must be aware of the possibilities of negligence or fraud. Because of these responsibilities it is necessary for quantity surveyors to prepare the final account in strict conformity with the conditions of the contract. They do not have the same latitude as with a private client, and cannot always use the 'give-and-take' methods of balancing trifling or obvious self-cancelling variations. Only an experienced quantity surveyor can judge the fairness of 'give-and-take' methods, and the auditor cannot be expected to have this skill. As public money is being spent, each account and payment must not only be correct but be seen to be correct.

Comparison of public and private practice

Surveyors in public service have different concerns and anxieties to those in private practice. They may not have the difficulties of securing the necessary capital, finding offices, ensuring a flow of work and avoiding financial losses. But the public sector is subject to the whims of government and its ministers, and ideas are some-times implemented for political rather than for practical or purposeful gain.

Though the variety of work in some public offices may be limited, there is some satisfaction in being concerned in a continuous programme of national or local public works and, though the 'client' may always be the same, at least the surveyor may have a better understanding of their requirements. In the larger offices the work is varied, but if the staff are specialised, surveyors may be confined to a limited range of duties and responsibilities. The surveyor's great responsibility in the public service is that of controlling the expenditure of public funds, money placed at the disposal of departments who must assume that this is being expended wisely and carefully.

Contracting surveying

The organisation of building and civil engineering companies varies considerably from firm to firm. Some of the larger firms, for example, may almost be general contractors, while other firms with comparable turnovers of work may do very little of the construction work themselves, relying almost entirely on subcontractors. The smaller firms will often expect a wider range of skills from almost everyone they employ. The contractor's surveyor may in some companies undertake a specialised range of tasks, but in other firms may be expected to undertake work that is nor-mally outside the periphery of quantity surveying. The size and type of the con-tracting firm is therefore a very important influence on the surveyor's work.

The second important factor affecting the surveyor's work is the management structure of the firm. In some companies quantity surveying is seen as a separate function and is under the direction of a surveying manager. In other firms the quantity surveyors work with other disciplines under the authority of a contract manager.

Conditions of employment

The conditions of employment for the majority of quantity surveyors working for contractors are different to those of their counterparts employed in professional practice or public service. The most obvious difference is that a contractor's surveyor is likely to spend more time on site. In many cases they may be resident surveyor on a single project, or using this as a base for a number of other smaller contracts.

Being a member of the contractor's project team also means more interdisciplinary working at all levels of the surveyor's career. They may, for example, share an office with someone from a different profession, and therefore have the opportunity of gaining a better understanding of other people's work in the industry.

Many contractors also expect their quantity surveyor to work the same hours as their other employees on site. This often means an earlier start and a longer working week than those surveyors employed by the professional consultants. Furthermore, it is unusual to receive overtime payments on those frequent occasions where it is necessary to work late in the evenings or at weekends.

Contractors' surveyors do, however, tend to enjoy a greater amount of freedom in which to do their work, and they may also receive extra responsibilities earlier in their careers. Some contractors do, in addition, pay a site allowance, especially in those circumstances where the surveyor may have to live away from home. Contractors' surveyors are generally thought to be financially better off than surveyors working in private practice. This will, however, vary at different times during a career and the economic cycle, and will need to take into account the various differences in the respective conditions of employment.

For a contractor's surveyor who works with one of the large national companies, there may also be the disadvantage or advantage (depending on how one looks at it) of needing to be more mobile. It will be necessary to move to where the work is located, and this could require the family moving house frequently throughout a career. The building industry tends to be more regional in nature and it may therefore be reasonable to expect only to work on sites within a certain location. Many of the small to medium sized firms tend to work only within a limited region. Civil engineering works tend to be more nationally organised, but such larger projects may tend to offset the otherwise more frequent mobility of staff. Contractors' surveyors working on industrial engineering projects are often confined to those areas where these industries are located.

Role

The role of the contractor's quantity surveyor is somewhat different from that of the professional or client's quantity surveyor. They are unashamedly more commercially minded, and sometimes the financial success or failure of a project or even a company is due in part at least to the work of the contractor's surveyor. While the client's quantity surveyor may claim impartiality between the client and the contractor, the contractor's surveyor will be representing their own employer's interests. Prudent contractors have always employed quantity surveyors to look after their commercial and financial interests, and have particularly relied upon them in the more controversial contractual areas.

Function

Contractors employ quantity surveyors to ensure that they receive the correct payment at the appropriate time for the work done on site. In practice the quantity surveyor's work may embrace estimating and the negotiation of new contracts, site measurement, subcontractor arrangement and accounts, profitability and forecasting, contractual disputes and claims, cost and bonus assessment, site costing and other matters of a management and administrative nature. The various activities of a contractor's surveyor may include estimating, financial management, site costing and bonusing, contract management, negotiation with suppliers and subcontractors, interim certificates and payments, contractual matters and the preparation and agreement of claims. The detailed aspects of the contractor's surveyor's work are further considered in the relevant chapters.

Design and build

There are now many construction companies who are actively engaged in design and build, speculative development and similar sorts of ventures. Contractor's surveyors in these companies are now more likely to become more involved in pre-contract work in some way or other. They may be responsible for advising the designer on the cost implications not only of the design but also of the methods which the firm may use for construction on site. They will also be consulted on the contractual methods to be used and the documentation that is required, and will recommend the ways in which the scheme might be financed, taking into account the speed of completion.

The importance and increased trend towards single point responsibility from clients in the construction industry have made everyone more aware of this aspect of procurement. Private practice quantity surveyors are often heavily involved, either directly on behalf of the client or through the engagement of contractors who request their services in documentation preparation for tendering purposes.

Work for subcontractors

Since a large amount of work is now undertaken by subcontractors, this section of the construction industry benefits from using quantity surveyors. The surveyors who work for these companies will undertake a far wider range of duties and may well be concerned with all aspects of a financial nature, including negotiating with the bank, the submission and agreement of insurance claims, VAT and dealing with the companies' accountants. In other capacities they may be responsible for general management, the allocation of labour to projects, planning their work and dealing with their wages.

Future prospects for contractors' surveyors

The future prospects for the contractors' surveyors are seen to be very good, for several reasons:

- The increase in the direct employment of contractors for design-and-build arrangements and management contracts places a greater onus and more emphasis on their work. In these circumstances it may become necessary for them to become more involved in precontract activities.

- The fact that contracting margins are very small. Companies tend to rely upon the surveyors for expertise to turn meagre profits into more acceptable sums, and loss-making contracts into profit. This is often achieved through the active management of cash flows.
- Because of their proven performance there are now wider opportunities for them amongst a whole new range of firms who can utilise their skills.

Contractors' surveyors are employed at the highest levels within firms, which in itself emphasises their importance and their potential for the future. Many contracting firms now have at least one director whose professional origins have been in quantity surveying. Such a director may be responsible for the surveying and estimating departments in a company or more commonly a role in commercial management, and is likely to be concerned with the financial decisions and management of the construction company. Many medium-sized firms have seen the importance of giving their chief quantity surveyors considerable responsibility in financial matters, with their importance overall in the firm being recognised by a directorship.

Relationships and the client's quantity surveyors

The contractor's surveyor's first contact with the client's quantity surveyor will probably be in raising queries on the contract documentation, particularly if it is the surveyor who has priced the work. Upon receipt of the acceptance of the contractor's tender, the contractor will inform the architect of the names of the contracts manager, agent, quantity surveyor and other senior personnel who will be responsible for the project.

The contractor's surveyor and the quantity surveyor would then normally meet to discuss points that were raised at tender stage or during a post-tender meeting. These would include matters relating to the correction of errors in pricing, the procedures to be followed regarding remeasurement and financial control, and the dates of interim valuations.

The relationship between the two surveyors will usually start amicably enough, and this is the way it usually ends, but there is always the possibility of deterioration in the relationship developing. This may occur because one party does not feel that the other is being as fair or as scrupulous as they ought to be.

It needs to be remembered that the two surveyors are trading in a very desirable commodity: money. While fairness is an excellent motive to follow and should help to reduce the number of potential problems arising, it must also be remembered that there is a contract to be considered, and auditors will need to be satisfied that correct procedures have been followed. Each party to the contract, the client and the contractor, has employed its respective surveyors to ensure that its own interests are properly protected. On matters of dispute the advice may well be to settle these differences on a give-and-take basis.

It is all too common nowadays for disputes to lead to adjudication or arbitration or indeed to litigation. Before this happens both surveyors should warn their employers of the considerable financial costs of legal disputes.

The professions

The professions are one of the fastest growing sectors of the occupational structure in Britain. At the turn of the twentieth century they represented about 4% of the employed population. One hundred years later, at the start of the twenty-first century, this had risen to 20% of the working population, although the definition had become much broader. Several reasons are given for the rapid growth of the professions, such as an increasing complexity of commerce and industry, the need for more scientific and technical knowledge and an improved desire for greater accountability.

However, the professions are vulnerable as information and communications technology continue to develop and advance. Some of their more repetitive functions have already been subsumed and much of their specialist knowledge can already be readily accessed. The understanding and application of this knowledge is more difficult, but progress continues to be made through the use of, for example, expert systems. Easy access to telecommunications allows their work to be carried out in the most economical parts of the world. In addition, the blurring of professional boundaries means that non-specialist disciplines are now competing for a share in the workload.

The concept of a professional person is continually evolving. It is generally meant to mean someone who offers competency and integrity of service at a high level coupled with the use of skilled intellectual techniques. The relationship between them and the people whom they serve is one of mutual trust. However, in recent years there has been a blurring of professional and business roles, but the distinction between professional and commercial roles still exists.

It is relatively easy to identify those occupations in society which constitute a profession but much less easier to explain why some groups are included and others are not. The mere fact that an individual may belong to a professional society does not automatically mean that one is a professional. There are four essential attributes which generally account for professional status.

- A body of systematic knowledge which can be applied to a variety of problems. For example, structural engineers have a body of knowledge which they can apply to projects to determine the structural behaviour of a part of a structure.
- Professionalism involves a concern for the interests of the community at large rather than just self-interest. Thus, the primary motivation of a professional is in the interests of the client rather than personal gain. On some occasions it may be necessary to advise a client not to go ahead with a project and thus lose a commission rather than to develop a project which is not worthwhile to a client.
- The behaviour of professionals is strictly controlled by a code of conduct or ethics which is maintained and updated by a professional association and understood by those in training as a requirement of qualification as a full professional member. If a member breaks the code, then the association can impose restrictions upon their activities or ban them from practising using the benefits of membership. In the case of an architect, expulsion from ARB (Architects' Registration Board) means that the individual can no longer be described as an architect.

- Professionals receive high rewards in terms of earnings and status within society. These represent the symbols of their achievements, reflect their contribution to society and the view held by society.

In terms of their market situation, the professionals can be divided into two groups: the higher professionals and the other professionals. The higher professionals include the chartered construction professionals as well as accountants, lawyers, doctors, university lecturers, etc. There are significant earning differences between the two groups. Measured in terms of earnings, the higher professionals as a group do extremely better than the other groups and in addition enjoy better fringe benefits such as company cars and health care schemes.

The built environment professions

The built environment professions in Britain are many and varied; they represent a distinctiveness about the industry and are a matter for much debate. There are in the order of 300 000 members and students amongst the seven main chartered professional bodies who work in the construction industry. In addition to these chartered bodies there are a further 20 non-chartered bodies competing for a share of the work. The RICS is the largest professional body in construction and property with over 110 000 members.

It is sometimes argued that the difficulties which arise in the industry are due, at least in part, to the many different professional groups which are involved. Others argue that the services which the construction industry provides are now so specialised that one or two different professional groups would be inadequate to cope with the complexities of practice. Whilst the UK is different in this respect to many countries around the world, there is no standardisation of practice and considerable differences exist, even across the different countries of western Europe.

The quantity surveying profession

There were approximately 35 000 members of the former Quantity Surveyors' Division of the RICS. It is difficult to estimate how many other quantity surveyors there are in the UK and around the world, but the figure is probably in excess of 300 000. These include members of other professional bodies, such as the Chartered Institute of Building in the UK, various professional bodies overseas and quantity surveyors who have no professional affiliation.

There has been a shift in employment patterns from the public sector towards contracting (Fig. 2.4). Whilst the profession weathered the economic storm of the early 1990s better than many other professions, it has nevertheless suffered in terms of both employment and remuneration.

In comparison with other professions, such as engineering, accountancy and law, quantity surveying private practices are relatively small. The largest quantity surveying practice has about 3000 staff world-wide, and this is exceptional. By comparison the largest British engineering consultant has almost 5000 staff world-wide.

	1981	1991	2001
Private practice	54	53	48
Public service	22	15	12
Contracting	17	20	26
Commercial	5	10	12
Education	2	2	2

Fig. 2.4   Employment patterns of quantity surveyors (%) (*Source:* RICS).

On an international scale some of the larger accountancy, law and management consultancies employ in excess of 10 000 staff. At the other extreme the smaller practices with fewer than ten staff account for between 50 and 60% of quantity surveying private practices. There has been a trend to develop multi-disciplinary practices over the past decade.

Quantity Surveying Institute (QSi)

QSi was formed in 2004 as a professional body solely for quantity surveyors working in the construction, engineering and ship-building industries. It is the only professional body in the UK that devotes all of its time and efforts to quantity surveying.

It was formed, to some extent, as a result of dissatisfaction in the RICS. Whilst the fees in the RICS had been increased substantially (about six times higher than QSi), the feeling that quantity surveyors were being neglected by the RICS was the main driver for its formation. When the RICS decided to scrap its skills based divisions, the quantity surveyors felt out on a limb. Six years later the RICS did an about turn and renamed the Construction Faculty the Quantity Surveying and Construction Faculty. There was a feeling amongst RICS officers prior to this that quantity surveyors should change their name. Engineers, architects and accountants don't change their names, even though their roles are changing (see Chapter 1). There was the danger that as the role of the quantity surveyor continued to evolve, any name change might soon become outdated.

QSi has about 1000 members, which is of course tiny compared to the RICS, and a fair sized group within the Chartered Institute of Building. The focus of QSi is not directly on private practice but includes those working in central and local government and with contractors and subcontractors.

The QSi's key objectives are to:

- Support and protect the character, status and interests of the profession of quantity surveying
- Promote good practice and high standards from its members
- Consider all questions affecting the interests of the profession and to petition Parliament to promote legislation affecting quantity surveyors or their profession

- Australian Institute of Quantity Surveyors
- New Zealand Institute of Quantity Surveyors
- Singapore Institute of Surveyors and Valuers
- Institution of Surveyors in Malaysia
- Japan Institute of Quantity Surveyors
- Sri Lankan Institute of Quantity Surveyors
- Hong Kong Institute of Surveyors

Fig. 2.5 Membership of PAQS.

- Disseminate to members information of a technical and contractual nature which relates to their work
- Publicise throughout the construction industry the benefits that can be derived from the employment of quantity surveyors
- Promote the employment of quantity surveying skills in helping to make the construction industry more efficient.

Pacific Association of Quantity Surveyors (PAQS)

The Pacific Association of Quantity Surveyors is an organisation of quantity surveyors located around the fast developing Pacific Rim countries. They share information and have intentions to develop common accreditation procedures that allow accreditation by one of the professional bodies to be accepted by the remainder. The membership of PAQS is shown above in Fig. 2.5. Other quantity surveying institutions around the world are associate members.

The future of the built environment professions

The built environment professional bodies grew steadily, both in number and in membership, throughout the twentieth century. It can easily be argued that there is a proliferation of professional bodies working in the construction industry in the UK. Their future is influenced by:

- The diversification and blurring of professional boundaries often including non-built environment professions such as those involved with the law and finance
- Their role as learned societies
- The education structure of courses in the built environment
- The pressure groups both within and outside of the construction industry
- The desire in some quarters for the formation of a single construction institute, to unify all professionals involved in construction and property
- The need to maintain standards and codes of professional practice.

Role of the RICS

The Royal Institution of Chartered Surveyors was formed in 1868 and offices were leased at 12 Great George Street, which is still part of the RICS Headquarters building today. The Institution of Surveyors which later became the Royal Institution of Chartered Surveyors (RICS) evolved into a renowned international organisation with approximately 110000 members working in 120 countries. It was granted its Royal Charter in 1881 and in 1922 the Quantity Surveyors Association amalgamated with it. In 1930 the then Institution of Surveyors became the Institution of Chartered Surveyors. In 1946 it was granted the title Royal to become Royal Institution of Chartered Surveyors. The RICS coat of arms with its motto, *Est modus in rebus* (There is measure in all things) was adopted. A number of different professional bodies have merged with the RICS over the years and in 1983 the Institute of Quantity Surveyors merged with the RICS. In 1998 the *Agenda for Change* initiative was launched by Richard Lay in his Presidential Address. The Harris Report on *The future role of the RICS* was launched later that year. In 2000 the RICS Foundation for leading-edge research was established, although this initiative was relatively short lived. In 2001 the seven divisions were replaced with the 16 faculties. Further detail on how the RICS developed can be found in Thompson (1968).

The RICS Medium Term Strategy from 2003 sets out RICS strategy up to July 2010 to achieve its vision: to be the mark of property professionalism worldwide. The Strategy covers:

- immediate recognition
- trust and confidence
- regulation and consumer protection
- promoting the public interest
- competence
- property-related information and knowledge
- good careers
- development worldwide
- governance, membership structures and relations
- resources management
- funding.

The RICS is accountable to both members and the public and has three main roles:

- To maintain the highest standards of education and training
- To protect consumers through strict regulation of professional standards
- To be the leading source of information and independent advice on land, property, construction and associated environmental issues.

It is organised around 16 faculties of which one is designated as Quantity Surveying and Construction. There are a number of other faculties to which quantity surveyors are also likely to belong. These include Dispute Resolution, Facilities

Management, Management Consultancy and Project Management. The current key objectives of the RICS are:

- Communicating and influencing
- Delivering professional knowledge
- Maintaining professional standards
- Growing membership
- Running effective operations.

Whilst these objectives are still being agreed in terms of corporate targets, the framework put in place aims to:

- Increase qualified membership, while upholding standards
- Increase profit delivery from business services
- Improve operational efficiency and effectiveness by ensuring operating costs and staff costs as a proportion of income are in line with other benchmarked professional bodies
- Enhance status – nationally and regionally across the UK as well as outside the UK
- Improve member satisfaction, in particular in respect of communications, local service delivery and professional knowledge/information.

Further information about the RICS can be found at www.rics.org.

The RICS remains the premier institution for quantity surveyors. This is due largely to history and also to size. The RICS remains one of the largest professional bodies around the world. In terms of construction professional bodies in the UK, the RICS is by far the largest and most influential with a total membership of over 110 000. Quantity surveying professional bodies also exist in several countries around the world, most notably those that were members of the former commonwealth. The RICS liaises with these bodies at different levels and there are many RICS members who hold dual qualifications. Powell (1998) identified the following as key requirements expected of the RICS and other local surveying bodies:

- Improving education and entry level to respond to challenges likely to be met by individuals and firms
- Change from protectionism to facilitator
- Provide leadership in a fragmented construction industry
- Accept that it is the commercial world that is the leading edge but use its position and authority to influence and create and spot trends
- Reflect the need for individuals to remain champions of propriety and fairness and uphold the charter
- Constantly upgrade professional skills, business skills and knowledge to a level which is envied by other professions
- Promote the benefits of employing a chartered surveyor so that the currency value of belonging to the RICS is as high as possible
- Improve communication and develop the single profession culture amongst its members.

Discussion topic

What is meant by ethical behaviour and how is this viewed in the construction industry?

Ethical behaviour initially and intrinsically stems from a person's own value system. This can be simply described as knowing the difference between right and wrong and having the courage to do what is right. There is never a right way to do the wrong thing. Simply put, that's what ethics is all about. In some ways, society has almost become desensitised to unethical behaviour and come to expect it, especially after the effects of the behaviour of notable companies such as Enron and WorldCom.

Personal ethics tell us that if we are going to get along with one another we shouldn't lie, steal and cheat. In business, the same principles apply. Business ethics are based upon a willingness to live up to our word and provide all the necessary information so that the other party can fulfil their obligations in a fair manner.

Smith (1992) in *Accounting for Growth* identifies some dubious practices in the world of accountancy. He identifies internationally well-known companies who were supposedly performing very well financially but yet, in reality, were almost in a state of bankruptcy. Whether the auditors collaborated with these practices is unknown. What was known was that they were not doing their jobs ethically. One of the former big five accountancy and management consultancies was virtually forced out of business because of unethical practices that were subsequently revealed.

James (2003) notes two types of integrity: personal and process. Personal integrity is when a person's words and actions are the same, and those same things are the right things. Process integrity is what companies trust to produce the right results. Unfortunately, many companies in the construction industry have lousy processes, and lousy processes allow bad things to happen.

According to Doran (2004) in his *Survey of Construction Industry Ethical Practices*, which was an in-depth study of ethics in the construction industry, he found that the perception of the industry was that it was riddled with unethical practices. It is clear from this survey that this behaviour is both recognised by most, accepted by some, but few want to do anything about it.

According to the responses from 270 owners, architects, engineers, consultants, construction managers, contractors, and subcontractors, Doran's theory is correct. When asked if they had experienced, encountered, or observed construction industry-related acts or transactions that they would consider unethical in the past year, 84% of respondents said they had, and 34% indicated they had experienced unethical acts many times. The construction industry needs to pay more attention to ethical issues since the industry has gained a poor and nasty reputation. The respondents to this survey stopped short of saying the industry was full of criminals. A large proportion agreed or strongly agreed that the construction industry is tainted by prevalent acts that are considered unethical (Fig. 2.6).

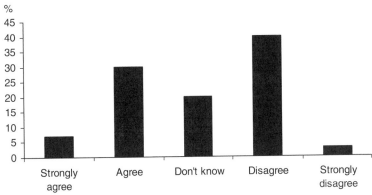

Fig. 2.6 Is the construction industry tainted by illegal acts? (*Source:* Doran 2004, reproduced with permission).

Loosely defined as the discipline dealing with what is good and bad and with moral duty and obligation, ethics, whether in one's personal life or in business, come down to doing the right thing, as governed by rules, guidelines, laws and principles for good moral conduct. In the construction industry, ethical behaviour is measured by the degree of trustworthiness and integrity with which companies and individuals conduct business. Professional bodies such as the RICS have their own Rules of Conduct (see page 91), to guide its members' behaviour.

When you look at the nature of the relationships between owners, general contractors and subcontractors in the construction industry, Doran suggests that it is not surprising that ethics get so complicated. The subcontractors, who were the largest stakeholder group that responded to the survey, are a little further down the line from the client and feel that they are subjected to more unethical acts.

The survey also found a large discrepancy between the value people place on ethics and what they actually do in practice to support their values. Although the large majority (85%) believe there should be an industry-wide code of ethics, only 30% agreed that adding regulations concerning ethical behaviour is necessarily a good idea. (See also Contracts of employment in Chapter 4.)

As far as enforceability of ethics goes, this endeavour is easier said than done. Cavill and Sohail (2006), when studying corruption in the construction industry on an international scale, indicate that the difficulty of eradication at all levels is considerable. Their general definition of corruption is the misuse of public office for personal gain either at one's own instigation (e.g. extortion) or in response to inducements (e.g. bribes). Nye's definition (1967) of corruption refers to behaviour that deviates from the formal duties of a public role. This is because of a private role regarding pecuniary or status gains. Ley (1965) refers to corruption as behaviour that breaks some rule, written or unwritten about the proper purpose to which a public office or institution has been placed. Cavill and Sohail (2006) identify specific forms of corruption that include:

- Bribes: payments made in order to gain an advantage or to avoid a disadvantage
- Fraud: theft through misrepresentation

- Embezzlement: misappropriation of corporate or public funds
- Kickbacks: sweeteners or rewards for favourable decisions.

All businesses have at least unwritten codes of behaviour that determine how they deal with others, both within a firm and those who are external to it. These represent a set of personal ethical standards that are going to be the reflection of the employees of the company. Employees need to be trained in these ethical standards, of what are acceptable and what are unacceptable practices.

In order to minimise the chances of unethical or illegal behaviour in the construction industry, respondents agreed that several initiatives should be implemented, including stiffer penalties for those caught in unethical or illegal acts, an industry-wide code of ethics, more emphasis placed on social responsibility in award criteria and increased levels of training.

References and bibliography

Ashworth A. *Education and Training of Quantity Surveyors*. Chartered Institute of Building. 1994.

Cavill S. and Sohail M. Accountability arrangements to combat corruption and improve sustainability in the delivery of infrastructure services. CIB Building Education and Research, Hong Kong. 2006.

Davis, Langdon and Everest. *QS2000*. RICS. 1991.

Doran D. *Ethical Practices Today: Survey of Construction Industry Ethical Practices*. FMI Consulting. 2004.

Hubbard C. *Investing in Futures*. The Education Task Force Final Report. RICS. 1999.

James R. *The Integrity Chain*. FMI Consulting. 2003.

Kolesar S. *The values of change*. Presidential address. RICS. 1999.

Ley C. What is the problem about corruption? *Journal of Modern African Studies*, 1965.

Nye J. Corruption and political development: a cost–benefit analysis. *American Political Science Review*, 1967.

Powell C. *The Challenge of Change*. RICS. 1998.

RICS. *Degrees and Diplomas Accredited by the RICS*. Royal Institution of Chartered Surveyors. 1999 (and subsequent updates).

RICS. *The Future of Surveying Education*. Royal Institution of Chartered Surveyors. 2006.

Smith T. *Accounting for Growth*. Business Books, London. 1992.

Thompson F.M.L. *Chartered Surveyors: the Growth of a Profession*. Routledge and Kegan Paul. 1968.

Venning P. *Future Education and Training Policies*. RICS. 1989.

3 Organisation and Management

Introduction

Setting up and expanding a practice or business, together with general management and finance, are all subjects adequately dealt with in the many textbooks on business management. However, there are certain specific aspects of organisation and management relating to the work of the quantity surveyor that are worthy of consideration in some detail. They include:

- Staffing
- Office organisation
- Marketing
- Management of quality, time and cost
- Education and training
- Finance and accounts.

Staffing

The staffing structure within a quantity surveying practice, public service or contracting organisation will comprise members of staff carrying out the quantity surveying or other specialist service and those providing the necessary support services such as administration, accounts and information technology. The overall staffing structure will differ from one organisation to another, but certain general principles will be similar.

Private practice

The way an office organises and carries out its surveying work will to some extent be determined by the size of the practice and the nature of its commissions. The larger the practice, the more specialised the duties of the individual surveyor are likely to be. There are essentially two modes of operation, as shown in Fig. 3.1. Type (a) separates the normal surveying work on a project into the cost planning, contract documentation, final account work and specialist services, e.g. value management, taxation advice. Some coordination will be necessary between the phases of a project, as different personnel will carry out the work in each. This type of

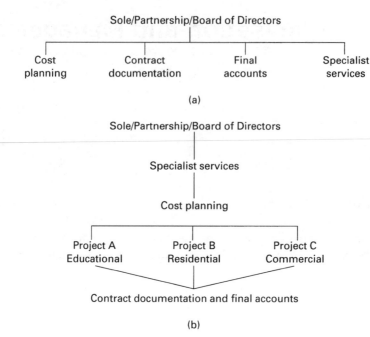

Fig. 3.1   Modes of operation of office organisation.

organisation gives staff the opportunity to develop their expertise in depth and, with regard to the growing range of specialist services in larger practices, this may be essential.

Type (b) allows the surveyors to undertake all aspects of the project from inception to final completion and provides the surveyor with a clearer understanding of the project. This alternative will often allocate the staff to teams, who then become specialists in certain types of project, depending of course on the type of work undertaken and the workload at any one time. This model does not easily accommodate the development of specialist service expertise, which requires concentration of activity rather than a more generalist approach.

Most practices today also carry out other work that may be incidental to their main work, possibly because of the interest or particular skills and knowledge of members of staff. In other instances specialist work represents a major source of work for the practice.

Whichever procedure is used by the office for carrying out its work, some programme of staff time will be necessary if the services are to be carried out efficiently to meet the project requirements. The first management aid needed to plan and control the work is a planning and progress chart. This is a linear timetable, which breaks down the project into its various parts, and to be complete it must take into account all aspects of the project, including the preparation of interim valuations and the final account. It may be desirable in the first instance to prepare the chart on the basis of pre-contract work alone, but for overall planning purposes some account must also be taken of the other duties that have to be performed. The type, size and complexity of the project will need to be taken into account in planning the work that has to be done.

The overall management of the practice and the well being of the employees are the responsibility of the partners/directors. Each project will be the responsibility of an associate partner/director, or at least one of the senior surveying staff who, having identified the work that needs to be carried out within a particular time, will determine the resource level required to perform the various tasks, either using staff within the office or calling on freelance staff as necessary. This is likely to be a necessary aspect of calculating the fee when bidding for work or negotiating with clients. It is also essential to pre-plan resources in order to budget costs against anticipated fee income. A major problem often facing the practice is the restricted time available to produce tender documentation and the like. Being at the end of the design process, they have to rely upon the other members of the design team to supply their information in good time. This emphasises the importance that should be attached to an adequate programme of work, by identifying the work that has to be done, who will do it, and the date by which it should be completed.

Support services

In addition to the surveying and other specialist staff within an organisation there will also be the support staff necessary for its efficient operation. These will include secretarial and administrative support as well as the finance and accounts and information technology departments (information technology is covered in Chapter 5 and elsewhere in this book). However, the development and influence of IT systems on the quantity surveying and office administrative functions means that the balance of the general staffing structure is continually changing.

Office organisation

General organisation

Quantity surveyors, whether in private practice or in a support department within a contracting organisation, provide a service, and therefore the direct staff costs and indirect costs (such as employers' National Insurance contributions and pension contributions) form the largest part of the cost of running the practice or department.

The general establishment, expanding and equipping of an office are outside the scope of this book. However in addition to staffing costs there are significant further costs associated with the running of an office, and it is worth noting the main items that constitute the general office overhead.

The major component of overhead expenditure is likely to be the office accommodation in terms of rent and the cost of furniture and general equipment, IT equipment (which has a very high rate of obsolescence) and communications necessary for the office to function efficiently. Other items of overhead expenditure are likely to include administrative costs, marketing and general expenses. There are several factors to be considered in selecting an office location and quality of premises and often, particularly to the new practitioner, there is a balance to be achieved between desired image, the realistic forecast of client base and fee income.

Specialist stationery

In addition to general stationery, there is specialist stationery for use by the surveying staff in performing various tasks. The extent and format of this will be dependent upon the level of use of information technology and the scale of activities. Pro-formas and specialist paper will be required for most activities including cost planning, measurement, final accounts and specialist services such as value management. When selecting such materials, thought should be given to the ability to photocopy, store and use it in IT systems.

Reference books and information

A characteristic of a quantity surveying office or department is the specific need for reference books and cost and technical information to assist the surveyors of specialists in performing their duties. Reference libraries will have been developed over a number of years and it is important that they be kept up to date to ensure that the staff can be fully aware of current developments in all aspects of the construction industry.

The core material that is required for reference, and which will need frequent updating, includes the following:

- Textbooks on the principal subjects encountered in the office
- British Standards Institution Handbooks
- Construction price books
- Building Cost Information Service (BCIS) cost information (also available on-line)
- Current editions of all forms of contract
- Wage agreements
- Plant hire rates
- EU directives/publications
- Professional and general construction industry journals
- Technical literature on products, materials and components
- Specifications.

Increasingly, much of this information is becoming available on-line although hard copy formats are preferred by many surveyors in practice. The management of information is a major and important task and should not be done casually.

Building costs records

A certain amount of keeping of records is essential in any office, though there is a tendency to delay preparing them because they are not directly productive.

They have a habit of never getting done when the information is fresh. The surveyor's own cost records will be of much greater value than any amount of published cost data. They relate to live projects and can be stored in a manner that easily facilitates retrieval. Some care, however, needs to be exercised when using them because there could have been special circumstances which may or may not need to be reflected in any re-use of the information. It is, however, more likely to reflect the pricing in the local area and therefore in this context it is more reliable.

It is good practice to prepare a cost analysis of every tender received, and these analyses will form a bank of useful cost records. If prepared in accordance with the principles of analysis prepared by the BCIS then some measure of comparability with their cost analyses can be achieved. Different clients may require different forms of cost analysis to suit their particular purpose. Supplementary information on market conditions and specification notes should also be included. These will be necessary in any future comparison of schemes. One of the outcomes of changing procurement trends, in particular the increased use of design and build in lieu of the traditional method, is the reduction in quality of information for analysis purposes.

It is important when compiling cost records to maintain a constant basis. The use of the BCIS principles will greatly assist in this. For instance, it may be desirable or necessary to omit external works, site clearance or preparatory alteration work from the price. The analysed tender sum must, however, make this omission clear.

Refurbishment projects do not lend themselves so well to records of this kind, and it is therefore preferable to provide an analysis in a rather different way, perhaps by identifying the cost centres for this type of project. Although these types of scheme are commonplace today in the UK, a standard form for their cost analysis has yet to be devised. Their unique nature, in terms of existing design, condition and constraints, is such that estimating by means of previous cost analysis will in any event require considerable adjustment and frequently reversion to resource based techniques.

When the building scheme is complete a final cost can be recorded, and this may also be calculated as a cost per functional unit or superficial floor area. No attempt should be made to recalculate the cost analysis, as the project costs by now will be largely historic and the allocation of the final accounts to elements is very time consuming.

Building Maintenance Information (BMI) have a form of cost analysis for the maintenance and running costs of buildings. Surveyors concerned with costs-in-use or the occupancy costs of buildings may need to refer to or prepare analyses in this way.

Employer's responsibilities

There are certain health and safety and insurance issues to be addressed by employers.

Health and safety

There are continuing amendments to the statutes concerned with health and safety matters that are relevant to offices, and employers should establish the current legislation and ensure adherence. In general terms, there is responsibility on the employer in respect of the safety, health and welfare of employees and this incorporates a wide range of obligations including the following:

- The employer must ensure that any plant or equipment is properly maintained
- The employer must ensure that the systems of work are safe
- The employer must ensure that the entrances and exits are easily accessible
- Relevant instruction or supervision should be provided where required to avoid misuse of equipment, etc.
- In addition to employees, the general public visiting the premises must also be properly safeguarded from any risks
- The need to provide adequate heating, lighting, ventilation, sanitary conveniences, washing facilities and drinking water
- The provision of minimum space standards to prevent overcrowding, making allowances for space occupied by furniture, fittings and equipment
- The need to provide reasonable means of escape in the event of fire for all office premises.

In some buildings some of the occupier's responsibilities are transferred to the owner of the premises, who is also responsible for complying with the provisions of legislation for the common areas, such as entrances, passages, stairways and lifts. Environmental health officers from the local authority are responsible for enforcing the legislation in private offices.

With regard to the current fire precaution related legislation, where premises fall within the relevant Act an application must be made to the fire authority for a fire certificate. The authority will then visit the premises to assess the means of escape and other fire prevention measures and make recommendations where necessary.

This is by no means a comprehensive review of health and safety related matters. The demands on employers are increasingly onerous and failure to comply with health and safety requirements may lead to both corporate and personal prosecution. There are also changing work practices, which need attention. For example, one aspect that is becoming a major concern relates to the extensive and concentrated use of workstations and VDUs. This practice may result in health problems to users requiring consideration of several factors including workstation ergonomics, light quality and background noise. There are several comprehensive web pages that may now assist in updating on health and safety matters, including those of the Health and Safety Executive and the HMSO (e.g. for details of the Fire Precautions (Workplace) Regulations 1997).

Fire insurance

Insurance of the building will usually be the landlord's responsibility, but the tenant will need to cover the contents of the office against damage by fire and burglary.

The general furniture and stationery will present no difficulty; however, a serious consideration in the event of a fire is the replacement of documents in the office, particularly those referring to contract documentation in the process of preparation. If everything is destroyed there may be no alternative to beginning again. The client of course will not pay a double fee. It is therefore advisable to insure the documents in the office under a special item. It will be very difficult to assess a figure but it should be substantial, the premises being a comparatively small matter when the contingent liability is considered. Only the cost of their replacement as it relates to salaries and overhead costs will be covered, unless a 'loss of profit' policy is taken out.

The impact of damage or theft relating to IT installations is likely to be very significant. Suitable provision should be made to minimise such risks including adequate back-up systems, preferably in separate and secure locations.

Employer's liability insurance

An employer is liable to pay compensation for injury to any employees arising out of and in the course of employment, caused by the employer's negligence or that of another member of staff. In order to be able to meet these obligations the employer is required by law to take out a specific insurance under approved policies with authorised insurers. Premiums are based upon the type of employee and the salaries that are paid.

Public liability insurance

An owner or lessee of premises may be liable for personal injury or damage to the property of third parties caused by their negligence or that of a member of staff. The insurance cover should be sufficient to cover the different status of individuals, and actions by both principals and employees not only on the premises but anywhere while on business, and in countries abroad if this is appropriate.

Professional indemnity insurance

A conventional public liability policy does not normally cover any liability where professional negligence is involved. There is a requirement for members of the RICS to maintain approved professional indemnity insurance.

Marketing

In the current climate of changing professional roles and status within the construction industry, the marketing of quantity surveying and other construction-related professions has taken on added importance. The marketing of professional services is a difficult task, however, as it involves selling a service rather than a product.

As quantity surveyors look to diversify the services they offer, it is necessary for them to recognise their specific strengths and weaknesses and those of their competitors, and to identify potential clients (in both existing and new market sectors) together with their needs and requirements. The services provided can then be tailored to meet these requirements and a marketing strategy can be targeted accordingly.

It is essential to have a clear marketing plan, to develop the marketing and presentational skills of key individuals within the organisation, and to maintain an awareness of potential business growth.

Corporate image

Quantity surveying practices, whether consciously or not, present an image of themselves, both to others employed in the construction industry and to new and existing clients. This image may be developed from the way the firm's partners or directors see the future role of the profession of quantity surveying. Some practices, particularly smaller ones, are therefore content to see themselves in a traditional role, offering a service that is still in demand. Others have developed new skills and services and offer a much wider portfolio of advice and surveying services to meet the expanding needs and expectations of their clients, which in turn are being changed by active marketing from within the industry.

The image of the firm is therefore reflected to a large extent by the aspirations of the partners or directors and their collective experience, together with the commissions that they have undertaken. The expertise of the other members of the practice must also not be underestimated, as they are able to contribute to and enforce this image. It is important to have available details of staff development and achievements to reinforce the image. The important qualities to be portrayed are those of integrity, reliability, thoroughness and consistency.

The types of project, services provided and methods of working are important aspects to consider, and ones in which future clients are likely to be interested. Clients will also want to know how the practice keeps to programmes and costs and how good generally its advice has been. They will wish to identify its various strengths and weaknesses, and will be interested to know how the practice has managed changes in technology and perhaps how this will affect the future method of working. Surveyors should recognise areas of service that may be used by clients as performance indicators and should ensure that their related output is to a high standard. For instance, accuracy with estimates will be seen by most clients as vital, as will work affecting the achievement of deadlines in design and construction programmes. Perhaps some practices have in the past concentrated on other issues of professional practice at the expense of those that are seen as more important by clients.

The firm's track record, future potential and ability to solve problems should be conveyed in such a way that the practice demonstrates that it is the most appropriate to undertake the client's work, will suit the client's method of working, and will provide the quality of service commensurate with the fee charged.

Public relations and marketing

The surveying profession generally has done little in the past to advertise its skills to the public. The majority of the public therefore still presume that surveying equates with land surveying, and among the other established professions outside the construction industry quantity surveying is still largely unknown. The increasing diversity of surveying activities has also made it difficult to clearly identify a common role. Some practitioners are involved in the traditional areas of the profession whilst a growing number are involved in the provision of a much wider range of services. In addition, the role of the contractor's surveyor, which in some respects has been previously neglected by the profession, is also diverse and a growth area.

The RICS has invested, and intends to continue to invest, in the promotion of all areas of surveying. The quantity surveyor is represented via the 'Quantity Surveying and Construction Faculty Board', one of sixteen such groupings. This faculty board aims to reflect the many functions of the surveyor within the construction sector. The Institution has adopted a more positive and proactive role in support of its members than its hitherto relatively passive position. It now seeks to ensure that the major financial institutions, development banks and aid agencies are continually aware of the surveyor's essential contribution to the development and construction process. It would be of even greater advantage to ensure that the quantity surveyor's services are properly understood and appreciated by those with whom contacts are first made. Client organisations must be encouraged to make full use of quantity surveying services and be convinced that they do provide value for money in all commercial and contractual aspects of construction.

In 1991, the RICS published *Quantity Surveying 2000: The Future Role of the Chartered Quantity Surveyor* (Davis, Langdon & Everest 1991), which through research studies reviewed the changing markets, the changing nature of the construction industry, the changing profession and concluded with how the profession should respond to these changes through diversification into new markets and provision of new services. The publication predicted that services such as facilities management and value management would be major surveying activities, by the year 2000 if taking the title of the publication literally. Whilst these predictions are far from reality, it is clear that the profession is developing at a pace with new markets (e.g. PFI) and services emerging continually.

The regulations on advertising and publicity for the function of quantity surveying have been significantly relaxed by the RICS over the years. However, care must still be taken to ensure that all statements made about a firm are an accurate representation. Advertisements must not seek to explicitly solicit instructions, nor compare a firm's services with those of others.

Public relations and marketing are seen as vitally important to the recruitment of able young people into the profession, including women and ethnic minorities, who, within the construction sector generally, are greatly under represented. The current levels of recruitment into the quantity surveying profession remain at a critical level although there has been a rise in the number of undergraduates undertaking quantity surveying programmes of study since 2002. Despite this, most major practices and contracting organisations are finding it difficult to recruit from a dwindling supply of graduates.

Practice brochures

The general easing of advertising restrictions presented both the quantity survey-ing profession and individual firms with new ways of marketing their services and seeking new clients. One of the most effective ways of achieving this is the prepa-ration of a brochure that describes the practice and the services that it has on offer. It can be sent out in response to enquiries or distributed when the opportunity arises. It may be necessary in the first instance to prepare the brochure in conjunc-tion with a marketing agency in order that the correct balance and effect is achieved.

The type of brochure available varies from something prepared internally by the practice and assembled into an appropriate folder, to the elaborate, professionally prepared booklet which comes complete with colour illustration. Some practices provide a variety of literature and other material, which seeks to emphasise the dif-ferent specialist services that they offer. Other practices, in addition, provide a quar-terly contract or cost review or newsletter, which aims to keep their name firmly in their client's mind.

The production of a brochure must be done with care and skill to create the correct corporate view of the practice. It should not be dull and uninteresting but should give the appearance of something that demands to be read. The informa-tion must be factual, but its presentation should be persuasive and emphasise the qualities and skills of the practice. The information may include the origins and development of the practice, partnership details, services provided, projects com-pleted, office addresses, lists of clients and photographs of partners and projects.

Presentations

A considerable proportion of new commissions are obtained through fee competi-tion, which often involves preparation of pre-qualification documents, details in support of a fee tender submission and in certain instances formal presentation to clients in person.

Although the format of presentations will differ depending upon the specific client, the project and the services to be provided, the main aim is to get across to the client general details of the practice, specific previous relevant project experi-ence and (most importantly) the capability to carry out the required role. Likewise, contractors are increasingly being required to prequalify for inclusion on tender lists. In addition to general company information and experience, they need to iden-tify their approach to the important aspects of management strategy and pro-gramming capability.

In recent years, the opportunity to improve the quality of presentations has greatly improved with the assistance of information technology. Use of presenta-tion software, for instance the ubiquitous Microsoft Powerpoint, in conjunction with a laptop computer and a compatible projector, will greatly enhance presenta-tion quality. Whilst cynics may frown upon the trend towards what they may see as over elaborate presentations, absence of such quality may reflect negatively upon an organisation. Notwithstanding these comments, there is an increasingly heard view that there is an over-use of powerpoint and certainly, good quality presentation now seem to be the rule rather than the exception. It is therefore suggested that some thought should be given to the presentation objectives and the audience before deciding on format.

Internet

Many organisations have now developed their own web pages, which may be used to market their services and provide general information to clients. How successful this device may be as a marketing tool within a very traditional industry is unknown; however, its importance will clearly grow in the coming years. In designing a web page, the care and attention required to portray the correct image, as in other areas of marketing, is very important. Other factors that should be considered include ease of use, links to other appropriate web sites and currency and quality of information provided.

Quality management

Quality management systems

Considerable attention is given nowadays to the aspects of quality of the service provided by consultants and contractors to their clients. Increasingly clients in all sectors of the construction industry are demanding that consultants and contractors operate a quality management system and obtain third-party certification to demonstrate compliance with ISO 9001.

Practices and companies wishing to obtain certification following assessment by one of the certification bodies need to demonstrate that their procedures comply with the Standard. The scope of services for which registration is sought has to be clearly identified on the application for assessment and evidence demonstrating this scope has to be presented at the time of assessment.

Quality of documentation

While quality management systems set procedures to be followed to ensure quality of service provided, another aspect that is worth considering is the quality of documentation leaving the office, as this is the chief means of communication with other organisations. There will be correspondence in the form of letters or reports issued on every project, from the first letter of instruction to a final letter sending in an account for fees. The science and art of letter and report writing constitute a subject well worth studying, but only a few points can be mentioned here.

The object of writing is to convey the ideas of one person to the mind of another, who is not present to be addressed verbally, and at the same time to make a permanent record of the communications. The writer must convey by words alone both the emphasis required and the tone in which the letter or report is written. Words and phrases must, therefore, be very carefully chosen.

Without going into the subject too deeply, a few suggestions may be made:

- Be sure that the points made are clear
- Be as brief and straightforward as possible and do not use two words where one will do
- Start a new paragraph with each new point
- If a long letter develops, consider whether it is not better to put the matter in the form of a schedule or report, with only a short covering letter
- Be sure to write with the reader in mind; do not use technical terms when writing to a non-technical client
- Avoid commercial clichés, journalese, Americanisms and slang
- Avoid spelling mistakes and bad grammar; they give a poor impression to an educated reader
- Avoid the impersonal.

Surveyors, particularly recent graduate surveyors, will find many letters appropriate to a variety of situations that have been prepared on past and present projects and are held on file. Use of these previous communications, either directly or as a basis for a letter in a new situation, will save much time and should normally reflect a mature and professional style.

E-mail

The benefits of e-mail are clear in providing the opportunity of communicating quickly and effectively with an individual or group of individuals almost anywhere in the world. Furthermore, it is possible to 'attach' information which can speed up the interchange of relevant documentation and thus improve the quality of service. It is difficult to imagine life without it and this is now reflected within the Standard Form of Building Contract, JCT 2005, which accommodates the use of specified electronic communications (clause 1.8). However, despite the opportunities provided, there are some downsides that need to be considered, for example;

- It is too easy to communicate to anyone and everyone, thus there can be an unnecessary amount of communication to deal with
- E-mail messages can be forwarded to third parties without the control or approval of the sender; this should be borne in mind when sending any e-mail
- E-mail messages can be a challenge to good time management; it is easy to let the contents of the inbox interfere with and gain priority over more urgent work plans
- The language and tone of an e-mail message can easily be misinterpreted; like any written communication, the rules outlined above need to be considered
- E-mail messages accommodate spontaneous replies which can be as binding as any other communication and therefore should be given the same amount of caution with regard to, for example, discussions which may have an impact on a contractual claim.

Benchmarking and the QS

Benchmarking within the construction industry is increasingly considered from the perspective of the construction company and members of its supply chain. This is a natural inclination since construction may be viewed as a 'manufacturing' process with an easily recognised and evaluated product (a building). It would be a mistake, however, to ignore the opportunity that benchmarking may offer to construction consultants who also operate within the same increasingly competitive and changing domain. For instance, Pickrell *et al.* (1997) present a case study of the Bucknall Group which describes how the company introduced benchmarking to combat the difficulties it faced in the early 1990s. One example of the success achieved related to the introduction of an electronic time sheet system that fed into the costing system and saved an estimated equivalent of £55 000 per annum. In addition to improving the profitability of a QS practice, benchmarking can also be used by clients to compare consultant performance in a similar fashion to that discussed above for contracting organisations. The principles of and rationale for benchmarking are discussed further in the discussion topic below.

Time and cost management

In the increasingly competitive fee and tender market the management of time and cost is of significant importance. There is a need for staff to be managed in as efficient a way as possible. To assist in this, detailed records of time expended and thereby costs incurred need to be monitored on a regular basis.

Resource allocation

Resources need to be allocated to specific tasks on one or more project depending upon their size and complexity. Teams may be assembled for document preparation or post-contract administration on large projects. The overall resource allocation needs to be regularly monitored. Certain work will be indeterminate or require immediate attention, which can cause problems in overall planning. Agreement of design and documentation production programmes with members of the design team can assist in preplanning the timing of resource requirements.

Individual time management

Each member of staff needs to manage his or her time as efficiently as possible. This is particularly important when they are involved on a number of projects at any one time. Effective time management, although achievable in theory, becomes more of a problem in practice.

Individual tasks should be prioritised, with the most urgent (not the most straightforward) being addressed first. Where deadlines are critical, a timescale should be applied to each task to provide a target for completion.

Staff time records

It is also necessary to plan and coordinate the individual staff members' time. This information can be presented on a bar chart for each member of staff. It needs to take staff holidays into account, to be revised at regular intervals, and to be reviewed weekly.

Whether or not an organisation has a full job costing system, it is necessary that detailed records be kept of the time worked by each member of staff. See the sample weekly time sheet relating to a traditional quantity surveying practice in Fig. 3.2. This may be required as a basis on which to build up an account for fees for services that will be charged for on a time basis, or to establish the cost for each particular project. Such costs will be used primarily to establish whether a particular project is making a profit or a loss, but may also be used to estimate a fee to be quoted for similar future work.

Each member of staff needs to keep a diary in which to record not just movements, meetings and other matters but also time spent on each project. It is essential that entries are made on a daily basis and entered up before leaving the office. In addition to identifying the project worked on, notes of specific activities, such as cost planning and preparation of interim valuations, need to be identified in order for time sheets to be comprehensively completed.

Progress charts

The keeping of charts showing the work in the office and its progress is useful for management purposes. It can also identify future commitments in workload, although these may only be tentative. The form of a progress chart may be similar to that used for construction works, different jobs taking the place of the stages in a building contract.

The progress chart may also be useful in helping to decide whether to look for additional staff or whether there is an under-utilisation of staff for which work in the long term will need to be forthcoming.

Education and training

The various methods of entry into the profession are described in publications by the RICS.

Whether staff are selected from school-leavers, who will then attend a university on a day-release basis, or directly from an undergraduate university degree course, will depend on availability and the practice's preference. There are advantages in both cases. Professional practice is really only something that can be learned by experience, and this argument favours the school-leaver/part-time student who has the opportunity to link learning with practice, and of course, the advantage of being paid throughout the training period. However, intensive study of theoretical knowledge may provide the young surveyor with a wider view of the profession. Full time study, albeit frequently interrupted by work placement opportunity

Weekly Time Sheet

Name	Grade	Location	Staff No.	Week Ending

Job Description	MON		TUES		WED		THUR		FRI		SAT		SUN		TOTAL		CODE		
	BASIC	O/T	BASIC	O/T	BASIC	O/T	BASIC	O/T	BASIC	O/T	BASIC	O/T	BASIC	O/T	BASIC	O/T	JOB NO.	WORK CODE	CH/NC
																	FOR ANALYSIS OF WORK CODES SEE BACK OF THIS DOCUMENT		
Office Administration																			
Training/Study Leave																			
Public Holidays																			
Annual Holidays																			
Sick																			
Other (Specify)																			
TOTAL																			
Signed (Staff)								Approved (Partner/Associate)											

WORK CODES

CODE	DESCRIPTION
A	PRE CONTRACT ESTIMATES AND COST CONTROL
B1	TENDER AND CONTRACT DOCUMENTATION INCLUDING REPORTS
B2	NEGOTIATIONS
C1	POST CONTRACT – VALUATIONS AND COST REPORTS
C2	POST CONTRACT – FINAL ACCOUNTS
D	CLAIMS
E	OTHER QS SERVICES eg Liquidations, attendance at additional meetings, etc
F	PROJECT MANAGEMENT
H	FACILITIES MANAGEMENT
I	OTHER eg Insurance valuations, expert witness work, arbitration, etc

Fig. 3.2   Sample weekly time sheet.

within the typical sandwich course structure, provides a different working environment which some students find preferable. In either case the surveyors of the future who wish to achieve chartered status are required to undertake some type of RICS exempting course, and professional membership of the Institution will be via the Assessment of Professional Competence (APC).

The underlying objective of academic study, therefore, should be to develop in the student an understanding of the principles and concepts relating to the process of quantity surveying. Although this should contain some technical training, this responsibility is largely left to the professional offices and contracting organisations. Recognition as a qualified surveyor will still be in the hands of the Institution.

Upon qualifying, the surveyor should not consider this to be the end of studies. Regular continuing professional development (CPD) must be undertaken as a requirement of RICS membership. This may mean attending courses or further study for specialisation in a particular aspect of the profession, such as management. It may also incorporate postgraduate courses and research. Education and training are more fully described in Chapter 2.

Finance and accounts

Generally, the subject of finance and accountancy is not seen as an essential part of some undergraduate surveying courses. However, it is a fundamental and important business skill and as such is important to the quantity surveyor in general. This has been recognised by some universities, which in response are expanding the business content of their courses. Larger practices and companies will have finance and accounts departments with specialist accountancy staff, whereas within smaller organisations senior management deals with all financial matters.

The accounts

The primary purpose of keeping accounts is to provide a record of all the financial transactions of the business, and to establish whether or not the business is making a profit. The accounts will also be used:

- In determining the distributions to be made to equity holders
- In determining the partners' or company's tax liabilities
- To support an application to a bank for funding
- To determine the value of the business in the event of a sale
- As a proof of financial stability to clients and suppliers.

All limited companies are required under Companies Acts to produce accounts and to file them annually with the Register of Companies in order that they are available for inspection by any interested party.

The principal accounting statements are the profit and loss account and the balance sheet.

Profit and loss account

The profit and loss account records the results of the business' trading income and expenditure over a period of time. For a surveying practice, income will represent fees receivable for the supply of surveying services; expenditure is likely to include such items as salaries, rent and insurance. After adjustments have been made for accruals (revenue earned or expenses incurred that have not been paid or received), and prepayments (advanced payments for goods or services not yet provided), an excess of income over expenditure indicates that a profit has been made. The reverse would indicate a loss.

The preparation of the profit and loss account will enable the business to:

- Compare actual performance against budget
- Analyse the performance of different sections within the business
- Assist in forecasting future performance
- Compare performance against other businesses
- Calculate the amount of tax due.

An example of a simple profit and loss account is shown in Fig. 3.3.

Balance sheet

The balance sheet (Fig. 3.4) gives a statement of a business' assets and liabilities as at a particular date. The balance sheet will include all or most of the following:

- *Fixed assets:* those assets held for long-term use by the business, including intangible assets
- *Current assets:* those assets held as part of the business' working capital
- *Liabilities:* amounts owed by the business to suppliers and banks
- *Owner's capital:* shareholder's funds (issued share capital plus reserves) in a limited company, or the partners' capital accounts in partnership.

Profit and loss account for the two months to 28 February 2007		
	£	£
Income		
Fees received		300 000
Expenditure		
Salaries	150 000	
Rent	50 000	
Others	30 000	
Depreciation	20 000	250 000
Profit for the period		£50 000

Fig. 3.3 Example profit and loss account for QS practice.

Balance sheet as at 28 February 2007

	£	£
Fixed assets		
Fixtures and fittings		200 000
Less: Depreciation		20 000
		180 000
Current assets		
Fees receivable	60 000	
Cash at bank	60 000	
	120 000	
Current liabilities	50 000	
Net current assets (working capital)		70 000
Net assets		£250 000
Capital		200 000
Retained profits		50 000
		£250 000

Fig. 3.4 Example balance sheet for QS practice.

The various types of asset and liability accounts are considered in more detail below.

Assets

The term assets covers the following:

- *Intangible assets,* which include goodwill, trademarks and licensing agreements, usually at original cost less any subsequent write-offs
- *Fixed assets,* which include land and buildings, fixtures and fittings, equipment and motor vehicles, shown at cost or valuation less accumulated depreciation
- *Current assets,* which are held at the lower cost or new realisable value. Current assets include cash, stock, work in progress, debtors and accruals in respect of payments made in advance.

Depreciation, referred to above, records the loss of value in an asset resulting from usage or age. Depreciation is charged as an expense to the profit and loss account, but is disallowed and therefore added back for tax purposes. Depreciation is recorded as a credit in the balance sheet, reducing the carrying value of the firm or company's fixed assets. For tax purposes, depreciation will be considered under the provisions of the Capital Allowances Act.

Receipts	£	Payments	£
Capital introduced	200 000	Salaries	150 000
Fees received	240 000	Rent	50 000
		Other expenses	30 000
		Fixtures and fittings	150 000
		Balance carried forward	60 000
	£440 000		£440 000

Fig. 3.5 Bank account summary statement.

Liabilities

Liabilities include amounts owing for goods and services supplied to the business and amounts due in respect of loans received. Strictly they also include amounts owed to the business' owners: the business' partners or shareholders. Note that contingent liabilities do not form part of the total liabilities, but will appear in the form of a note on the balance sheet as supplementary information.

Capital

Sources of capital may include proprietors' or partners' capital or, for a limited company, proceeds from shares issued. Capital is required to fund the start-up and subsequent operation of the business for the period prior to that period in which sufficient funds are received as payment for work undertaken by the business. The example in Fig. 3.4. identifies an initial capital investment of £200 000; however, this investment is soon represented not by the cash invested but by the business.

To illustrate the movement of cash in terms of receipts and payments, a simple example of a summarised bank account statement is included in Fig. 3.5. This statement is produced periodically, usually monthly, and is the source of the cash postings to the other books of account.

Finance

There are several ways in which a business can supplement its finances. The most common way is by borrowing from a bank on an overdraft facility. The lender will be interested in securing both the repayment of the capital lent and the interest accruing on the loan. The lender will therefore require copies of the business' profit and loss account, balance sheet and details of its projected cash flow.

Cash forecasting and budgeting

It is necessary for a business to predict how well it is likely to perform in financial terms in the future. Budgets are therefore prepared, usually on an annual basis, based on projected income and expenditure. Once a business is established, future projections can be based to a certain extent on the previous year's results.

	January	February	March	April	May	June
Capital introduced	200 000	—	—	—	—	—
Fees received	60 000	180 000	50 000	110 000	100 000	200 000
Asset sales	—	—	—	—	—	—
Receipts	260 000	180 000	50 000	110 000	100 000	200 000
Salaries	75 000	75 000	75 000	75 000	75 000	75 000
Rent	25 000	25 000	25 000	25 000	25 000	25 000
Equipment	100 000	50 000	20 000	30 000	—	—
Others	15 000	15 000	15 000	15 000	15 000	15 000
Payments	215 000	165 000	135 000	145 000	115 000	115 000
Movement in cash	45 000	15 000	(85 000)	(35 000)	(15 000)	85 000
Balance brought forward	—	45 000	60 000	(25 000)	(60 000)	(75 000)
Balance carried forward	45 000	60 000	(25 000)	(60 000)	(75 000)	10 000
Borrowing facility	50 000	50 000	50 000	50 000	50 000	50 000
Additional requirement	—	—	—	10 000	25 000	—

Fig. 3.6 Example cash-flow forecast.

As mentioned above, in the event that it is the intention to borrow money from a bank then the bank is likely to request a cash-flow forecast for the next six or twelve months. The preparation of cash-flow forecast is a relatively easy process and in practice a computerised spreadsheet package or accounting software will be used to project the likely phasing of receipts and payments.

The example cash flow forecast in Fig. 3.6 illustrates the starting-up of a professional business. It identifies the initial introduction of capital, the borrowing facility requested and the projected effect of expenditure and receipts over the period. It can be seen from the forecast that an additional £10 000 of funding will be required in April and a further (£25 000 – £10 000) = £15 000 in May.

This cash flow is typical of a business start-up, when substantial sums are spent in advance of income being received. Provided the business is run profitably the outflow should be reversed before too long.

Books of account

The underlying books of account are likely to comprise the general ledger (which will include all general items such as salaries and rents) and totals from the subsidiary ledgers such as the sales ledger (fees or other income receivable), the bought ledger (accounts payable), the cash book (a record of the bank transactions) and the petty cash account. Other books, such as fee and expenses books, may also be kept.

In addition to the books of account, businesses must retain vouchers such as receipts, invoices, fee accounts and bank statements to support the accounting records. These are required by businesses' auditors and for VAT purposes.

Use of IT in accounting

Accounting functions have become less time-consuming through the use of computers for the regular and routine entries and calculations that are necessary. Entries can be allocated to different accounts, and up-to-date information can be retrieved quickly and efficiently in a variety of formats to meet particular needs. This greatly assists in the financial management of a business.

Annual accounts/auditing

At the end of a business' financial year a set of 'end of year' accounts is prepared bringing together all the previous year's financial information in the form of a profit and loss account and balance sheet as described earlier. In most circumstances an independent accountant will audit accounts; indeed this is a requirement for all larger limited companies under the Companies Acts. Audited accounts will carry more authority with the Inspector of Taxes and are also useful to prove to third parties, including prospective clients, that the financial status of the business has been independently scrutinised.

Discussion topic

Discuss the origins and rationale for benchmarking, and what it may bring to the construction sector.

Introduction

Benchmarking is a technique that, like other management practices discussed in this book, developed in the USA and is now successfully applied in many industry sectors around the world. It owes much of its relatively recent development to the success achieved by the Xerox Corporation. In the late 1970s, Rank Xerox was severely challenged by Japanese competitors who were able to manufacture photocopiers of a better quality and at a lower cost. In response, Xerox successfully implemented a programme of benchmarking which incorporated the comprehensive examination of their organisation throughout. Better practice, in any relevant process used by another company, was identified and translated into new Xerox practice. This facilitated a significant improvement in company performance and an improvement in their competitive position (Pickrell *et al.* 1997).

Benchmarking can be applied across the business sector and throughout the entire hierarchy of a company from strategic level to operational. The aim of benchmarking is to improve the performance of an organisation by:

- Identifying best known practices relevant to the fulfilment of a company's mission
- Utilising the information obtained from an analysis of best practice to design and expedite a programme of changes to improve company performance. (It is important to stress that benchmarking is not about blindly copying the processes of another company, since in many situations this would fail. It requires an analysis of why performance is better elsewhere, and the translation of the resultant information into an action suited to the company under consideration.)

Despite the acknowledged benefits of benchmarking in assessing company performance and prompting its improvement, its use in practice within the UK appears to be low. Research carried out in 1997 by the Confederation of British Industry (CBI) concluded that 'benchmarking is weak overall in industry' and that there was a 'need to raise awareness and practice of benchmarking across British industry' (CBI 1997). Any benchmarking practice at all appears in the main to be restricted to large companies and in many instances isn't carried out adequately. The level of use of benchmarking within the construction industry is consistent with this view. Some large construction organisations and construction clients have begun to use the tool successfully, however, the extent of use is very limited and is, in many companies, in its infancy.

The importance of benchmarking to the construction industry has recently been acclaimed at a very high level. In the report of the Construction Task Force (1998) to the Deputy Prime Minister, John Prescott, *Rethinking Construction* (The Egan Report) stated that:

'Benchmarking is a management tool which can help construction firms to understand how their performance measures up to their competitors' and drive improvement up to 'world class standards . . .'

The report makes clear the under-performance of the construction industry and the need for fairly dramatic improvement, much of which is considered to be reliant upon successful benchmarking practice. It also highlighted the significance of the construction industry to the British economy which, in 1998, had an output of £58 billion (10% of GDP) and a workforce of approximately 1.4 million, further emphasising the need for change.

In other sections of this book, recognition has been given to the poor rate of implementation of 'new' management practices. This introduction suggests that benchmarking practice in the UK may also currently be seen to fall into this group of 'fashionable' but little used activities. In the case of benchmarking however, business survival may be dependent upon it and therefore its wide acceptance may become a natural consequence. It is likely that the use of benchmarking practice will accelerate and, as companies improve by it, the need for other organisations to follow will be apparent.

The need for benchmarking

It is clear from the above that the purpose of benchmarking is to improve a company's performance with the aim of maintaining or achieving a competitive edge. The UK construction industry is witnessing a rapid change in many areas, some of which are of great significance, for example: methods of procurement; client expectations; client/contractor relationships; professional appointments; the Private Finance Initiative; government focus; warranties and guarantees; payment provisions; safety regulations. Consideration of this incomplete list should otherwise convince anyone who believes our industry stands still. In addition, there is an increasing level of competition from Europe and elsewhere – including the USA and Japan (where benchmarking practice is part of the culture). There are good incentives for the use of benchmarking. The advantages that it may bring include:

- An opportunity to avoid complacency which may be nurtured by monopolistic situations resulting in the loss of market leadership
- An improvement in the efficiency of a company leading to an increase in profits
- An improvement in client satisfaction levels resulting in an improved reputation, leading to more work, better work and an increase in profits
- An improvement in employee satisfaction in working in a more efficient, profitable and rewarding environment.

Many organisations that have successfully used benchmarking in order to achieve a specific objective continue to use it thereafter as part of a total quality management programme. In this way it can contribute to the long term objectives of the company which primarily may be to achieve a high level of customer satisfaction, recognised excellence in their sector and, thereby, a high return.

What can benchmarking bring to the construction industry?

The relatively low standing of the construction industry within the UK suggests that there is significant opportunity for benchmarking to be used with great effect. As discussed in other chapters of this book, clients too frequently criticise the industry in that buildings are delivered late, exceed budget costs and are of poor quality. This provides three broad areas for improvement. This view has recently been enforced by the publication of the 'Egan Report' (1998) which stated that:

'. . . In construction, the need to improve is clear. Clients need better value from their projects, and construction companies need reasonable profits to assure their long-term future. Both points of view increasingly recognise that not only is there plenty of scope to improve, but they also have a powerful mutual interest in doing so.'

The inefficiencies and waste in construction are demonstrated by the following examples of poor performance which have been identified by recent studies in the USA, Scandinavia and the UK (Egan Report 1998):

- As much as 30% of construction work is carried out to remedy problems of first time construction
- The output of the labour force is at 40–60% of optimum
- Poor site safety is demonstrated by an accident rate which can result in 3–6% of total project costs
- Material waste is high with wastage being 10% or more.

As stated above, there may be difficulties to overcome in carrying out benchmarking studies. One of the major problems relates to the accessibility of information that may prove difficult to obtain in all studies but particularly where external sources are being used. There are benchmarking organisations which have been established to assist with this; the recent proposal in the Egan Report suggests that an information centre should be set up whereby those involved in the construction industry can have access to a bank of data which will prove helpful in benchmarking exercises.

In recent years, initiatives promoted by Constructing Excellence have increased the use of benchmarking practice within the sector and have identified case study examples highlighting the benefits it can bring. They have utilised a set of Key Performance Indicators which have allowed construction organisations to compare their performance with competitors in the rest of the industry. Participation with Constructing Excellence provides access to a range of business assessment tools:

'enabling you to compare your business' performance from project to project, and with the wider industry: identify areas for improvement; and plan for the future.

- Measure against your peers
- Identify strengths and weaknesses
- Self assessment
- Drive business improvement'

(constructingexcellence.org.uk; accessed 17 July 2006)

Benchmarking and partnering

To justify the adoption of partnering, clients may need to validate the benefits of such an arrangement by establishing that good value and continued performance improvements exist. Since the usual assurance of 'value' – traditionally dependent upon the receipt of competitive bids – is absent, benchmarking may be required to provide an alternative method of 'measuring' the relative performance of a partner contractor, to demonstrate the added value arising from a partnering arrangement.

References and bibliography

Barrett P. *Profitable Practice Management for the Construction Professional.* E & F N Spon. 1993.

British Standards Institution. ISO 9001, *Quality Management. Systems: requirements,* 3rd edition. BSI 2000.

Confederation of British Industry. *Benchmarking the Supply Chain.* Partnership Sourcing Ltd. 1997.

Construction Task Force. *Rethinking Construction* (The Egan Report). Department of the Environment, Transport and the Regions. 1998.

Davis, Langdon and Everest. *Quantity Surveying 2000: The Future Role of the Chartered Surveyor.* RICS. 1991.

Jennings A. R. *Accounting and Finance for Building and Surveying.* Macmillan Press. 1995.

Kolesar S. *The Values of Change.* Presidential address; Royal Institution of Chartered Surveyors. 1999.

Park A. *Facilities Management: An Explanation.* Macmillan. 1998.

Pickrell S., Garnett N. and Baldwin J. *Measuring Up; a Practical Guide to Benchmarking in Construction.* Construction Research Communications Ltd (by permission of Building Research Establishment Ltd). 1997.

Powell C. *The Challenge of Change.* RICS. 1998.

Websites

Constructing Excellence available at http://www.constructingexcellence.org.uk (accessed 17/06/2006).

Health and Safety Executive (HSE) [online] available at http://www.hse.gov.uk (accessed 17/07/2006).

Her Majesty's Stationary Office (HMSO) [online] available at http://www.hmso.gov.uk/legislation/uk.htm (accessed 17/06/2006).

4 The Quantity Surveyor and the Law

Introduction

The purpose of this chapter is to describe in general terms how the law affects the quantity surveyor in practice, concentrating on the form of agreement between the quantity surveyor and the client, the impact of the demand for collateral warranties and performance bonds, the requirement to maintain professional indemnity insurance cover and the Employment Acts as they relate to the employment of staff. The wider issue of construction law is a complex and ever-changing area, the detail of which falls outside the scope of this book.

The quantity surveyor and the client

The quantity surveyor in private practice provides professional services for a client. The legal relationship existing between them is therefore a contract for services. The nature of this contract controls the respective rights and obligations of the parties. So far as the quantity surveyor is concerned it determines the duties to be performed, powers and remuneration for the particular work undertaken.

Agreement for appointment

There is no legal requirement that the agreement or contract between the surveyor and the client should consist of a formal document nor even, indeed, that it be in writing. Nevertheless, given that it may be crucial to establish the precise nature of the relationship in the event of a dispute, it is desirable that the understanding reached be confirmed in writing. If differences subsequently arise and proceed to litigation, the court will be faced with the problem of ascertaining the true intentions of the parties from the available evidence. Clearly a written record will, in these circumstances, be a great deal more persuasive than the possibly disputed recall of the contending parties.

The existence and nature of the agreement can be established by an exchange of correspondence or by the use of a standard form of appointment, such as the Form of Agreement, Terms and Conditions for the Appointment of a Quantity Surveyor which is published by the RICS. It cannot be overstated that whatever practice is adopted, there is a need to ensure that a valid, comprehensive and adequately evidenced contract exists between the respective parties.

Professionals can sometimes find themselves in difficulty if they undertake work relying only upon an incomplete agreement, in which important items remain to be settled. However, the law does not recognise the validity of a contract to make a contract and, where any essential element is left for later negotiation, the existing arrangements are unlikely to be recognised as a binding agreement. Given the real pressures to acquire business, it is easy to succumb to this temptation without adequately weighing the risks involved.

In *Courtney & Fairbairn Ltd* v. *Tolaini Bros (Hotels) Ltd* (1975) 1 AER 716; 2 BLR 100, the defendants, wishing to develop a site, agreed with the plaintiffs, a firm of contractors, that if a satisfactory source of finance could be found, they would award the contract for the work to the plaintiffs. No price was fixed for the work but the defendants agreed that they would instruct their quantity surveyor to negotiate fair and reasonable contract sums for the work. Suitable finance was introduced, and the quantity surveyor was instructed as agreed. The quantity surveyor was unable, in the event, to negotiate acceptable prices, resulting in the contract being awarded elsewhere. The plaintiffs sued for damages, claiming that an enforceable contract had been made. It was held that the price was a fundamental element in a construction contract and the absence of agreement in that regard rendered the agreement too uncertain to enforce. This case was, of course, not directly concerned with the provision of professional services but the legal principle illustrated is of general application.

A persuasive incentive to take care arises from the fact that the ability to recover payment for work done will usually depend on the existence of an appropriate contract. Performance alone does not automatically confer a right to remuneration, although where a benefit is conferred, the court will normally require the beneficiary to make some recompense, possibly by way of a *quantum meruit* payment, meaning, as much as is deserved.

In *William Lacey Ltd* v. *Davies* (1957) 2 AER 712, the plaintiff performed certain preliminary work for the defendant, connected with the proposed rebuilding of war-damaged premises, in the expectation of being awarded the contract for the work. The defendant subsequently decided to place the contract elsewhere, and eventually sold the site without rebuilding. The plaintiff sued for payment for work already done. In this case it was held that in respect of the work done, no contract had ever come into existence but, nevertheless, as payment for the work had always been in the contemplation of the parties, an entitlement to some payment on a *quantum meruit* arose.

In this context it is reassuring, from the quantity surveyor's point of view, to note that, where professional services are provided, there is a general presumption that payment was intended. In *H.M. Key & Partners* v. *M.S. Gourgey and Others* (1984) I CLD-02-26, it was said: 'The ordinary presumption is that a professional man does not expect to go unpaid for his services. Before it can be held that he is not to be remunerated there must be an unequivocal and legally enforceable agreement that he will not make a charge'. However, while some recovery of fees may be possible without the formation of a binding contract, the lack of such an agreement enhances the possibility of disputes and litigation. Moreover, if the basis of enforced payment is to be *quantum meruit*, there is no guarantee that the court's evaluation of the services provided will correspond with the practitioner's expectations.

Given that all relevant terms are settled and agreed, and incorporated in a formal contract, the intentions of the parties may still be frustrated by a failure to express the terms clearly.

In *Bushwall Properties Ltd* v. *Vortex Properties Ltd* (1976) 2 AER 283, a contract for the transfer of a substantial site provided for staged payments and corresponding partial legal completions. At each such completion 'a proportionate portion of the land' was to be transferred to the buyer. A dispute arose as to the meaning of this phrase. It was held that in the circumstances, no certain meaning could be attributed to the phrase; that this represented a substantial element in the contract; hence the entire agreement was too vague to enforce.

However, if a valid contract exists in unambiguous terms, the court will enforce it. It is therefore vital to ensure that the terms are not merely clear but do in fact represent the true understanding of the parties, both at the outset and as the work progresses. When the actual work is in hand with all attendant pressures, it is all too easy to overwork the fact that the obligations undertaken and remuneration involved are controlled by the contract terms. Departure from or misunderstanding of the original intentions, unless covered by suitable amendments of those terms, may have very undesirable consequences as the following case illustrates.

In *Gilbert & Partners* v. *R. Knight* (1968) 2 AER 248; 4 BLR 9, the plaintiffs, a firm of quantity surveyors, agreed for a fee of £30 to arrange tenders, obtain consents for, settle accounts and supervise certain alterations to a dwelling house on behalf of the defendant. Initially work to the value of some £600 was envisaged, but in the course of the alterations the defendant changed her mind and ordered additional work. In the end, work valued at almost four times the amount originally intended was carried out; the plaintiffs continued to supervise throughout and then submitted a bill for £135. This was met with a claim that a fee of £30 only had been agreed. It was held that the original agreement was for an all-in fee covering all work to be done; the plaintiff was entitled to only £30.

The moral is clear: avoidance of difficulty and financial loss is best ensured by accepting engagement only on precise, mutually agreed and recorded terms, setting out unequivocally what the quantity surveyor is expected to do and what the payment is to be for so doing. If, as often happens, circumstances dictate development and expansion of the initial obligations, the changes must be covered by fresh, legally enforceable agreements. Oral transactions, relating to either the original agreement or later amendment of it, should always be recorded and confirmed in writing. This is more than an elementary precaution, for it should be borne in mind that what is known as the parol evidence rule will normally preclude any variation of an apparently complete and enforceable existing written contract, by evidence of contrary or additional oral agreement.

The agreement for appointment of a quantity surveyor, whether a standard or non-standard document, will encompass certain general provisions including the following:

- Form of agreement/particulars of appointment
- Scope of services to be provided
- Fee details
- Payment procedures
- Professional indemnity insurance requirements
- Assignment
- Suspension
- Copyright
- Duty of care
- Dispute procedures.

Other provisions such as limitation of the quantity surveyor's authority, communications and duration of appointment, might be included.

The agreement can be executed either as a simple contract or as a deed. There are important differences, two of which are the most significant as far as the quantity surveyor is concerned: the need for consideration and the limitation period.

In a simple contract there must be consideration. This is a benefit accruing to one party or detriment to the other, must commonly payment of money, provision of goods or performance of work. The period in which an action for breach of contract can be brought by one party against the other is limited to six years. In a speciality contract, a contract executed as a deed, however, there is no need for consideration and the limitation period is 12 years. The significance of the latter is that a quantity surveyor who enters into an agreement as a deed, doubles the period of exposure to actions for breach of contract.

Responsibility for appointment

What has been written so far assumes that the quantity surveyor's appointment arose from direct contact with the client. Additional problems may occur where the appointment arises indirectly from the retained architect or project manager. In such cases the power to appoint on behalf of the client may subsequently be called into question. There is no general solution to this problem. The actual position will depend on the express and implied terms of the other consultant's contract with the client. If they have express power to appoint, then, of course, no problem arises and the appointment is as valid as if made by the client in person. However, reliance on their possessing an implied power to appoint would be very unwise. It is clear that the courts do not recognise any general power of appointment or delegation as inherent in an architect's or other consultant's contract with a client.

In *Moresk Cleaners Ltd* v. *T.H. Hicks* (1966) 4 BLR 50, the Official Referee stated bluntly that 'The architect has no power whatever to delegate his duty to anyone else'. That case concerned the delegation of design work but it would seem equally applicable to other unauthorised appointments.

This absence of a general implied power to appoint does not preclude the possibility that in particular circumstances it may be held to exist. There is, for example, some rather dated authority for the proposition that, where tenders are to be invited on a bill of quantities basis, such implied authority may be present.

Potential difficulties in the matter can be easily avoided, by the simple expedient of ensuring that, where the employment of the quantity surveyor is negotiated by another consultant, the terms of the appointment are conveyed in writing to the client and the client's acceptance thereof similarly secured. Ratification by the client will then have overcome any deficiencies in the consultant's authority.

Responsibility for payment of fees

Where an effective contract exists between the quantity surveyor and the client, provision will be contained relating to the payment of the professional fees involved. No real difficulty should be experienced where the appointment has been made by a duly authorised agent, or the client has ratified an appointment purported to have been made on their behalf. The position where no valid agreement exists has already been mentioned, and it was suggested that even in such unfortunate circumstances some remuneration, probably by way of a *quantum meruit*, will usually be forthcoming.

However, if the authorised agent has made the appointment in excess of their powers, the quantity surveyor will have to look elsewhere than to the client for payment. In such circumstances it will usually be possible to bring an action against the agent personally. Such a claim would normally lie for warranty of authority, either where the agent had misunderstood or exceeded the authority granted by the client, or had not actually purported to act on the client's behalf. The existence of a legal remedy is nevertheless of doubtful consolation if the debtor is unable to pay.

If there are any reservations regarding the financial standing of a potential client, the surveyor must make enquiries, perhaps by taking up bank references, and then trust to commercial judgement. In this connection, it is vital to ensure that the documentation accurately reflects the true identity of the client. This may seem too obvious to mention but misunderstandings can and do occur, particularly in dealings with smaller companies controlled by a sole individual. It is easy to confuse the individual acting on behalf of a company and acting in a personal capacity. The unhappy result may be dependence for payment on a company of doubtful solvency, having imagined that one was acting for an individual of undoubted substance. Finally, it is chastening to reflect that monies owned in respect of professional fees are in no way preferred in the event of insolvency.

Amount and method of payment

Actual fees and fee rates are now a matter for negotiation, since recommended scales do not exist. Entitlement depends on the terms of agreement under which the services are provided, and any negotiations are constrained by practical rather than legal considerations. However, where the work involves advising on matters connected with litigation or arbitration, for example in respect of claims, it is not permissible to link the fee to the amount recovered. In *J. Pickering* v. *Sogex Services Ltd* (1982) 20 BLR 66, arrangements of that nature were said to savour of champerty

– that is, trafficking in litigation – and as such to be unenforceable as contrary to public policy.

Where possible it is prudent to make provision for the payment of fees by instalments at appropriate intervals. The payment of a series of smaller amounts while services are being provided tends to be more readily accepted than the settlement of a substantial bill when the work has been completed. Moreover, failure to pay on time may be a useful guide to the state of the client's finances. If the worst happens and the client is rendered insolvent, there is some comfort in being an ordinary creditor for only the balance and not the entirety of the fee. Failure to pay on time should be treated seriously and, if necessary, legal steps taken promptly for recovery. In these matters patience and understanding are more likely to lead to disappointment than to reward.

Negligence

Where the law is concerned, negligence usually consists either of a careless course of conduct or such conduct, coupled with further circumstances, sufficient to transform it into the tort of negligence itself. As stated earlier the extent and nature of the duties owed to the client by the quantity surveyor, as well as the powers and authority granted to the client, will be determined by the contract for services between them.

It has always been implied into a professional engagement that the professional person will perform duties with due skill and care. This requirement is reiterated by provisions in the Supply of Goods and Services Act 1982. Lack of care in discharging contractual duties is, and always has been, an actionable breach of contract.

As late as the mid-1960s, when it was so held in *Bagot* v. *Stevens Scanlan & Co* (1964) 3 AER 577, the existence of a contractual link between the parties was believed to confine liability to that existing in contract and to exclude any additional liability in tort. Since then, the position gradually changed and the courts appeared to recognise virtually concurrent liability in both contract and tort. Thus an aggrieved contracting party was able to sue the other contracting party or parties both in contract and tort. This was illustrated by the decisions in *Midland Bank Trust Co Ltd* v. *Hett, Stubbs & Kemp* (1978) 2 AER 571 and, more immediately relevant to the construction industry, in *Batty* v. *Metropolitan Property Realisations Ltd* (1978) 7 BLR 1.

Liability was also considered to exist independently, where there is no contractual link between the parties, enabling a third party to sue in the tort of negligence. A plaintiff suing in negligence must show that:

- The defendant had a duty of care to the plaintiff, and
- The defendant was in breach of that duty, and
- As a result of the breach the plaintiff suffered damage of the kind that is recoverable.

In the first place, the plaintiff would try to show that the defendant owed a duty of care. From the principles established in *Donoghue* v. *Stevenson* (1932) AER I (the celebrated 'snail in the bottle' case), the courts tended to find the presence of a duty of care in an ever-increasing number of circumstances.

The tentacles of the tort of negligence even extended well beyond normal commercial relationships, at least so far as the professional person was concerned. Since the well-known case of *Hedley Byrne & Co Ltd* v. *Heller & Partners Ltd* (1963) 2 AER 575, any negligent statement or advice, even if given gratuitously, seemed in certain circumstances to afford grounds for action.

In more recent times in the case of *Junior Books Ltd* v. *The Veitchi Co. Ltd* (1982) 21 BLR 66, the House of Lords held that a specialist flooring subcontractor was liable in negligence for defective flooring to the employer with whom the subcontractor had no contractual relationship. Almost immediately, however, the courts began to retreat from the position by means of a long string of cases which culminated in *Murphy* v. *Brentwood District Council* (1990) 50 BLR 1 which, among other things, overturned the 12-year-old decision in *Anns* v. *London Borough of Merton* (1978) 5 BLR 1.

The tortious liability for negligence is therefore reduced, which in itself has led to the growth in the use of collateral warranties.

There is little case law concerning the negligence of a quantity surveyor. However, it is always advisable to limit the risk involved. The best safeguard is discretion and a reluctance to express opinions or proffer advice on professional matters, unless one is reasonably acquainted with the relevant facts and has had the opportunity to give them proper consideration. Indeed, it is one of the requirements of the RICS Code of Conduct that a surveyor should never accept work on some ad hoc basis, but should always seek to formalise the process in some way.

Death of the quantity surveyor

Whether the liability to carry out a contract passes to the representatives of a deceased person depends on whether the contract is a personal one. That is, one in which the other party relied on the 'individual skill, competency or other personal qualifications' of the deceased. This is a matter to be decided in each particular case.

In the case of a quantity surveyor with no partner, the appointment must be regarded as personal, and the executors could not nominate an assistant to carry on with business unless the respective clients agreed. With a firm of two or more partners the appointment may be that of the firm, in which case the death of one partner would not affect existing contracts. But the appointment may be of one individual partner, as an arbitrator for example, where another partner in the firm could not take over, even though he or she may be entitled to a share of the profits earned by the partner in the arbitration.

The fact that a contract between a quantity surveyor and the client is a personal contract, if that is the case, does not mean that the quantity surveyor must personally carry out all the work under the contract; this is unless it is obvious from the nature of the contract, for example a contract to act as arbitrator, that the quantity surveyor must act personally in all matters. In other cases, such as the preparation of a bill of quantities and general duties, the quantity surveyor may make use of the skill and labour of others, but takes ultimate responsibility for the accuracy of the work.

Death of the client

The rule referred to in the previous paragraph as to a contract being personal applies equally in the case of the death of the client. Here the contract is unlikely to be a personal one, and the executors of the client must discharge the client's liabilities under the building contract and for the fees of the professional people employed. The fact that the appointment of the surveyor was a personal one will not be material in the case of the death of the client.

Collateral warranties

A collateral warranty, or duty of care agreement, is a contract that operates alongside another contract and is subsidiary to it. In its simplest form, it provides a contractual undertaking to exercise due skill and care in the performance of certain duties which are the subject of a separate contract. An example of this would be a warranty given by a quantity surveyor to a funding institution to exercise due skill and care when performing professional services under a separate agreement between the quantity surveyor and the client. The purpose of a collateral warranty is to enable the beneficiary to take legal action against the party giving the warranty, for breach of contract if the warrantor fails to exercise the requisite level of skill and care in the performance of the duties.

It used to be the view that such an agreement was not very important, because it merely stated in contractual terms the duties that were owed by the quantity surveyor to a third party in tort. That view is no longer tenable.

Until relatively recently it was always a fundamental contract principle that only parties to a contract had any rights or duties under that contract. The principle was called privity of contract. For example, in a contract between a client and quantity surveyor each had duties towards one another. The quantity surveyor had a duty to carry out specific duties but had no duty to any third party, even if the contract stated that there was such a duty.

However, this situation has been radically changed by the passing of the Contract (Rights of Third Parties) Act 1999, which allows reference to be made to the Act, and provision written into an agreement stating certain extended liabilities as/if appropriate. This will not necessarily affect the quantity surveyor in their work for the client, but similar provisions within contracts between contractors and clients (such as JCT 2005 Clause 7A and B) may have widespread consequences, as between contractors and purchasers, tenants or funders, and the quantity surveyor should be aware of such possibilities when advising their clients.

At one time the third party would have been able to overcome problems of the above type by suing in the tort of negligence if there was no contractual relationship. However, this is no longer the case. This has been stated earlier in connection with negligence and the law in this regard has changed over recent years, culminating in the case of *Murphy* v. *Brentwood District Council* (1990) 50 BLR 1.

As a result of the fundamental changes in the law, collateral warranties are now of significant importance to clients. These have therefore proliferated in recent years. It is now common for contractors, subcontractors and suppliers and all the consultants to execute a collateral warranty in favour of the client, the company providing the finance for the project and/or prospective purchasers/tenants.

There are standard forms of warranty to be given to funders and purchasers or tenants published by the British Property Federation (BPF) which have been approved by the Royal Institute of British Architects (RIBA), RICS, ACE (Association of Consulting Engineers) and the BPF. There are also a great many other forms of warranty, some of which have been especially drafted by solicitors with a greater or lesser experience of the construction profession and the construction industry generally.

The party requiring collateral warranties in connection with a project would probably gain maximum benefit from those provided by the design consultants. It is more likely that a claim under a warranty will be related to design matters. However, it is common for warranties to be requested from all consultants on a project, including the quantity surveyor. It is important to ensure that all consultants enter into a warranty on the same terms, with the exception of clauses relating to the selection of materials (see below).

Usually, as detailed above, the requirement for a collateral warranty will be independent of the form of contract being used. However, in the case of parties to the Partnering Contract (PPC 2000 – clause 22) which could include the client's quantity surveyor, all those concerned are contractually required to obtain collateral warranties for each of the (other) parties to the contract, if and when requested by the client.

The following are specific issues that should be considered by quantity surveyors when called upon to sign a warranty.

Relationship to terms of appointment

Prior to agreeing any warranty terms, it is essential to have full written terms of appointment as described earlier. Unless the duties to be undertaken by the quantity surveyor are fully detailed, the standard warranty term that 'all reasonable skill and care be taken in the performance of those duties' leaves it open to argument as to the definition of such duties if a claim is brought under the warranty in the future.

The warranty should refer specifically to the terms of appointment – and it is important that the terms and conditions of the warranty are no more onerous than those contained in the appointment.

Materials

The standard forms of warranty include provisions regarding taking reasonable skill and care to ensure that certain materials are not specified. This clause should be deleted from quantity surveyors' warranties as it relates primarily to design consultants.

Assignment

The warranty is likely to make provision for the warrantee to be able to assign the benefits to other parties. The more restrictive the assignment clause, the better as far as the quantity surveyor's potential liability is concerned. Much of the value of the warranty is the ability to assign, and therefore if a quantity surveyor agrees to an assignment clause it should be limited in terms of number of assignments and time-scale: for example, assignment only once within a limit of three years subject to the quantity surveyor's consent.

Professional indemnity

The warrantee will be concerned with the level of professional indemnity insurance cover carried by the quantity surveyor, and the required level of cover will be stated in the warranty. It is important that all warranties are passed to the insurers before being signed or the insured could be at risk.

A requirement to maintain professional indemnity insurance cover at a level for a specific number of years is impractical. A requirement to maintain cover should be limited to using best endeavours to maintain cover as long as it is available at commercially reasonable rates.

Complete records should be kept of all warranties given, as it is necessary to disclose these annually to the insurers at the time of renewal of the policy.

Execution

The essential differences between a simple contract and a deed have been highlighted earlier. If the quantity surveyor is requested to enter into a warranty agreement as a deed, whereas the terms of appointment are executed as a simple contract, the quantity surveyor's period of liability to the third party will be twice as long.

Performance bonds

There is an increasing demand for consultants to provide performance bonds, particularly on major projects. Although it is a concern that clients consider it necessary to require such bonds in the pursuit of work, the quantity surveyor may not be in a position to object.

The conditions of the bond are likely to be similar to those required from a contractor. The value is calculated as a percentage of the total fee, 10%, for example, and the conditions under which it can be called upon are stated. In certain instances 'on demand' bonds are being requested, whereby payment by the surety can be demanded without the need to prove breach of contract or damages incurred as a consequence. It is therefore important to check the conditions in detail and to ascertain the cost of providing the bond prior to agreeing fee levels and terms of appointment.

The common sources of protection are bank guarantees and surety bonds. The terms bonds and guarantees have similar meanings and are used synonymously within the construction industry. Guarantees are documents that indemnify a beneficiary should a default occur. They are usually provided by banks. Performance bonds are usually issued by insurance companies. A performance bond assures the beneficiary of the performance of the work involved up to the amount stated. Performance bonds are three-party agreements between the bondsman, beneficiary and the principal debtor.

Professional indemnity insurance

It has always been prudent for quantity surveyors to protect themselves against possible claims from their clients for negligence, for which they may be used. Such mistakes may not necessarily be those of a principal's own making but those of an employee. The RICS by-laws and regulations make it compulsory for practices, firms or companies to be properly insured against claims for professional negligence. Minimum levels of indemnity are specified. Premiums are calculated according to the limit of indemnity selected, the number of partners or directors, the type of work that is undertaken and the fee income. The policy chosen must be no less comprehensive than the form of the RICS Professional Indemnity Collective Policy as issued by RICS Insurance Services Ltd.

The policy covers claims that are made during the period when the policy is effective, regardless of when the alleged negligence took place. Claims that occur once the policy has expired, even though the alleged event took place some time previously, will not be covered. A sole practitioner is therefore well advised to maintain such a policy for some time after retirement. The RICS requires its members to maintain run-off cover for a period of six years after they retire, to cover just such eventualities. Recent court cases suggest that a professional person may be held legally liable for actions for a much longer period than the normal statutory limitation period would otherwise suggest.

Contracts of employment

There are certain legal requirements relating to the employment of staff. The basic relationship between the employer and the individual employee is defined by the contract of employment. This is a starting point for determining the rights and liabilities of the parties. Although these rights originated from different statutes, they are now consolidated in the Employment Protection (Consolidation) Act 1978, though further amendments were introduced in the Employment Act of 1990. An Employment Relations Act was introduced in 1999. An important feature of these rights is that they are not normally enforced in the courts but in employment tribunals.

The Department of Trade and Industry has an Employment Relations Directorate that seeks to develop legislation on:

- Hours of work
- Pay entitlement
- Public holidays
- Employment agency standards
- Individual employment rights
- Redundancy arrangements
- Employee consultation
- Trade unions and collective rights
- European employment directives.

A contract of employment must be given to each employee within 14 days of commencing employment. It should cover matters regarding the conditions of employment, including hours of work, salary, holiday entitlement, sick leave, termination, and the procedures to be followed in the event of any grievance arising.

Other Acts worthy of note are the Sex Discrimination Act 1975, whereby a person cannot be discriminated against because of his or her sexual orientation or marital status, and the Race Relations Act 1976, which considers discrimination on the grounds of colour, ethnic or national origins or nationality. In addition, from October 2006 the Employment Equality (Age) Regulations 2006 makes it illegal to discriminate against a person on the grounds of their age. These Acts cover not only recruitment but also promotion and other non-contractual aspects of employment. The Equal Pay Act 1970 covers contractual terms *and* conditions of employment in addition to pay, making it unlawful for an employer to treat someone differently because of their sex. There is also the Disability Discrimination Act 2004, which has consequences for the QS at both a personal and professional level. This Act is addressed separately below.

An employer must give the employee the amount of notice to which he or she is entitled under the contract of employment. This will relate to the employee's length of service up to a maximum of 12 weeks, although the contract may specify a longer period. Employees may be dismissed for acts of misconduct, but the employment tribunal must be satisfied that the employer acted reasonably should a complaint be brought to them. Some of the following might be considered as misconduct:

- Absenteeism
- Abusive language
- Disloyalty
- Disobedience
- Drinking
- Using drugs and smoking
- Attitude
- Personal appearance
- Theft or dishonesty
- Violence or fighting.

In order for an employee to bring a case for unfair dismissal, an employee must be able to show at least one of the following:

- Employer ends employment without notice
- A fixed term contract ends without being renewed
- An employer forces an employee to resign; this is known as constructive dismissal
- An employer refuses to take back a woman returning to work after pregnancy
- An employer gives a choice of resignation or being dismissed.

Am Employment Tribunal will investigate whether the employer was acting reasonably after investigating all of the facts of the case. It will want to establish that:

- Warnings were given
- Adequate notice was provided for a disciplinary hearing
- The employee had the opportunity to comment on the evidence
- A decision was not made by someone who had not heard the employee's view
- An appeal was decided by someone who was not already involved.

Sometimes a job comes to an end because the firm has no more work or because the kind of work undertaken by the employee has ceased or diminished. In these circumstances the employee will normally be entitled to redundancy payments. In order to establish a claim the employee must have at least one year's service, be between the ages of 20 and 65 (modified in some cases by the 2006 Act), and have been working for a minimum number of hours per week, depending upon the length of service.

The Disability Discrimination Act 2004

Latest in a line of legislation seeking to protect and enhance the rights and opportunities of disabled persons, this Act has consequences for the quantity surveyor both as an employer/employee and as a practising professional.

In the field of employment, employers must make facilities available for those of their staff who are registered disabled, as true for the quantity surveyor as in any other field. Whilst *occupational exceptions* may be appropriate for certain situations on site, in the office setting an employer will be expected to construct or adapt their premises accordingly.

In their work with and for clients, quantity surveyors should be aware of the latest legislation, particularly that regarding buildings to which the public have access. There will surely be cost consequences arising out of the current legal requirements concerning access, lighting, signage and the like. As the UK moves steadily towards a growing proportion of elderly people this will present further challenges in the future in the designing and costing of construction works, challenges for which the quantity surveyor must be prepared.

Discussion topic

What is conduct that is unbefitting of a chartered surveyor and what processes does the RICS use to deal with them?

The public are increasingly sceptical of self-regulated professions. There are too many examples of where professional bodies have failed to properly investigate a member who on the surface at least has failed to properly safeguard either a client or public interest. The RICS, in common with other professional bodies, has a set of bye-laws, regulations and rules of conduct to which every chartered surveyor must adhere. The RICS describes itself as a global professional body that represents, regulates and promotes chartered surveyors.

Of course the vast majority of chartered surveyors go through their professional careers upholding the highest standards of professional conduct. Under the Royal Charter and bye-laws, all RICS members are expected to comply with regulations governing their conduct. These include:

- Provisions relating to the keeping of members' accounts
- Professional indemnity insurance
- General standards of behaviour.

On 1 January 2006 some minor changes to the Rules of Conduct came into effect. The changes relate to life long learning and the UK Schedule Part II Professional Indemnity Insurance. In every other way the Rules of Conduct 2006 are in keeping with the Rules of Conduct 2004.

The RICS bye-law 19(1) states that every member shall conduct themselves in a manner befitting membership of the Institution. However, in common with many of the other professional bodies and, despite various attempts to do so, there is no precise definition of what this means in practice. The lack of such definition allows a professional body to move with the times as practices and procedures employed change. For example, unacceptable practices in the past may no longer be unacceptable to the profession today where standards have changed. In the case of discrimination, this might not have been much of a consideration fifty years ago, but today this is high on the agenda.

Unbefitting conduct is in practice judged by, in the case of the RICS, its Professional Conduct Panel or Disciplinary Board. Such conduct is likely to be judged against the following kinds of criteria:

- Acting with integrity
- Honesty
- Transparency
- Objectivity
- Accountability for one's actions
- Know one's limitations
- Treating others with respect

- Setting the right example
- Having the courage to make a stand for what is right and decent.

In summary, *doing to others what you expect others to do to you* (Matthew's Gospel chapter 7 verse 12) is the measure which is to be applied.

The above provides a framework for membership and as Williams (2004) states, this calls for honesty, trustworthiness, proper standards of work, respectful behaviour and independence of mind.

Rule 3 provides that chartered surveyors or anyone acting on their behalf should not act in a manner that either compromises or impairs the:

- Integrity of the member
- Reputation of the Institution, the surveying profession or other members
- Compliance with any code, statute or statement of the Institution
- Member's duty to act in a legitimate way.

The obligation to behave in a manner befitting membership of the Institution does not cease when the office is closed (Williams 2004) but also extends into the member's private life.

The RICS deals with cases that arise in the following ways.

Professional Conduct Panel (PCP)

These hearings take place when a member is alleged to have breached a bye-law, rule of conduct or regulation. The hearings take place in private in front of an independently appointed panel consisting of five people, at least one of which must be a lay member and the others RICS members.

A Disciplinary Board (DB)

These take place when a member is alleged to have breached a bye-law, rule of conduct or regulation but the Chief Executive believes that the PCP does not have sufficient powers to deal with a particular case.

However, only with the approval of the chairmen or vice-chairmen of the PCP can the case be referred to a DB. A DB is heard in public, although the member subject to the complaint can request for the hearing to be made in private. The Board is independently appointed, has a chartered surveyor majority and sits with a legal assessor.

Procedures

On receipt of a complaint, allegation or information about an alleged contravention, regardless of the originator or source, RICS staff are empowered under the disciplinary rules to investigate as appropriate. Where they are satisfied that there is evidence of a breach of the bye-laws, rules or other regulations, and that further action is appropriate, the case can be referred to either a PCP, or DB; or the Chief

Executive may impose a penalty. The rules do not require a complaint to be made to initiate an investigation. It is a fundamental aspect of self-regulation that the RICS itself is able to investigate matters based on information from a variety of sources.

Decisions and appeals

The decision on whether a breach has taken place is based on the balance of probabilities. There is no right of appeal following the findings of a PCP, however, members are always informed of this matter, in writing, before a hearing and given the option of having their case heard by a DB from which there is a right of appeal.

Peer review and decision making

Members can be assured that any cases brought as a result of a staff investigation are carefully considered, either by a panel or board, prior to any decision. Peer involvement in this process is vital and is another essential plank in any self-regulatory regime.

Other than in exceptional circumstances, the results of all cases are publicised via RICS.org and RICS Business and sometimes in a local newspaper or trade magazine where the public interest demands it.

Indemnities and immunities

The bye-laws, rules and regulations apply to all members of the RICS regardless of any official position they may hold or have been elected to within the Institution. There are no exceptions, immunities or indemnities for officers of the RICS.

All members are treated equally. The only indemnity applicable to an officer of the Institution is when he/she is acting in an authorised, official capacity.

There is no question of any member of a RICS committee being able to claim any form of immunity in a professional conduct process or of any provision for indemnity against third party claims being either relevant or appropriate in such circumstances.

All members have an opportunity to have their cases heard and that investigating staff follow proper procedures and rules. RICS Panels and Boards hear fewer than 200 cases a year, which is very positive in the context of an organisation regulating 110 000 members and 12 000 firms.

References and bibliography

Cornes D. and Winward R. *Winward Fearon on Collateral Warranties*. Blackwell Publishing. 2002.

Dugdale A.M. and Stanton K.M. *Professional Negligence*. Butterworth. 1989.

Griffiths M. *Partnership and Legal Guide*. CLT Professional Publishing. 2000.

James P.S. *Introduction to English Law*. Butterworth. 1994.

Jess D.C. *Insurance of Professional Negligence Risks*. Butterworth. 1989.

Patten B. *Professional Negligence in Construction*. Spon Press. 2003.

RICS. *Caveat Surveyor: Negligence Claims Handled by the RICS Insurance Services*. RICS Books. 1986.

RICS. *Direct Professional Access to Barristers*. RICS Books. 1989.

RICS. *Caveat Surveyor II*. RICS Books. 1990.

RICS. *Compulsory Professional Indemnity Regulations*. RICS. 1993.

Ryley M. and Goodwyn E. *Employment Law for the Construction Industry*. Thomas Telford. 2000.

Williams G. *Professional Conduct for Chartered Surveyors*. RICS Books. 2004.

5 Research and Innovation

Introduction

The wider role of quantity surveying is concerned with best use of scarce and available resources. Traditionally many of its practices have been applied to the construction industry, although the techniques and skills have a much wider application. A majority of the practices and procedures that are used have been developed in practice from a pragmatic approach to meet the needs and requirements of clients and the construction and property industries. Whilst there is much that can be described as research and development, little has been carried out in a structured manner and then only by a relatively few members of the profession. There is a fundamental need to verify that the practices being used are the most effective for the job. At the same time the profession does not stand still and there is the need to develop its full potential. These two statements are both valid but are frequently seen as a dichotomy.

One of the most frequent criticisms of quantity surveyors has been their narrowness of vision. Problems are considered in the context of construction costs with only a limited appreciation of the other project attributes. Until recently the context of the quantity surveyor's work was restricted to costs, with hardly any consideration of value. Davis, Langdon and Everest (1991) suggested that it was simpler to cost engineer a design solution to a predetermined cost target, than to ensure a proper balance of expenditure throughout the building to maximise the client's benefit or to add value. However, the history and development of the quantity surveyor (Ashworth and Hogg 2000) clearly shows that there has been a paradigm shift in thinking and practice from cost to value. Also, as new techniques and methods have been added to the quantity surveyor's portfolio, there has been an incremental shift in the development of new tools.

Research and development work is now seen by some of the larger professional practices to be an important part of their activities. In some practices it forms a distinctive feature of the practice's profile and portfolio of work and as a source for income generation. It achieves this in two ways: by supporting the work of the practice and through providing high quality publications for business and industry. These also provide a high profile for such practices. Work within the profession:

- Provides a framework within which quantity surveying research and development is encouraged to take place

- Raises an awareness amongst quantity surveyors of the importance and role of research and innovation
- Further develops a dynamic research and development community
- Seeks to persuade government and other agencies of the importance of such research and development and the need for its proper resourcing
- Stimulates debate on the future direction of the profession and the role of research and development within it.

RICS

The RICS in 1999 launched a Research Foundation recognising a rapidly changing world and work environment. The Foundation, whilst short-lived, was to support and enhance the profession through an accurate and rigorous understanding of the way in which the natural and built environments behave. Research is critical to this understanding. The objectives that were set still apply to:

- Act as the leading international body in the understanding of all matters relating to the development, management and use of the built and natural environments
- Provide leadership on the major debates that are taking place
- Contribute to the more effective and efficient use of built and natural environment, to the benefit of everyone involved
- Disseminate this to the widest audience.

All research, development and innovation should:

Set an agenda for debate
- Using focused and rigorous research to bring about a clearer understanding and analysis of the major policy issues
- Acting to bring about a consensus on the key issues to be addressed by policy and research development
- Influencing policy and strategy at the highest levels.

Promote the latest thinking
- Identifying innovative solutions and ideas
- Putting forward best practices in the implementation of new techniques and processes
- Developing a clearer understanding of the operation of the property, construction and development markets.

Support an active research community
- Developing world class research capabilities in the built and natural environments
- Encouraging the exploitation of new thinking and approaches to the key issues that we face in the effective management and development of our built and natural environments.

The RICS remains committed to a research and development programme which combines long term strategic studies aimed at determining the future shape and direction of the profession in the construction and property industries and targeting those projects which support Institution policy developments. It awards a number of small research grants, often, but not necessarily, to those employed as academics. It publishes and distributes summaries of these research findings, so that debate and policy can be informed by the latest research and analysis. A full list of these is available from the RICS.

In 1991, the RICS undertook a survey to help determine the extent of these activities in the universities and published a report, *The research and development strengths of the chartered surveying profession: the academic base.* The information included in the RICS report provided an overall profile of research activities, general areas of capability and specific research expertise, research links with the profession and details of external research contracts.

The RICS sponsors annually the COBRA conference that is attended by academics and practitioners.

Classification of research and development

The research process is illustrated in Fig. 5.1 and usually commences with a desire or need to find out or verify existing knowledge. The initial starting point is to set out the objectives in the form of a research proposal. Often such proposals are too broad resulting in a superficial analysis and understanding and a failure to achieve the expected outcomes. The better proposals are those which relate to a problem, where the researcher already has some prior knowledge and experience. Whilst the objectives that are set at the start of the process may be clearly understood, it is not

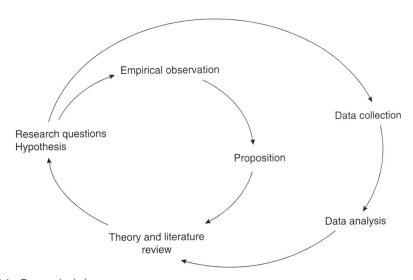

Fig. 5.1   Research circle.

uncommon for these to be altered as the research proceeds. It may be necessary to reappraise the problem after an initial investigation and once a thorough literature survey has been completed. Academic research that can be carried out in collaboration with practice is a useful combination to be pursued. Research may be classified under the following headings:

Basic research   Experimental or theoretical work undertaken primarily to acquire new knowledge of the underlying foundation of phenomena and observable facts. Undertaken without any particular application in mind.

Strategic research   Applied research in a subject area which has not yet advanced to the stage where eventual applications can be clearly specified.

Applied research   The acquisition of new knowledge which is primarily directed towards specific practical aims or objectives.

Scholarship   Work which is intended to expound the boundaries of knowledge within and across disciplines by in-depth analysis, synthesis and interpretation of ideas and information and by making use of rigorous and documented methodology.

Creative work   The invention and generation of ideas, images and artifacts including design. Usually applied to the pursuit of knowledge in the arts.

Consultancy   The development of existing knowledge for the resolution of specific problems presented by clients often within an industrial or commercial context.

Research and development in the construction and property industries

Research and development takes place in all industries including the construction and property industries and its associated professions and is important for the following reasons:

- Technical change is accelerating and progressive businesses tend to quickly adopt new techniques and applications
- Research and development is inseparable from the well-being and prosperity of a country and of the separate business within the country
- Research and innovation are inseparable
- Research and development are necessary to maintain international competitiveness and success, particularly as the craft based traditions of construction diminish and the technological base expands
- This background of constant change and challenges demands an effective research and development base to introduce change effectively and efficiently
- The expenditure on research and development, within the construction industry, is small when compared with our competitors overseas and with other British industries
- Whilst the expenditure on research in the construction industry is lagging behind that in other industries and countries, it is far ahead of that in surveying.

The expenditure on research and development in the construction industry in Britain amounts to about 0.65% of the construction output. Construction companies, for example, contribute about 10% towards this sum, which is approximately one-third of that of our competitors in France, Germany and Japan.

The RICS (1991) Report stated that for research and development activities to fulfil their potential within an area of activity, it was necessary to stimulate a 'virtuous circle' of research. This is typified by the following features:

- There is sufficient awareness and confidence in the capabilities of the research base to encourage the profession to be prepared to commit funds to research
- The flow of funds into research is perceived by the profession as being of great benefit to them
- The profession sees the value in supporting research activities and is willing to continue to invest
- There is a sufficient flow of funds into research in order to support and maintain a high-quality research base.

When a profession achieves this virtuous circle, then it is very likely that it will be extremely well supported by research and development. It will be proactive, dynamic and forward looking. In order to bring this about, one of the most important issues is how to overcome the barrier of lack of knowledge of the research and development capabilities that already exist. For those who still doubt the wider benefits to be achieved from research and development, it is worth noting what the Centre Scientifique et Technique du Bâtiment said in 1990:

> 'The stronger the research and development effort of a sector, the better its image; even in a fragmented sector. Just look at the image of doctors!'

Rethinking construction innovation and research

This Report was prepared under the chairmanship of Sir John Fairclough and has thus become commonly known as the Fairclough Report (2002).

The government spends around £23m each year on construction related research largely to underpin regulation, safety and health and to support the competitiveness and sustainability of the construction industry.

The industry and public interest are inextricably linked as the construction sector has a profound impact over the quality of life of its citizens. The innovative capacity of an industry influences its long-term competitiveness but it is not given the same priority in the construction industry as in some other sectors. Whilst there is a great deal of project based innovation in construction, there is no mechanism for capturing this for future projects.

There is an acknowledgement that a carefully focused research and development programme driven from clear strategic perspectives will be needed to support its work. Research and development has a pivotal role but the effort needs to be carefully focused on activities that the industry will pursue out of self-interest or that of its clients or government policy.

Further information can be obtained from the Construction Research and Industry Strategy Panel (CRISP) at www.ncrisp.org.uk.

Government should:

- Safeguard current investment in construction research and development
- Increase investment in support of productivity, value for public sector clients and strategic issues
- Assist industry in the production of a mission statement covering its contribution to the quality of life agenda
- Provide pump priming to help industry set research and development priorities based on a strategic analysis of the issues it faces
- Help the best innovators maintain their position in the international forefront
- Challenge the professional institutions to institute arrangements for consideration of near market competitiveness issues, e.g. setting of industry standards.

In commissioning research and development, government should:

- Procure on merit, avoiding monopolies but encouraging centres of excellence, which will encourage deeper skills and help retain staff
- Encourage collaboration between research centres
- Ensure direct relevance to industry needs.

The Strategic Forum should:

- Engage with key industry leaders in setting a strategic vision for industry, including key issues requiring action and a mission statement on construction's contribution to quality of life
- Prioritise research and development and define long term programmes based on the work of an enhanced CRISP.

Whilst direct construction activity is responsible for only 5.2% of GDP, it is responsible for 60% of all fixed capital investment in the form of buildings, structures and the infrastructure on which other industries depend. The efficiency of construction therefore has a bearing on the competitiveness of a large part of the UK economy and affects the quality of life of all.

Total public funding for construction related research is between £50m and £70m per annum, the bulk of which comes from the EPSRC (Environment and Physical Sciences Research Council), the DTI (Department of Trade and Industry) and the DTLR (Department for Transport, Local Government and the Regions). This needs to be set in the context of a total national expenditure on construction of some £90bn per annum.

The links between the construction industry and universities have traditionally been poor, although through the EPSRC's Innovative Manufacturing Initiative (IMI), contact has improved. The university construction research base is spread over 85 universities, focusing on longer-term problems and new ideas. However, there is little interdisciplinary working or incentive to collaborate with industry leading to poor exploitation of research outputs by the industry.

The industry has a very large number of firms of which few are of sufficient size to employ people with higher qualifications.

Government supports CRISP with a small secretariat and funds to commission small consultancy studies aimed at defining the research and development needs of the industry and providing input to government. However, it is not well known across the industry and is dependent upon a small number of very dedicated individuals to provide voluntary input.

Government support

The four main categories of government support are as regulator, sponsor for improved competitiveness, client and policy maker. In its role as a client, it is responsible for 40% of the industry's turnover, about £25bn. It is very much in government's interest to press for improvements in the industry, a strong research base and collaboration between research providers.

Building regulations/code and standards

The regulatory role in terms of Building Regulations means that government needs to have research competencies available as new technologies or design.

Government as industry sponsor (*Rethinking Construction*)

The government's aim as a sponsor of the industry is to improve its productivity, competitiveness and innovation. The report *Rethinking Construction* (Egan 1998) provides much of the policy framework for construction research, alongside the push for a more sustainable approach. The Movement for Innovation (M4i) demonstration projects set up under the Egan policy framework have been very successful in demonstrating improved performance and the value of learning from other sectors and applying technology transfer.

However, much more time and effort is needed to widen the lessons learned from the demonstration projects. Also, the industry still lacks industry standards which would provide economy of scale for suppliers and act as powerful productivity drivers.

Identification of key issues, long-term programmes and industry engagement

Government support for research needs to be informed by the identification of the key issues facing the industry. However, project by project funding is unlikely to make real inroads into these critical issues and they will need to be tackled by larger research programmes with integrated, multidisciplinary teams. In addition, industry engagement will be required to provide at least half of the resource. The government's research management contractors who provide additional resources to help manage the large number of research projects could be used to help define a more strategic programme in support of the key issues referred to above.

Government as client

In the public sector (the Ministry of Defence's *Building Down Barriers* excepted) government clients have not engaged with the construction research agenda. Government clients therefore need to be much more actively engaged in research collaboration to embed learning and innovation in their own organisations.

Strategic thinking

Strategic forum

The industry has not been able to develop longer term or strategic thinking and there is insufficient intellectual capacity devoted to its future. The recently established Strategic Forum is taking forward the Egan agenda and it is well placed to consider other, longer term issues.

Foresight

The recent Foresight exercise concentrated on the needs of the industry and operational effectiveness rather than a wider vision and strategy.

Defining long-term needs

It is critically important that a better mechanism is developed for defining the industry's long-term research needs, within a process of wider strategy setting. There is no mechanism at present to procure long-term research. There needs to be greater commitment from industry leaders to a strategic vision for the industry and its place in society. This would lead to better communication of the research and development agenda. Current input to CRISP relies on voluntary effort but the strategic thinking is that the industry needs cannot be undertaken on this basis in any sensible timescale. A strategic research and development agenda would enable industry leaders to engage with research funders to better prioritise their own investment and help define public funding.

Changing role of the quantity surveyor

The role and work of the quantity surveyor have changed considerably, particularly over the past two decades (see Chapter 2). These changes in direction and practice are expected to be overshadowed by the accelerated developments which are likely to take place in the immediate future. A major theme of the report *QS 2000* (Davis, Langdon & Everest 1991) is the changes facing the profession. The much later report by Powell (1998) was more alarmist but sent the same message.

Perhaps the quantity surveying profession had failed to heed the warning of *QS2000*? The RICS presidential address in 1999 (Kolesar 1999) further reiterated the importance and inevitability of change. Heraclitus (c.540–480 BC) stated: 'Nothing endures but change!' The RICS presidential address in 2000 (Harris 2000) continued with a similar theme, with the desire to elevate the status of the profession. Research and development have important roles in this vision of the future. The following represent some of the issues which, in the absence of appropriate and relevant research and development, may allow opportunities to be missed or to be ineffectively undertaken:

- Blurring of professional disciplines, both within the surveying profession generally but also with other professional groupings
- Wider range of services offered to present clients
- Application of quantity surveying expertise to new markets
- More extensive and intensive use of information and communications technology to improve efficiency and effectiveness
- Changes in the professional structure
- Multi-discipline working and development
- Increased emphasis on continuing professional development
- Geographical dispersion of work to allow for the most economical methods of working
- Forecasted shift between professional and technician activities.

Harris (2000) has further argued that to persist in prescribing tight definitions of our expertise in an attempt to safeguard our work from our competitors will not be successful. He further argues that we will also not succeed through our technical superiority alone.

As with all professions, quantity surveying has evolved and will continue to do so for the foreseeable future. This evolution has been a response to changing demands and services expected from clients and the developing skills and knowledge base of practitioners coupled with the wide implications of information technology. Fig. 5.2 indicates some of the changes to the profession which have occurred since the middle of this century. The majority of these changes have happened as a result of the pragmatic needs of practice in response to changes in the needs of clients and technology, rather than through any formal development or research.

Research and development in quantity surveying practice

Several of the larger quantity surveying practices have now established research and development sections as an integral part of their practices. This has been done both in an attempt to diversify and also to be at the leading edge of the profession. Some practices have been able to recoup income from work that is broadly described as research and development. Others have joined in collaborative ventures with universities, become members of research advisory teams or have allowed researchers access to non-sensitive data and information. Such activities

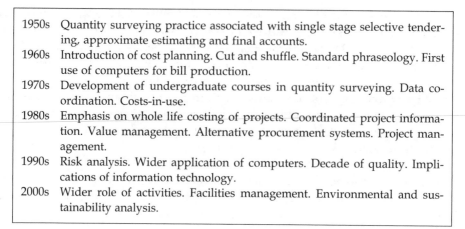

1950s   Quantity surveying practice associated with single stage selective tendering, approximate estimating and final accounts.

1960s   Introduction of cost planning. Cut and shuffle. Standard phraseology. First use of computers for bill production.

1970s   Development of undergraduate courses in quantity surveying. Data co-ordination. Costs-in-use.

1980s   Emphasis on whole life costing of projects. Coordinated project information. Value management. Alternative procurement systems. Project management.

1990s   Risk analysis. Wider application of computers. Decade of quality. Implications of information technology.

2000s   Wider role of activities. Facilities management. Environmental and sustainability analysis.

Fig. 5.2   Trends in quantity surveying practice.

are also able to provide a useful spin-off for public relations and publicity. Research and development is therefore seen as being important for the following reasons:

- Improving the quality of the service provided to clients
- Increasing the efficiency of work practice
- Extending the services which can be provided
- Developing a greater awareness of new technologies
- Providing a fee earning capability from research and development contracts
- Enhancing public relations and practice promotion.

A search of the websites of the leading firms of quantity surveyors indicates just how diverse a proportion of their work has become. All are inter-disciplinary and many use descriptions for their activities as management consultants to the construction industry.

Several of the websites refer to research that they are carrying out or supporting, in anticipation of being able to provide improved or extended services to their clients. Some have funded projects with universities that are specifically aimed at gaining an edge over their competitors and other firms who would like to enter these markets.

The type of research being undertaken relates directly to their own practice interests and includes:

- Building costs
- Cost reports
- Economic forecasts
- Country analysis
- Global market reviews
- Construction industry analysis
- Sector information.

A majority of these firms have developed a series of publications based on the above that can be downloaded from their websites.

Some of these practices publish an extensive array of economic information including commodity price movements, currency fluctuations and macroeconomic data that are all collated and analysed by full-time research analysts that enables them to improve the quality and accuracy of their forecasts of future construction cost trends. When these are achieved they enable clients to accurately plan ahead with the assurance that they will be one step ahead of the competition.

Many of these practices also produce purpose-designed reports to meet with specific client requirements.

A number of practices provide a range of construction price books both for the UK and for other countries around the world. These are now often available as a CD-Rom.

Data collection and analysis

In some practices, research and development departments evolved from the routine collection of cost and contract information for their own regular use in cost planning and cost forecasting associated with new commissions. It is well understood by all surveyors that the best information available is that which is collected by the practice itself, relying only upon other sources, such as the BCIS, price books, etc. as secondary sources of data.

Market trends

Since construction costs and prices are associated with productivity and market conditions it is important for surveyors to be informed of current trends in practice in order that clients can be properly advised. Whilst the national published data is again valuable, this will form only a secondary source to the specific market information and trends which are retained by the surveying practice.

Practice expertise

A practice that has an extensive involvement with a particular type of construction project is able to provide detailed analyses on the different project components, so that the full extent of any changes in design or specification can easily be assessed. Such a practice is then able to develop a detailed expertise with a particular type of construction or procurement arrangement.

Objective and speculative research and development

It is also important to develop a research database prompted by the needs of the individual surveyors in order to allow for possible future surveying services to be developed. On this basis clients can be provided with information which in the future might generate new commissions for the surveyor. Development of this type may be undertaken with specific objectives in mind or be speculative, attempting to forecast changes in the market or in the demand for surveying services. It may also be done to meet identified client needs for specific surveying services.

Fee earning capability

In addition to the more usual surveying services, research and development activities are able to generate their own contracts for commissioned research. Some practices, for example, have developed a part of their own research and development expertise to publish the systematic collection and analysis of their data. Surveying practices are also able to submit research and development ideas or bids to government or research bodies for research and development contracts. This has in the past been done in association with a university or college as a collaborative venture.

Academic research

Academic research specifically in quantity surveying and in the wider context of the construction industry that is carried out in universities is a relatively young discipline of no more than about fifty years old. Research by members of the RICS in a professional setting, such as the Wilderness Group, pre-dates this and was then aimed very much towards developing the profession and in advancing practice. This was in the profession for the profession. Many will suggest that for research to have any impact it is yet too early. Others will also argue that a large amount of research has to be carried out for some of it to have any real impact. It must also be accepted that there is a considerable amount of research undertaken in other, more scientific disciplines, much of which does not yield any positive outcomes. That is the reality of the research agenda. It is about finding out and testing ideas and hypotheses. If we knew the outcome at the outset then there would probably be no need for research at all, but simply the development and future application of new ideas. But that is not what the world is like. Students undertaking dissertations as a part of an undergraduate programme choose a topic for study, without knowing the outcomes that this might achieve.

However, there seems to be a real dilemma when trying to link research and practice. There are tensions between these as noted by Fairclough (2003).

Practice, unfortunately, appears to see little benefit from academic research. It is also difficult, for example, to identify any research that has changed quantity surveying, even in the rapidly changing environment. Virtually all of the changes in practice have been due to the way that clients and quantity surveyors themselves have sought to improve what they do, rather than because research has influenced these activities. This is unfortunate. There is an apparent gulf between what practice sees as research and what academic research wants to deliver. These tensions are exacerbated by the research councils and their own definition of what they believe constitutes research. There is also the danger of a widening gap where some academics have never practised.

Research papers are not read by those in practice. This is unlike the physical science disciplines, yet this is the model that is used. Also the complexities of some academic papers are not easily understood by those who are in key positions in practice. The possibility of any influence or enhancement is therefore negligible. There needs to be a better way of presenting research findings to those who are working in practice if researchers are to move into a position of influence. It has been suggested that, on average, no more than seven people will read a research paper. These papers require a large amount of investment and intellect in their preparation, yet it would appear that very few people are anxious to find out what has been discovered and to apply it in practice. There is an important and urgent need to provide better joined up thinking between research and practice and its potential for application for the benefit of both researchers and those in practice.

An important feature of the Research Assessment Exercise in 2008 is the need for researchers to identify what impact their research has had on, for example, other researchers or the practice of the discipline. This provides an important focus for a researcher's work.

Research dissemination

The publication of academic research papers is now a part of an academic's role for those working in universities, however, so much of this goes unnoticed, possibly because much of it is only disseminated through the academic mediums of journals and conference proceedings. Some of these include:

- *Construction Management and Economics* (Taylor and Francis Journals)
- *Engineering, Construction and Architectural Management* (Emerald)
- *Journal of Construction Procurement*
- *Journal for Education in the Built Environment* (JEBE). The journal of the Centre for Education in the Built Environment (CEBE) (www.cebe.heacademy.ac.uk)
- *Journal of Construction Engineering and Management* (American Society of Civil Engineers)
- CIB (International Council for Research and Innovation in Building and Construction) Conferences and Networks (www.cibworld.nl) that include:
  - W55 *Building Economics*
  - W65 *Organisation of Management of Construction*
  - W70 *Facilities Management and Maintenance*
  - W89 *Building Education and Research*
  - W92 *Procurement Systems*
- Association of Researchers in Construction Management (ARCOM) (www. arcom.ac.uk)
- RICS Construction Research Conference (COBRA).

Information and communication technologies (ICTs)

'Information technology' is a relatively recent addition to the English language. It has its counterparts in the French, 'Informatique', and the Russian, 'informatika'.

For many, information technology is synonymous with new technology such as microcomputers, telecommunications, computer controlled machines and associated equipment.

The use of computers has had a dramatic influence upon human behaviour and development since the early 1990s. Computers have also had a major impact upon the profession of quantity surveying, in respect of the role and function of the professional activities. Whilst the capability of computers and their associated software continues to increase, their relative and real price decreases. Reliability is now generally good and their use has become easier as simplified and user-friendly procedures have been introduced. The use of information and communication technologies has also created wide ranging implications. From a social point of view ICT has changed the way in which we communicate and reach decisions, manage our work and store information.

Computer literacy

Computer literacy requires an understanding of the following two related areas of computer knowledge:

- *Knowing computer capabilities and limitations:* General understanding of the organisation, capabilities and limitations of the various machines, i.e. the hardware.
- *Knowing how to use computer:* Familiarity with the common uses or applications of computers. Comfortable working with pre-written software.

Additional competence is gained by mastery of the following two additional areas:

- *Knowing how computer software is acquired:* General idea on how individuals and organisations develop custom made programs and information systems.
- *Understanding the computer's impact:* Aware of the impact that computers and information systems are having on people and organisations.

Information technology continues to develop at an exponential rate. Virtually everyone involved in the construction industry now has extensive access to this technology. What can be imagined will be achievable, if this is desired. Many aspects currently not imaginable will also be achieved, probably in a shorter term than envisaged. There is a tendency to over-estimate what will happen in the next couple of years, but to under-estimate what may take place in the medium term.

Quantity surveying practice

Information technology has been shown to be an effective tool for a wide range of applications in the construction industry. These have included computer aided design (CAD) and drafting, and assisting manufactures of building materials. The use of CAD by design professionals has been able to demonstrate considerable success in the modelling of design solutions. Although information and communication technology applications are capable of achieving high work levels and have been reported to offer time savings of up to 40%, they sometimes do not always meet the expectations for increased productivity and product quality in the construction industry. The opportunities are substantial. The following areas lend themselves to the development of information technology in construction applications:

- Design and production techniques which incorporate design aids, virtual reality, robots, energy management, commissioning of buildings and education and training
- Information systems which employ databases, quantities, drawings and models, specifications, property data and electronic data interchange
- Hardware and software which include interfaces, expert systems, standards, integration of applications and software techniques
- Communications which apply to intelligent buildings, wide area networks, local area networks, integrated services digital networks, optical fibres and wiring, radio technology and security.

Survey findings amongst quantity surveyors suggest that a majority are ICT literate. It is impossible to complete a quantity surveying degree course without the use of ICT for a wide range of applications. These will include the general applications such as world-processing and spreadsheets and the more specialist software such as digitised measuring packages, BCIS on-line, estimating and tendering software and integrated databases. Quantity surveying consultants use a complete range of software, but there is often a lack of an overall strategy for its implementation and investment and for on-going training.

In order to integrate the client, team and process more fully, supporting ICT networks will become increasingly adopted. However, the continuing problem with the construction industry remains its lack of standardised data and information and the lack of compatibility between different programs. EDICON and CITE are standards that are readily available and quantity surveyors should take full advantage of these. The multi-disciplinary and design and build environments are the most conducive to the development of integrated IT facilities.

Figure 5.3 provides an indication of the use of information technology in quantity surveying practices. Whilst the use of IT continues to grow owing to its efficiency and effectiveness of operation, in some areas of work manual methods still remain competitive in terms of familiarity and the speed of application. In other cases, computerised systems have been able to improve the quality of service provided to clients and to produce information that was not previously available or was too difficult to obtain from manual systems. In some quantity surveying practices IT has progressed so far that they no longer see the necessity for extensive secretarial support, since their surveyors access most of their information and data directly from the computer. However, typically only about 60% of practices currently have staff with some formal computer training.

The likely future methods of working according to Powell (1998) suggest the following in respect of information and communication technologies:

```
Application                     Percentages
                                10  20  30  40  50  60  70  80  90  100

Development appraisal           *****************#########
Early cost advice               ****************************######
Measurement                     *************************######
Digitisers                      ************************####
Documentation production        *********************************************####
Post contract                   ****************************########
Cash flow forecasting           *****************************########
Project management              ***********************####
Word-processing                 ****************************************************
Spreadsheets                    ***************************************************
Databases                       *********************#####
Desktop publishing              *********#####
Internet                        ***********************************####
Electronic mail                 *************************************************
```

***** Current use
Projected use

Fig. 5.3   Extent of use of information technology by quantity surveyors.

- Greater access to international and non-local markets and projects
- Use of low cost third world resources
- Cooperative team working
- Shared contributions to managed process
- Information becomes a commodity
- Concurrent engineering
- Tele-working
- Hot-desking.

Major ICT issues

Machiavelli made this comment in the sixteenth century:

'It should be borne in mind that there is nothing more difficult to handle, more doubtful of success, and more dangerous to carry through than initiating changes. The innovator makes enemies of all those who prospered under the old order, and only lukewarm support is forthcoming from those who would prosper under the new. Their support is lukewarm partly from fear and partly because men are generally incredulous, never really trusting new things unless they have tested them by experience.'

It is not surprising therefore that the use of the computer moved slowly in the early days. The removal of a measure of control from people and handing this over to a

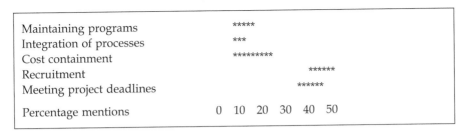

Fig. 5.4 Top five ICT problems.

machine is one of the biggest changes that humankind can make. The 1980s saw these attitudinal changes amongst surveyors.

Figure 5.4 identifies the top five problems of using computers in practice. Meeting project deadlines remains a major problem area. Some difficulties in the past have occurred where a computer system has failed with the surveyors' data locked inside. This provides an embarrassment and frightens surveyors from their possible future use. The reliability of computers has now largely overcome this problem. However, the importance of making back-up copies of work in progress cannot be over-stressed. Meeting project deadlines has always been a particular problem for quantity surveyors due to their position of being the last line in the design process and the possible cause of over-runs on the other consultants' activities.

Impact upon the quantity surveying profession

Information technology has its origins in the technologies related to restricted view of information: the generation, processing and distribution of representations of information. Examples are the telecommunication and computer engineering, the data processing and office machinery industries. The products of these industries still form the bulk of information technology products. Progress in recent years has been towards the extension of this data engineering or telematics, to an increasing range of areas of application. This has brought with it an active interest in the human aspects of information such as its quality, value, utilisation, etc.

Revolutions of this type, although they appear to take place dramatically quickly, do not happen overnight. A full working lifetime has expired since the advent of the first computer. Also many of the predications about the implications of computers such as paperless offices and the non-necessity for office space because we will all be working from home, have yet to take place. Even the demise of the bill of quantities has yet to occur. As with most forecasts of the future, individuals and organisations are often wide of the mark. Working life continues to evolve, perhaps gaining in momentum with future change now being the order of the day. The key words in all of the professions is adaptability and the development of transferable skills, rather than skills which will be left behind and of no future use.

The impact of computers on the quantity surveying profession can be summarised as follows:

- Reduction in the amount of time spent on repetitive processes
- Improvements in methods of communications, particularly world-wide
- Enhancement in the quality of the services provided (emphasis is already being placed upon this during this decade of quality)
- Development of a broader range of services, sometimes encroaching on other professional disciplines
- Speed in the execution of tasks.

The future of ICT

In *Information Horizons* (Miles, *et al.* 1988), three perspectives are provided on the nature of how information technology may change society in the future, as follows.

The continuists

Information technology is merely the current stage in a long term process of developing technological capacities. The so-called revolutionary claims are over-stated. The rate of diffusion of information technology will be much slower than claimed by interested parties. There will be many mistakes and failures and discouraging experiences. The main features of society are liable to remain unchanged by the use of information technology. The future will largely be based upon an extrapolation of the past.

The transformists

Information technology is based upon revolutionary technology with unprecedented progress made in computers and telecommunications. The positive demonstration, effects and proven success of information technology in meeting new social and economic needs will promote rapid acceptance of change. Information technology will foster a major shift in industrial and agricultural practices through the introduction of new technologies.

The structuralists

Information technology has revolutionary implications for the economic and social structure in society. Some countries, industries and professions will become far more able at capitalising on the possible potential of information technology. Change will appear in waves demanding new styles, structures and skills. In practice the future cannot be adequately forecasted; personal development is therefore required which will provide for flexible and adaptable solutions.

The importance of change

Change is the natural order of events. When we refuse to change, we become outdated and eventually this means unemployable. The reasons for change in the construction industry are not difficult to find, and include, for example, the following:

- Government intervention in the construction industry through privatisation, PFI, Construction (Design and Management) (CDM) Regulations, compulsory competitive tendering, European legislation
- Recent reports on the state of the industry:
  - *Constructing the Team* (Latham 1994)
  - *Improving Value for Money in Construction* (Atkin and Flanagan 1995)
  - *Towards a 30% productivity improvement in construction* (Construction Industry Board 1996)
  - *A Statement on the Construction Industry* (Barlow 1996)
  - *Value for Money: Helping the UK afford the buildings it likes* (Gray 1996)
  - *The Challenge of Change* (Powell 1998)
  - *Rethinking Construction* (Egan 1998)
  - *Modernising Construction* (National Audit Office 2001)
- Pressure groups formed to encourage change and improvement
- International comparisons, particularly the USA, Japan and the Single European market
- The apparent failure of the construction industry to satisfy the perceived needs of its customers, particularly in the way that it organises and executes projects
- The influence of patterns of education and research on these processes
- Trends generally in society towards greater efficiency, effectiveness and economy
- Rapid changes expected from information technology in respect of design, management and manufacturing processes and practices
- A desire to make the construction industry more high-technology orientated
- The varying attitudes amongst the professions
- Developments occurring in other similar and different industries and the need for the construction industry to catch up
- A desire to reduce the adversarial nature associated with the construction of buildings
- The awareness of quality assurance mechanisms in other industries
- A desire to establish best practices in our work
- The over-riding wish of clients for single point responsibility
- Changes in culture and work practices.

The construction industry is perceived as dirty, dangerous, exposed to bad weather, unhealthy, insecure, underpaid, of low status and with poor career prospects for educated people (Latham 1994). It is widely agreed within the industry that it is too easy to set up in business as a general contractor. No qualifications are required, no experience and virtually no capital. Whilst market forces ultimately remove incompetent firms by depriving them of work, the existence of such unskilled producers is bad for clients and damages the wider reputation of the industry. Women are seriously under-represented at all levels in the industry (Latham 1994). The need for change in the construction industry is clear.

The challenge of change

The British construction industry has a long and honourable tradition and records of achievement, both within the UK and overseas. A few years ago the same could be said of the British manufacturing industry. Whilst some notable parts of the latter have survived, many have ceased to exist or have declined at an alarming rate. Those that have managed to survive made radical changes to their practices. There are already sufficient signs and indicators to suggest that the construction industry of Great Britain is facing similar challenges (Barlow 1996).

The changes required first are mainly organisational and cultural rather than of a technical nature. Whilst increased resources are necessary in information technology and research and development, the underlying cultural changes remain of paramount importance.

The issues facing the future of the construction industry are many and varied and include:

- Under capitalisation exacerbated by fragmentation and the large numbers of small firms
- Low technology, labour intensiveness and traditional craft base
- Litigious basis for settling differences and disputes
- Lack of prototype development resulting in untested and ill-specified components and technologies
- Low level of computing and use of information technology
- The use by government of the construction industry as an economic regulator contributes towards a cyclical workload
- High number of company business failures
- Poor image, working practices and employment conditions
- Difficulties of recruiting, training and retaining a skilled and committed workforce.

There is a general lack of understanding of best practices or even an awareness in some situations that change is the order of the day and urgent. There is a determination to survive but a lack of realisation of international competition.

Innovation

Innovation is about the introduction of the new in place of the old, especially changes in customary practices. Innovation involves developing a strategy which involves people, practices, processes and technology to deliver high added value.

A great deal of technological change passes unnoticed. It consists of the small-scale progressive modification of products and processes. Such has been the description of the construction industry. Freeman (1987) has described such changes as incremental innovations. They are important, but their effects or shock-waves are only felt within the immediate vicinity. More important are the radical innovations or discontinuous events which can have a drastic effect upon products and processes. A single radical innovation will not have a widespread effect on the economic system. Its economic impact remains relatively small and localised unless a whole cluster of radical events are linked together in the rise of new industries or services, such as the semiconductor business. These are the more significant changes. The following five generic technologies have created new technology systems:

- Information technology
- Biotechnology
- Materials technology
- Energy technology
- Space technology

These represent new technology systems that change the style of production and management throughout the system. The introduction of the electronic computer is an example of such transformations.

The concept of long-ware developments, each of less than 50 years duration, is generally associated with the work of the Russian economist, N.D. Kondratiev. Figure 5.5 is a highly simplified picture of the sequence that might be commonly envisaged. Four complete K-waves are identified with the implication that we are currently entering a fifth. Each wave has lasted approximately 50 years and appears to be subdivided into four phases: prosperity, recession, depression and recovery. Each wave seems to be associated with significant technological innovation associated with production, distribution and organisation.

The fifth Kondratriev cycle, which appears to have begun in the early 1990s, is associated primarily with the first of the five generic technologies listed above. Information technology is around which the next wave of technological and economic changes will cluster (Freeman 1987). The development of information technology originates from communications technology and computer technology.

Innovation in the quantity surveying profession should seek to:

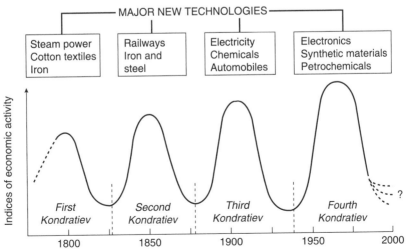

Fig. 5.5 Kondratiev long waves of economic activity and associated major technologies (*Source:* Freeman 1987).

- Identify important innovations which have contributed to construction quality cost effectiveness and added value
- Educate industry leadership in the importance of innovation within their respective companies and associations
- Encourage architects, engineers, contractors and surveyors to develop and implement processes for innovation to occur
- Recognise important innovations within the industry, once achieved, through awards, publications, etc.
- Develop financial support for further innovations within the construction industry and its varied services.

The Construction Productivity Network (CPN) was formed as a result of the Latham Report (1994) and is under the auspices of the Construction Industry Board (now superseded by the Strategic Forum for Construction). This is a partnership of the construction industry, its clients and government working together to improve efficiency and effectiveness in construction. The Board aims to secure a culture of co-operation, teamwork and continuous improvement in the industry's performance. CPN exists to promote the sharing of knowledge and the benefits of innovation across all sectors of the construction industry through workshops, visits, conferences and publications.

Conclusions

The Department of the Environment's Construction Sponsorship Directorate (1995) has provided a new focus for change through its research and technology support programme. In this it is seeking partnerships within the industry to develop a whole industry research strategy. Representatives will come together in the Construction Research and Innovation Strategy Panel (CPISP). The strategy seeks to build on existing industry improvement initiatives.

Anything that does not add value to the process or the product is unlikely to be around for much of the future. Powell (1998) has suggested that quantity surveyors add value in the following ways:

- Facilitating earlier trading by helping clients to gain market advantage by selecting the correct procurement route
- Reducing maintenance costs through the application of whole life costing
- Minimising the disruption of a business during repairs, adaptations or maintenance programmes
- Offering single point responsibility for the project
- Reducing the focus on the costs of components and increasing the benefits to a client's business
- Adding value to the client's business of any proposed alternatives
- Understanding how the client's profits might be increased or business costs reduced
- Being prepared to challenge designs and suggest new ways of executing work

- Suggesting, developing and advising on procurement options
- Understanding how work may be alternatively sourced, procured, supplied and installed
- Gaining an understanding of how a client's competitors source work elements and execute their projects
- Establishing sector benchmark data to facilitate project comparison and evaluation.

The future for the quantity surveying profession is intrinsically linked with a past analysis of its activities, i.e. the principle of reaping tomorrow what we have sowed today, together with a response to the changing forces in industrial and commercial society. The importance attached to education and training and continuing professional development is now well recognised if a profession is to be vibrant and dynamic. The importance of research and development within this, forms an integral part in acquiring capability for the future. An effective research and development programme can only strengthen and assure the quantity surveyor's long term future.

Discussion topic

What are the main barriers facing the transfer of new knowledge into the construction industry?

The construction industry sees itself as separate to other types of industry. It has thus remained somewhat isolated from other industries in the development of management theories and practices. Some might even suggest that it is aloof. Whilst construction management has borrowed much of its practices from general management, it attempts to remain distinctive and also to be seen as a separate academic discipline. For example, most of the management taught on courses is delivered by construction professionals rather than by the array of the different management disciplines. It is often argued that to teach this subject effectively requires someone from the discipline who can relate and use examples from their experiences whilst working in the construction industry. Also, in some cases new knowledge such as total quality management (TQM) or benchmarking is sometimes introduced into the construction industry from other sectors, without a proper evaluation of its transferability in a construction context. In other cases there is sometimes a piecemeal adoption of management techniques that results in a loss of impact or diminished understanding and application.

It is only by understanding the way that knowledge is embedded within institutional and cultural contexts that it is possible to understand more fully the obstacles and practical solutions to the effective transfer of knowledge between different sectors or industries.

Logically, there are a number of possible strategies and outcomes of the process of transferring knowledge between the different industrial sectors, across organisations and between practice and education. They are dependent upon two main factors. The extent to which they are accepted as relevant and applicable to the construction industry and the extent to which their original logic is retained when they are implemented. These can be summarised as follows (Bresnen 1999):

		Retention of original logic	
		High	Low
Perceived relevance and adaptability	High	Radical change	Pragmatism
	Low	Reformism	Rejection

Radical change involves the whole-hearted acceptance of new approaches as both appropriate and workable. This is based upon the universal principles that they embody. The result is an attempt at wholesale cultural transformation.

Reformism is a less radical approach and involves a modification of circumstances to suit new approaches and practices by adapting existing practices or regulations.

Pragmatism involves the adaptation of new approaches or approaches to practical situations, but only where these are likely to be accepted.

Rejection involves regarding approaches as largely inappropriate or unworkable and leads to a greater preference for solutions developed in-house.

For some, the problems facing the construction industry or quantity surveying practice are viewed as inherently different to those in other sectors, industries or professions. Practices, the market and local regulations are seen as being so different that ideas from elsewhere are not capable of either being adapted or transferred. Parts of the proposals may be considered, but are not found to be workable within the complexity of an existing framework. Internal conditions also affect reactions to change and the manner in which change is best handled (Kotter and Schlesinger 1979).

In more recent years there has been a huge interest within the construction industry in exploring the possibilities for transferring and applying the knowledge and practices from manufacturing industry. EPSRC's (Environment and Physical Sciences Research Council) Innovative Manufacturing Initiative explicitly sets out to explore, 'Construction as a Manufacturing Process'. In so doing it is attempting to transfer knowledge and to encourage the construction industry to capture and learn from experiences in manufacturing engineering. Some of the processes identified in recent industry reports have identified lean construction, for example, as a process that has direct application in construction. However, not everyone in the industry is convinced, and, amongst those that are, there remain a number of reservations because a construction site, unlike an engineering factory, is temporary. There are also questions raised by some as to whether certain of the new ideas and principles involved can apply throughout the industry rather than on just the large multi-million pound projects.

General management theories such as total quality management (TQM), business process re-engineering, continuous improvement, benchmarking, lean production, supply chain management and strategic alliances are all modern ideas that have been borrowed from elsewhere and applied to the construction industry with varying degrees of success, but, more successfully, where there is greater scope to do so in the larger organisations and on the larger projects.

The industry also has a habit of attempting to reinvent itself every few years (Murray and Langford 2002) and whilst similar messages are often repeated about a number of issues and practices, it appears that in general industry and practice take little notice or the proposals are too difficult to implement or dissemination to the appropriate people does not take place.

The construction industry does have a particular set of characteristics (see Chapter 1) that sets it apart from other industries. But are these sufficient to separate it from the rest of industry and especially manufacturing such as automobiles and aeroplanes?

Fong (2005) makes reference to the different ways in which quantity surveying practices manage knowledge in respect of their work. He selected quantity surveying firms because:

- They are knowledge-intensive organisations
- Quantity surveying firms function in a project-based industry where knowledge is regarded as a vital resource for delivering the client's requirements
- Quantity surveying firms are project organisations in which everyone in the project team is dependent on the other's work both within the firm and external to it
- A thorough review of the literature suggests that no previous studies on how knowledge is managed in quantity surveying firms have been carried out.

References and bibliography

Ashworth A. and Hogg K. *Added Value in Design and Construction*. Pearson. 2000.

Atkin B. and Flanagan R. *Improving Value for Money in Construction*. Royal Institution of Chartered Surveyors. 1995.

Barlow J. *A statement on the construction industry*. The Royal Academy of Engineering. 1996.

Brandon P.S. (ed.) *Quantity Surveying Techniques: New directions*. Blackwell Science. 1992.

Bresnen M. *Institutional and Cultural Barriers to Effective Knowledge Transfer*. ThinkPiece, CRISP. 1999.

Construction Industry Board. *Towards a 30% productivity improvement in construction*. Thomas Telford. 1996.

Construction Sponsorship Directorate *An Introduction to the Whole Industry Research Strategy*. Department of the Environment. 1995.

Davis, Langdon and Everest. *QS 2000*. RICS. 1991.

Egan J. *Rethinking Construction*. Department of the Environment, Transport and the Regions. The Stationery Office. 1998.

Fairclough J. *Rethinking Construction Innovation and Research*. Department of Trade and Industry and Department for Transport, Local Government and the Regions. 2002.

Fong P.S.W. Managing knowledge in project-based professional services firms: an international comparison. In: *Management of Knowledge in Project Environments* (edited by P. Love, P.S.W. Fong and Z. Irani). Elsevier. 2005.

Freeman C. The challenge of new technologies. In: *Interdependence and Co-operation in Tomorrow's World*. OECD, Paris. 1987.

Gray C. *Value for Money: Helping the UK afford the buildings it likes*. Reading Construction Forum. 1996.

Harris J. *Victorian to Virtual*. RICS. 2000.

IPRA Ltd. *Future skills needs of the Construction Industries*. Report prepared for the Employment Department. 1991.

Kolesar S. *The Values of Change*. RICS. 1999.

Kotter J. and Schlesinger L. *Choosing Strategies for Change*. Harvard Business Review. 1979.

Latham M. *Constructing the Team*. The Stationery Office. 1994.

Miles I. *et al. Information Horizons: The long term social implications of New Information Technologies*. Edward Elgar. 1988.

Murray M. and Langford D. *Construction Reports 1944–98*. Blackwell Publishing. 2002.

National Audit Office. *Modernising Construction*. The Stationery Office. 2001.

Nisbet J. *Called to Account: Quantity Surveying 1936–86*. Stoke Publications. 1989.

Powell C. *The Challenge of Change*. RICS. 1998.

Powell J. *Towards a new construction culture*. Chartered Institute of Building. 1995.

RICS. *The research and development strengths of the chartered surveying profession: the academic base*. RICS. 1991.

RICS. *The core skills and knowledge base of the quantity surveyor*. RICS Research Paper 19. 1992.

RICS. *New RICS foundation set to transform built environment agenda*. RICS. 2000.

Runeson G. and Skitmore M. *Writing Research Reports*. Deakin University Press. 1999.

Seymour D. and Rooke J. The culture of industry and the culture of research. *Construction Management and Economics*, 13 (6). 1995.

Thompson F.M.L. *Chartered Surveyors: The growth of a profession*. Routledge, Kegan Paul. 1968.

Websites

www.comdig.org
www.ncrisp.org.uk

6 Cost Control

Introduction

Anyone proposing to construct a building or engineering structure will need to know in advance the probable costs involved in the works. These costs include the cost of the works carried out on site by the contractor, professional fees, and any taxes that may be due to the government. In addition to these sums the client, or promoter on civil engineering projects, will also need to make provision for the costs of the site and other development costs, together with the fitting out and furnishings that are required in the completed project. Many of these costs are often excluded from the normal process of construction project cost control. However, within the emerging discipline of facilities management, all of these costs need to be accounted for. It is also one of the duties of the client's quantity surveyors to ensure that the building to be constructed is carefully controlled, in terms of costs arising throughout the entire design and construction process. The focus of cost control must be balanced with the importance of value in terms of what is being provided for a client. This chapter is concerned only with project cost control; other aspects are considered elsewhere in this book.

Project cost control

Cost control in the construction industry is a term that is used to cover the whole spectrum of activities required to meet this end (RICS 1982). The process starts at inception, when guide prices or indicative costs will be required, through the stage when an early price estimate is prepared and the tender process undertaken by the contractors through to the final completion and agreement of the final account for the project. If the estimate is within the budget and approved by the client, the project then moves forward to the design stage and the various processes that are involved (see Fig. 6.1). During the various stages of design, the architect or engineer will want to consider alternative solutions that meet the client's overall aims and objectives.

The quantity surveyor will offer cost advice for the comparative design solutions of the alternative materials to be used or the form of construction to be adopted. The quantity surveyor will also provide advice on the cost implications of the design morphology and procurement. This stage, known as cost planning, has been

Stage	Phase
Appraisal	Inception
Strategic briefing	Feasibility
Viability	
Outline proposals	Sketch design
Detailed proposals	Detail design
Final proposals	
Production information	Contract documentation
Tender documentation	Procurement
Tender action	
Mobilisation	
Construction to practical completion	
	Project planning
	Installation
	Commissioning
After practical completion	
In-use	Maintenance
	Repairs
	Modification
Demolition	Replacement

Fig. 6.1   The RIBA work stages.

developed in further detail and may be described as a system of relating the design of buildings and other structures to their costs. This takes into account quality, utility and appearance. The cost is planned within a combination of the budget provided by the client, and the design and construction considerations determined by the design team.

The good practices of cost planning further require the quantity surveyor to allocate the estimated costs into subdivisions, defined as elements, of the building. These element costs can be compared against the element costs of other similar projects from the quantity surveyor's cost library records. The contract documents may also be prepared on this basis to facilitate easier preparation of the cost analysis. This is uncommon, but the extensive use of computers allows for the appropriate elemental information to be easily retrieved where this is required.

Cost control does not stop at tender stage but continues up to the agreement of the final account, and the issue of the final certificate for the works. Post-contract cost control and procedures are described later in this chapter and in more detail in Chapters 12–14.

Cost advice

Quantity surveyors throughout the design and construction process are required to advise the client on any cost implications that may arise. Such advice will be necessary irrespective of the procurement method used for contractor selection or tendering purposes. However, the advice will be especially crucial during the project's inception. During this time major decisions are taken affecting the size of the project and the quality of the works, if only in outline form. The cost advice given must therefore be as reliable as possible, so that clients can proceed with the greatest amount of confidence. Where the advice is inaccurate, it may cause a client to proceed with a project that cannot be subsequently afforded or, because of a too high forecast of probable costs, may result in a project being prematurely aborted.

Quantity surveyors are the recognised professionals within the construction industry as cost and value consultants. Their skills in measurement and valuation are without equal. It should be recognised that clients and designers who are either unable or unwilling to provide proper information by way of brief, quality or budget, must therefore also expect the cost advice to be equally imprecise! Quantity surveyors recognise the importance of providing realistic cost information that will contribute to the overall success of the project. In this context it is important to have an awareness of both design method and construction organisation and management. Quantity surveyors are the industry's experts on building costs and must perform their duties to appropriate professional standards. The client when paying for a professional opinion requires this to be sound and reliable. Failure to carry out these duties properly, in the context of an expert, could provide grounds for liability for negligence. Surveyors must therefore avoid giving estimates 'off the cuff'. It is also preferable during the early stages of a project to suggest a range of prices rather than a single lump sum. The quantity surveyor should also remember the maxim that the first estimate of cost is always the one figure that the client remembers.

Precontract methods

The methods that can be used for pre-tender price estimating are shown in Fig. 6.2. Although they are often referred to as approximate estimating methods, this needs to be read in the context of the way in which they are calculated rather than in terms of the intended accuracy of price alone. The degree of accuracy will depend upon the type of information provided, the quality of relevant available pricing information and the skills and experience of the quantity surveyor who prepares the estimate of cost. Familiarity with the type of project and the location of the site are important factors to consider.

Most of the methods are reasonably well known, but only those that are current or have possible future use have been described.

Unit method

The unit method of approximate estimating consists of choosing a standard unit of accommodation and multiplying this by an appropriate cost per unit. The technique is based upon the fact that there is usually some close relationship between the cost of a construction project and the number of functional units that it accommodates. The standard units may, for example, represent the cost per theatre seat, hospital

Method	Notes
Conference	Based on a consensus viewpoint, often of the design team.
Financial methods	Used to determine cost limits or the building costs in a developer's budgets.
Unit	Applicable to projects having standard units of accommodation. Used as a basis to fix cost limits for public sector building projects.
Superficial	Still widely used, and the most popular method of approximate estimating. Can be applied to virtually all types of buildings and is easily understood by clients and designers.
Superficial perimeter	Never used in practice. Uses a combination of floor areas and building perimeters.
Cube	This used to be a popular method amongst architects, but is now in disuse.
Storey-enclosure	Is largely unused in practice.
Approximate quantities	Still a popular method on difficult and awkward contracts and where time permits.
Elemental estimating	Not strictly a method of approximate estimating, but more associated with cost planning; used widely in both the public and private sectors for controlling costs.
Resource analysis	Used mainly by contractors for contract estimating and tendering purposes. Requires more detailed information on which to base costs.
Cost engineering	Mainly used for petrochemical engineering projects.
Cost models	Mathematical methods which continue to be developed.

Fig. 6.2 Methods of pre-tender estimating (*Source:* Ashworth, 2004).

bed space or car park space. Such estimates can only be very approximate and vary according to the type of construction and standard of finish, but in the very earliest stages of a design they offer a guide to a board or building committee. Apparently the cost per prison cell is higher than a bedroom unit in a five star hotel!

Superficial method

This type of estimate is fairly straightforward to calculate and costs are expressed in a way that can be readily understood by a typical client. The area of each of the floors is measured and then multiplied by a cost per square metre. In order to provide some measure of comparability between various schemes, the floor areas are calculated from the internal dimensions of the building: that is, within the enclosing walls.

The best records for use in any form of approximate estimating are those that have been derived from the quantity surveyor's own previous projects and past experience. However, extensive use is also made of cost databases to provide, if nothing else, a second opinion on the estimated costs. These are also especially useful where a quantity surveyor has no previous records on which to base an estimate. In these circumstances the Building Cost Information Service (BCIS) database may need to be consulted. The BCIS *Quarterly Review of Building Prices* is also a useful guide to which further reference can be made. This type of information must, however, be treated with considerable caution since it can easily produce misleading results. It can also rarely ever be used without some form of adjustment. Buildings within the same category, such as schools or offices, have an obvious basic similarity, which should enable costs within each category to be more comparable than for buildings from different categories.

It is important to consider varying site conditions. A steeply sloping site must make the cost of a building greater than for the same building on a flat site. The nature of the ground conditions, and whether they necessitate expensive foundations or difficult methods of working, must also be considered. The construction design and details to be used will also have an important influence. A single storey garage of normal height will merit a different rate per square metre from one that is constructed for double-decker buses. Again, a requirement for, say, a 20 m clear span is a different matter from allowing stanchions at 5 m or 7 m intervals. An overall price per square metre will be affected by the number of storeys. A two-storey building of the same plan area has the same roof and probably much the same foundations and drains as a single storey of that area, but has double the floor area. However, the prices for high buildings are increased by the extra time involved in hoisting materials to the upper floors, the danger and reduced outputs from working at heights and the use of expensive plant such as tower cranes.

The shape of a building on plan also has an important bearing on cost. A little experimental comparison of the length of enclosing walls for different shapes of the same floor area will show that a square plan shape is more economical than a long and narrow rectangle, and that such a rectangle is less expensive than an L-shaped plan. There are, however, exceptions to these general rules.

The standard of finish naturally affects price. There will be clients who require office blocks with the simplest of finish and there will be others to whom, perhaps, more lavish treatment has advertisement value: they may want expensive murals or sculpture.

In projects offering different standards of accommodation it will be preferable to price these independently. A variety of rates may therefore be required, depending upon the different functions or uses of the parts of the building. There may also be the possibility of the need to include items of work that do not relate to the floor area, and these will have to be priced separately.

Approximate quantities

These provide for a more detailed approximate estimate than any of the other methods. They represent composite items, which are measured very broadly by combining or grouping typical bill measured items. In practice only items of major cost significance are measured. For example, strip foundations are measured per lineal metre and include excavation, concrete and brickwork items up to damp proof course level. A unit rate for this work can be readily built up on a basis obtained from a cost database price book or priced bills of quantities from previous projects. Walls may include the internal and external finish, and the total costs of the windows and doors can be enumerated as extra over.

This method does provide a more reliable means of approximate estimating, but it also involves more time and effort than the alternative methods. No specific rules exist, but the composite items result from the experience of the individual quantity surveyor. In order for the quantities to be realistically measured, more information is required from the designer. Specially ruled estimating paper is available, which is designed particularly for approximate estimating purposes.

Approximate quantities should not be confused with a bill of approximate quantities. The latter would usually be based upon an agreed method of measurement, but since the design is not at a sufficiently advanced stage, the quantities must therefore only be approximate. Also, the approximate quantities used for precontract estimating are simplified, since several of the bill items are grouped together to form a single composite description.

The use of approximate quantities for precontract cost control can create some costing and forecasting difficulties, as by the time the drawings have reached the required stage many of the matters of principle have already been settled. Often these cannot be altered without a major disturbance of the whole scheme.

Cost planning

Cost planning is not simply a method of pre-tender estimating, but seeks also to offer a controlling mechanism during the design stage. Its aims in providing cost advice are to control expenditure and to offer to the client better value for money. It attempts to keep the designer fully informed of all the cost implications of the design. Full cost planning will incorporate the attributes of whole life costing and value management.

Two alternative forms of cost planning (RICS 1982) have been developed, although in practice they have been combined into a single method. Elemental cost planning was devised by the then Ministry of Education (subsequently Department for Education and Skills (DfES)) in the early 1950s. It was developed largely in response to the extensive school building programme in which there was found to be a wide variation in cost throughout the country. Also costs needed to be monitored and controlled more effectively than had been the case previously. It is referred to as 'designing to a cost'. At about the same time, the RICS Cost Research Panel was established and introduced comparative cost planning, which became known as 'costing a design'. The Cost Research Panel was instrumental in developing the Building Cost Advisory Service, later renamed the Building Cost Information Service (BCIS).

The cost planning process commences with the preparation of an approximate estimate and then the setting of cost targets for each element. As the design evolves, these cost targets are checked against the developing design and details for any changes in their financial allocations. The prudent quantity surveyor will also always be looking for ways of simplifying the details without altering the design, in an attempt to reduce the expected costs. Not only will the building construction be considered, but also the ease or otherwise with which the design can actually be built. This process should also result in fewer abortive designs, and the seeking of value for money should not cease at tender stage, but should continue throughout the post-contract cost-control procedure. A more detailed description of cost planning can be found in Ashworth (2004) and Ferry *et al.* (1999) and in the interactive learning text, Flanagan and Tate (1997). A sample summary cost plan is shown in Fig. 6.3.

In addition to providing a full cost planning service the quantity surveyor can assist in evaluating the comparative costs for alternative systems of construction or finishings. This can involve, for example, comparing different plan shapes in terms of cost, or external cladding. The comparisons must be comprehensive in ensuring that all costs are fully taken into account. Reducing for example, the weight of external cladding may also result in savings in the substructure element.

Cost limits

These are sometimes referred to as financial methods of pre-contract cost control. A cost limit is fixed on the building design, based upon either a unit of accommodation or rental value. The architect must then ensure that the design can be constructed within this cost limit. These are common procedures to be adopted in connection with public buildings such as houses, schools and hospitals. They are also used in the private sector where a developer may need to place a limit on the building cost, based upon the other costs involved, such as the land cost and the selling price of a house or the commercial value of the project.

Cost modelling

These are usually computer based techniques that are used for forecasting the estimated costs of a proposed construction project. Although cost models were first investigated in the early 1970s, there is only scant evidence of their use in practice (Ashworth 1986, 1999). A statistical model or formula is constructed which best describes the building in terms of cost. The development of cost model building can be a lengthy process requiring the collection and analysis of large quantities of data. Prior to using the model in practice it needs to be tested against the more conventional methods. Cost modelling is a radical approach to pre-contract estimating and cost control, and whilst further research is being carried out it is not yet at the stage of replacing the more conventional methods that have been described (Fortune and Lees 1996).

General considerations

The selection of appropriate rates for pre-tender estimating and cost control depends upon a large variety of factors. Some of these can be considered

JOB No: 123   Date: January 2006

New Church of England Primary School

Blankchester Diocesan Board of Finance

COST PLAN

(Fixed price 65 weeks)   Floor area   1200 m²

Summary	Cost per m²	Elemental cost	
	£	£	£
1. SUBSTRUCTURE	54.00	64 800	64 800
2. SUPERSTRUCTURE			
2.1 Frame	32.67	39 200	
2.2 Roof	43.08	51 700	
2.3 External walls	20.08	24 100	
2.4 Windows and external doors	20.75	24 900	
2.5 Internal walls and partitions	15.67	18 800	
2.6 Internal doors	11.17	13 400	172 100
	143.42		
3. INTERNAL FINISHINGS			
3.1 Wall finishings	15.17	18 200	
3.2 Floor finishings	19.75	23 700	
3.3 Ceiling finishings	11.00	13 200	55 100
	45.92		
4. FITTINGS AND FURNISHINGS	21.25	25 500	25 500
5. SERVICE INSTALLATIONS			
5.1 Sanitary appliances	8.33	10 000	
5.2 Disposal installation	11.67	14 000	
5.3 Hot/cold water installation	20.83	25 000	
5.4 Heating installation	52.50	63 000	
5.5 Electrical installation	26.67	32 000	
5.6 Gas installation	2.50	3 000	
5.7 Communication installation	3.33	4 000	
5.8 Builders work in connection	7.50	9 000	160 000
	133.33		
6. EXTERNAL WORKS			
6.1 Site works	104.16	125 000	
6.2 Drainage	19.76	23 700	
6.3 External services	8.33	10 000	158 700
	132.25		
7. PRELIMINARIES	54.17	65 000	65 000
8. CONTINGENCIES			35 000
TOTAL ESTIMATED COST	584.33 per m²		**736 200**

Exclusions: Professional fees and site supervision   Abnormal ground conditions
            Building control fees   Value Added Tax
            Site investigation costs

Fig. 6.3 Sample cost plan.

objectively, but in other circumstances only experience or expert judgement will be sufficient. Some of these considerations are as follows.

Market conditions

The quantity surveyor, in order to get the price as near as possible to the tender sum, must be able to interpret trends in prices based upon past data and current circumstances. This demands a great deal of skill and some measure of luck! Allowances may also need to be made for changes in contractual conditions, type of client, labour availability, workloads, and the general state of the industry. When work is plentiful contractors' tender prices will be higher.

Design economics

When using previous costs or cost analyses, changes in these costs for design variables such as shape, height and size will need to be taken into account.

Quality factors

Cost information is deemed to be based upon defined or assumed standards of quality. Where the quality in the proposed project is considered to be different from this, then changes in the proposed estimate rates will be required. The surveyor should always provide quality indications with any cost advice.

Engineering services

These have become an ever-increasing proportion of building project costs. Their cost importance is such that they now need to be considered in detail. On large schemes it is now usual to employ specialist engineering services quantity surveyors.

External works

There are considerable differences between building sites and hence few established cost relationships for this building element. The size of the site and the nature of the work to be carried out will be important factors to consider.

Exclusions

The proposed estimate of cost should clearly identify what has been included, by way of the outline specification and, equally important, what has been excluded. Clients may well assume that a tender sum includes all of their costs and will be concerned should they find that items were excluded. The obvious items to identify include fees, VAT, land costs, loose furniture and fittings. In some projects the whole of the fitting-out work may form part of a separate contract.

Price and design risk

During the design and construction process aspects of the design will still be evolving and a contingency sum will need to be provided for possible additional costs that were not envisaged. This is design risk. The contingency sum allowed is therefore likely to be larger at the early stages of the design than at the tender stage. The price risk factor is related to market conditions. A market that is more volatile than usual will result in a larger percentage being allowed in the estimate for this factor.

Accuracy of approximate estimates

Whether the process of cost planning at the design stage is used or not, an analysis of tenders will be valuable for future approximate estimating purposes. Quantity surveyors must attempt to anticipate matters that will affect the level of future tender sums. They cannot, however, forecast exceptional swings in existing tendencies, just as they cannot forecast the future. They must, however, be aware of all current published trends. These publications will include those specifically for the construction industry as well as publications of a more general nature.

The accuracy or reliability of the estimate is of prime importance. On average, ±10% is the typical sort of estimating accuracy that is expected in the construction industry. This percentage will be much higher at inception, when only vague information is available (Ashworth and Skitmore 1986). Forecasts are unlikely to be consistently spot-on, as by definition estimates will always be subject to some degree of error. Estimates should be shown as a range of values rather than as a single lump sum. Alternatively, confidence limits could be offered as a measure of estimates' reliability. The following factors are said to have some influence upon the accuracy of estimating:

- Quality of the design information
- Amount, type, quality and accessibility of cost data
- Type of project, as some schemes are easier to estimate than others
- Project size, as accuracy increases marginally with size
- Stability of market conditions
- Familiarity with a particular type of project or client.

Proficiency in estimating is the combination of many factors, including skill, experience, judgement, knowledge, intuition and personality (Ashworth and Skitmore 1986).

Preparing the approximate estimate

The method to be used for the preparation of an approximate estimate will depend to some extend on the type of project and amount of information provided by the designer and the client. The more vague the information, the less precise will be the estimate.

The estimate for a complicated refurbishment project could be prepared on the basis of the gross internal floor area, but this would require a large quantity of other calculations on the part of the quantity surveyor. Of course, if the surveyor was familiar with the type of work, the designer and the client, then this approach would be satisfactory. A more useful method would be to use some form of approximate quantities. Once the estimate is acceptable, it will form the budget for the project. *Precontract Cost Control and Cost Planning* (RICS 1982) provides useful information on preparing approximate estimates.

The first part of the process is to quantify the project using one of the methods described earlier and then to price these quantities using current cost information. The cost information can be obtained from previous projects, from cost analyses, or from some published source. It is usual to add on a contingency amount to cover unforeseen items of work. The amount of this is highest at inception and it gradually reduces as the design becomes more certain. The contingency amount is not removed entirely until the completion of the final account.

As the project is priced at current prices some addition is required to allow for possible increased costs. This is normally added as two separate amounts – the first up to tender stage in order to allow for comparison with tenders, and a second to allow for increased costs during construction. The forecasting of these sums in periods of high inflation is difficult.

Approximate estimates of construction costs normally exclude VAT, even on those projects where VAT will be charged. This can represent a considerable item so this should be clear to the client. Even this total sum will not represent the full costs of construction to the client. Professional fees for the architect, engineer and quantity surveyor, and other charges for planning approval, must also be added. The professional fees will always attract VAT, unless the practice is very small. The estimate should also be clear as to items that have been excluded altogether.

Whole life costing

It has long been recognised that it is unsatisfactory to evaluate the costs of buildings on the basis of their initial costs alone (Ashworth 1999). Some consideration must also be given to the costs-in-use that will be necessary during the lifetime of the building. Whole life costing is an obvious idea, in that *all* costs arising from an investment decision are relevant to that decision. The primary use of whole life costing in construction is in the evaluation of alternative solutions to specific design problems. The whole life cost plan is a combination of initial, maintenance, replacement, energy, cleaning and management costs. Whole life costing must take into account the building's life, the life and costs of its components, inflation, interest charges, taxation and any consideration that may have a financial consequence on the design (Flanagan and Norman, 1993). Many of these are indeterminate. Whole life costing is considered in more detail in Chapter 7.

Value management

Value analysis or value engineering was developed as a specific technique during the 1940s, and has been extensively used for a variety of purposes, particularly in the USA. It has been described as a system for trying to remove unnecessary costs before, during and after construction. It is an organised way of challenging these costs and is based upon a functional analysis that requires the answer to the six basic 'what if' questions of value analysis. In essence, the technique seeks to improve the value for money in construction projects by improving their usefulness at no extra cost, by retaining their utility for less cost, or by combining their improved utility with a decrease in cost. This technique has been renamed value management (Kelly and Male 2000), with an emphasis on it being a more proactive tool. It is considered in more detail in Chapter 8.

During the 1950s value engineering started to penetrate European manufacturing industries and it is today recognised as a means of efficiency-oriented management. It has only in recent years begun to be applied in the UK construction industry. Value management is seen to complement current quantity surveying practice and procedure.

Risk analysis

Quantity surveyors have traditionally presented their clients with single-price estimates, even though it was apparent that on virtually every project differences would occur between this sum and the final account.

The construction industry is subject to a greater amount of risk and uncertainty than most other industries. Also, unlike other major capital investment, construction projects are not developed from prototypes. Many schemes represent a bespoke solution, often involving untried aspects of design and construction, in order to meet an individual client's requirements. All construction projects include aspects of risk and uncertainty (Flanagan and Norman 1993).

Risk is measurable and can therefore be accounted for within an estimate. For example, quantity surveyors involved in forecasting the costs of a new project will have access to different sorts of cost data and these, coupled with their expertise, will enable a budget price range to be calculated within specified confidence limits. It is desirable to offer an estimate in this way rather than to suggest a single price. Uncertainty is more difficult to assess, as it represents unknown events that cannot be even assessed or costed. Different techniques can be applied, such as Monte Carlo simulation, to assess the risk involved. The risk will of course not be eliminated but at least it can be managed rather than ignored. Risk is considered in more detail in Chapter 9.

Best value

In examining the evolving role of the quantity surveyor there is no doubt that there has been a paradigm shift from thoroughly assessing the costs of construction towards adding value (Ashworth and Hogg 2000). The importance of value for money in building and engineering design is not new. Cheapness in itself is of no virtue. Added value is the new watchword. In its simplest sense it is a term that is used to describe the contribution a process makes to the development of its products.

Best value (DETR 2001) is a concept that has come out of the Local Government Act 1999 which sets out the requirements that are expected. The key sentence in the Act is, 'A Best Value authority must make arrangements to secure continuous improvement in the way in which its functions are exercised, having regard to a combination of economy, efficiency and effectiveness'. The so-called 3Es (efficiency, effectiveness and economy) concept has been in existence for some time. The concept of best value applies equally as well in the private as in the public sector. Best value aims to achieve a cost-effective service, ensuring competitiveness and keeping up with the best that others have to offer. It embraces a cyclical review process with regular monitoring as an essential part of its ethos. The best value concept for local authorities is being managed through the Audit Commission.

Best value extends the concepts of value for money that have been identified by Gray (1996), Egan (1998), Powell (1998) and Ashworth and Hogg (2000). Egan for example, defines value in terms of zero defects and delivery on time, to budget and with a maximum elimination of waste. In order to show that best value and added value are being achieved, it becomes essential to benchmark performance including costs. It is also necessary to benchmark the overall cost of the scheme so that improved performance in the design can be assessed against its cost. The sharing of information underpins the whole best practice process. Even the leaders in an industry need to benchmark against their competitors in order to maintain that leading edge. Whilst the aspiration of best value is both admirable and essential, its demonstration in practice presents the challenge.

Taxation

The implications of taxation can have an important effect on construction projects. Whilst it is unlikely to be the main aim of a client, tax efficient design must be considered. It has implications on the construction, fitting out, repairing, running and maintenance costs of buildings. The designing of buildings to be optimally tax efficient can yield substantial benefits for owners and occupiers. Several years ago a House of Lords judgment stated, 'No man in this country is under the smallest obligation, moral or otherwise, to arrange his affairs as to enable the Inland Revenue to put the largest possible shovel into his stores'. Taxation avoidance, rather than evasion, is within the law, and is a necessity bearing in mind the high and diverse incidence of taxation.

The influence of taxation and its effects on buildings and property are constantly changing owing to revisions in taxation principles and the introduction of new measures or rates by the Chancellor of the Exchequer. This is often incorporated as part of the annual budget presentation. Clients cannot expect full guidance on these matters from their accountants, since their technical knowledge of the construction process and its products is limited. In this respect quantity surveyors are well placed to advise the client on these matters, perhaps working in conjunction with an accountant. Every project has its own peculiarities and will require an individual analysis.

All taxes have some bearing on buildings and property. Stamp duty, for example, is a tax on documents and is charged on the transfer of ownership of all land and buildings above a certain value. Council tax and non-domestic rates (business tax) are annual charges relating to the ownership of buildings. With each of these taxes there is little room for reducing the taxation burden. In connection with the latter, clients considering building in a general location may be advised to build in an area that has lower local taxes. This will have to be balanced with the need to be in a particular location, the availability of sites and their respective charges.

Value added tax (VAT)

VAT is charged on the supply of goods and services in the UK, and on the import of certain goods and services into the UK. It applies where the supplies are taxable supplies made in the course of business by a taxable person. This tax was introduced to the construction and property industries through the Finance Act 1972. It is a direct result of the UK joining the European Union. The principal legislation is now embodied in the Value Added Tax Act 1974.

Building work is either standard-rated, currently 17.5%, or zero-rated. Examples of zero-rated works include residential buildings, which include children's homes, old people's homes, homes for rehabilitation purposes, hospices, student living accommodation, armed forces' living accommodation, religious community dwellings and other accommodation which is used for residential purposes. Certain buildings intended for use by registered charities may also be zero-rated. Buildings which are specifically excluded from zero-rating include hospitals, hotels, inns and similar establishments. The conversion, reconstruction, alteration or enlargement of any existing building are always standard-rated. All services which are merely incidental to the construction of a qualifying building are standard-rated. These include architects', surveyors' and other consultants' fees and much of the temporary work associated with a project. Items which may be typically described as 'furnishings and fittings', e.g. fitted furniture, domestic appliances, carpets, freestanding equipment, etc. are always standard-rated irrespective of whether the project may be classified as zero-rated. The VAT guides provide examples, but throughout these documents, individuals are advised to check their respective liability with the local VAT office. The ratings of some items are arbitrary whereas others will need to be tested by the courts.

Corporation tax

Profits, gains and income accruing to companies who are resident in the UK incur liability to corporation tax. The level of such liability is governed by the profits, gains or income for an accounting period.

Capital and revenue

Capital expenditure is money that is expended in acquiring assets, or for the permanent improvement of, addition to or extension of an existing asset. Such assets must generally have a useful life beyond one year and include items of buildings, machinery and plant.

Revenue expenditure is concerned with the maintenance of such an asset whilst it is in use. It is, by definition, those costs which cannot be classified as capital expenditure. It includes local taxes, annual water and sewage charges, energy, cleaning, insurances and minor repairs.

Capital expenditure will result in increased amounts for fixed assets on a balance sheet, whereas the revenue expenditure is chargeable to the trading or profit and loss account.

Capital allowances

The taxable profits of a company may be reduced through applying allowances against capital cost expenditure. The law on capital allowances is contained in the Capital Allowances Act 1990, together with provisions for future amendment through subsequent Finance Acts. The allowances are calculated on the basis of the following:

- *Initial allowance*   This is an initial sum that is allowed against the expenditure of an item in any financial year.
- *Writing down allowances*   These are sums that can be offset against taxable income on an annual basis for a specified term of years. These may represent, for example, the theoretical depreciation of an asset.
- *Balancing allowances*   There are charges or additional allowances made at the end of an asset's life. Their intention is to balance the allowances against the actual amounts.

Industrial buildings

Industrial buildings are treated differently to the majority of other types of buildings, being broadly defined as buildings used for the processing or manufacture of goods. They include buildings that are used for the storage of materials before manufacture and for goods after production. They must have a direct link with production and as such, wholesale warehouses are excluded. The offices that form a part of the factory are included if they do not exceed 25% of the total cost.

The full costs of construction including professional fees are allowed. Whilst land costs are excluded, the costs of any site preparation may be included. The full costs of the purchase of a building from a builder are allowed. Costs expended on existing buildings, plant and equipment are also included. There is an allowance of 4% per annum, calculated on a straight line basis, until the costs have been fully written down (i.e. 25 years). In the event of a sale, a balancing adjustment is applied.

Where the location of the building has been designated as an Enterprise Zone, expenditure incurred on buildings can qualify for an initial allowance of 100%. Where fixed plant and machinery are an integral part of the building, this is treated as a part of the building in respect of claiming Enterprise Zone allowances.

Plant and machinery

The treatment of capital expenditure on plant and machinery is very complex. Whilst the definition regarding machinery is generally understood, plant is not and has come before the courts on many occasions. It may include 'whatever apparatus is used by a businessman for carrying on a business'. This will exclude stock-in-trade which is bought or made for sale, but will include 'all goods and chattels, fixed or moveable, alive or dead, which are kept for permanent employment in a business'. This legal opinion was stated in *Yarmouth* v. *France* (1887). The main problem lies in distinguishing the apparatus with which a business is carried out from the setting in which it is carried on. Lifts and central heating systems are treated as plant, but plumbing and electricity systems are not. Specific lighting to create an atmosphere in a hotel and special lighting in fast food restaurants have been held as plant.

In *Jarrold* v. *Johngood and Sons* (1963), it was stated that items need not be subject to wear and tear, in this case movable metal office partitioning. The maintenance of plant is always an allowable revenue expense. Builder's work specifically required for the installation of plant items is deemed to be part of the capital cost item.

Computer hardware is a capital expenditure item and allowances are usually claimed under the short life asset rules. Where software is purchased at the same time the Inland Revenue have suggested that this should also be treated as part of the capital cost. Licences to operate software are treated as a revenue expense.

Plant and machinery costs, for taxation purposes, are not treated in the same way as the buildings that house them. First year allowances are from time to time 100% of their capital cost. The full amount of taxation relief can therefore be gained immediately on these items. However, it is more usual to allow a first year allowance and thereafter a writing down allowance of 25% in subsequent years.

Financial assistance for development

In building development some consideration must also be given towards the financial assistance that may be available.

Whilst the various planning regulations are able to prevent undesirable development from taking place, they are unable to encourage socially desirable development to be undertaken. As a part of the planning process the government and its agencies can suggest to a developer that certain specified works are carried out as a part of the approval for development. These might include, for example, leaving part of proposed housing development as a public open space. However, in order to encourage desirable developments to take place in unattractive locations, some form of financial assistance may be necessary.

The intention of such financial assistance is to support projects in areas where they might otherwise not take place, or in circumstances where there may be little obvious economic benefit to a developer. Such assistance may be offered from a local or regional authority or from a central government department. Major considerations are given to the effective co-operation between the public sector, local and central government and private business and voluntary organisations. Financial assistance may arise for one of the following reasons:

- Urban renewal programmes
- Regeneration of industrial areas
- Investing in jobs to benefit areas of high unemployment
- Land reclamation schemes
- Property improvement, such as housing improvements
- Slum clearances and derelict land clearance.

The aim is to encourage private companies or public managing agents to develop areas either as a means of improving the standards and amenities, or through investing in projects that will help in wealth creation. At the same time unemployment in a region may be able to be reduced. Financial assistance is therefore targeted in areas where it may otherwise be difficult to encourage companies to invest. Whilst financial assistance is available for a variety of different purposes, it is in the designated areas where the size of grants is the largest.

Investment grants are made by the government to manufacturing and extractive industries only, in respect of new buildings or adaptations and plant and machinery. The grants are treated as non-taxable capital receipts. Loans are treated in a similar way and may be free of interest. Loans are sometimes offered to companies who for a variety of reasons cannot secure finance in the normal conventional ways through commercial banks. Financial assistance may be a combination of:

- Taxation allowances on capital expenditure for buildings and plant. The company in receipt of such an allowance must in the first instance make a profit to secure the benefit.
- Low rents or business rates, offered by a local authority as inducements to locating in their area. These may be offered only for a limited number of years.
- Grants of up to 50% for capital items to assist firms in new developments.
- Extending, converting and improving industrial and commercial property.
- Amenity grants of up to 100% of the costs associated with providing access roads, car parking and other amenities.
- Bridging finance to close the gap between developing a building and its market value.
- Interest relief grants to offset some of the costs of borrowing finance.
- Building loans. As well as acting as a guarantor for bank finance for building, up to 90% of the market value of land and buildings may be met through loans charged at preferential rates of interest.

- Enterprise Zone benefits, which include 100% tax allowances for money invested in commercial and industrial buildings and exemption from local property taxes.
- Enterprise Zone status providing simplified planning procedures for developments.
- Subsidies paid to companies who employ additional employees in specified occupations.

Regional initiatives

Regional industrial policy operates within a general economic framework designed to encourage enterprise and economic growth in all areas of Britain. However, in some areas specific additional help is provided under the Regional Initiative. Help is thus focused on the assisted areas, which are designated intermediate and development areas.

The areas are those that have high unemployment caused by the demise of traditional industries or the loss of major employers. In order to obtain assistance, projects must create or in exceptional circumstances safeguard jobs within a designated area. Projects must have a good chance of long-term viability. In addition, the greater part of the project costs must be financed by the applicant or from private sector sources. Applicants must also be able to show that without this assistance the project would not take place at all. A further criterion is that improved economic efficiency and greater security of employment should result. Grants are based on the fixed capital costs of new buildings or adaptations of existing buildings, plant, equipment, machinery, vehicles, etc. Grants are available under a number of schemes such as Regional Selective Assistance, Regional Enterprise Grants, Enterprise Initiative, Training and Research Grants and the European Community.

Post-contract methods

The cost control of a construction project commences at inception and ends with the agreement of the final account. A variety of methods are used to control the costs of construction during the post-contract stage. If the cost control is to be effective then any changes that might affect the contract should be costed *prior* to instructions being issued to the contractor.

Budgetary control

Budgets are used for planning and controlling the income and expenditure. It is through the budget that a company's plans and objectives can be converted into quantitative and monetary terms. Without these a company has little control. A budget for a construction project represents the contract sum divided between a number of different subheadings or work packages.

The contractor will have a costed work programme for the project, although this can be disrupted through changes (variations) to the scheme or the acceleration or the deceleration of activities.

The client's budget represents the time-scale of payments and the availability of funds for honouring the contractor's certificates. Clients with several projects under construction will need to aggregate the amounts of interim certificates from different projects to obtain the total funding requirements. In addition, clients are concerned with the total forecasted project expenditure. The ability to control this depends upon the sufficiency of the pre-contract design, the need for subsequent variations, the steps taken to avoid unforeseen circumstances, and matters which are beyond their control, such as strikes.

The contractor's budget provides a rate of expenditure and a rate of income throughout the project. The contractor's funding requirements represent the difference between these two items, and the amount of capital required at the different times can then be calculated. Contractors also need to aggregate this information from all their current projects in order to determine the company position. Budgetary control compares the budgets with the actual sums incurred, explaining the variances that arise. In common with other control techniques, budgetary control is a continuous process undertaken throughout the contract duration (see Ashworth 2002).

Client's financial reports

Financial reports are prepared at frequent intervals throughout the contract period, depending upon the size and complexity of the project, to advise clients on any expected changes to the contract sum. An example of a typical financial statement is shown in Chapter 13, where such reporting is covered in more detail.

Client's cash flow

In addition to the client's prime concern with the total project costs, the timing of cash flow is also important, since this will affect borrowing requirements. The client's quantity surveyor will prepare an expenditure cash flow profile based on the contractor's programme of activities, and any subsequent changes or revisions to this programme. On large and complex projects, and in periods of high inflation, the timing of payments, based upon different constructional techniques and methods, might result in a different contract sum representing a better economic choice for the project as a whole.

Contractor's cost control

The contractor, having priced the project successfully enough to win the contract through tendering, must now ensure that the work can be completed for the estimated costs. One of the duties of the contractor's quantity surveyor is to monitor the expenditure, and advise management of action that should be taken. This process also includes the cost of subcontractors, as these are likely to form a significant part of the main contractor's total expenditure. The contractor's quantity surveyor will also comment on the profitability of different site operations. Wherever a site instruction suggests a different construction process from that originally envisaged, then details of the costs of the site operations are recorded. The contractor's quantity surveyor will also advise on the cost implications of the alternative construction methods that might be employed.

Discounting the fact that estimators can sometimes be wide of the mark when estimating, even with common work items, contractors need to satisfy themselves if wide variation between costs and prices arise. This will be done for two reasons: first, in an attempt to recoup, where possible, some of the loss; and secondly to remedy such estimating or procedural errors in any future work. There are various reasons why such discrepancies may arise:

- Character of the work is different from that envisaged at the time of tender
- Conditions for executing the work have changed
- Adverse weather conditions severely disrupted the work
- Inefficient use of resources
- Excessive wastage of materials
- Plant standing idle for long periods
- Plant being incorrectly selected
- Delays due to a lack of accurate design information.

Often when the project is disrupted by the client or designer this can have a knock-on effect on the overall efficiency and output of the contractor's resources. Contractors may sometimes suggest that they always work to a high level of efficiency. This is not always the case, and losses occur due to their own inefficiency. Costing systems that indicate that a project or site operation has lost money are of limited use if a contractor is unable to remedy the situation. A contractor needs to be able to ascertain which part of the job is inefficient and to know as soon as it begins to lose money. The objectives therefore of a contractor's cost control system are to:

- Carry out the works so that the planned profits are achieved
- Provide feedback for use in future estimating
- Cost each stage or building operation, with information being available in sufficient time so that possible corrective action can be taken
- Achieve the benefits suggested within a reasonable level of administration charges.

Contractor's cash flow

Contractors are not, as is sometimes supposed, singularly concerned with profit or turnover. Other factors also need to be considered in assessing the worth of a company or the viability of a new project. Shareholders, for example, are primarily concerned with rate of their return on the capital invested. Contractors have become more acutely aware of the need to maintain a flow of cash through the company. Cash is important for day-to-day existence, and some contractors have suffered liquidation or bankruptcy not because their work was unprofitable but because of cash flow problems in the short term. In periods of high inflation, poor cash flows can result in reduced profits, which in their turn reduce shareholders' return. A correct balance between the objectives of cash flow, profit, return and turnover is required. In addition, inflation and interest charges will also have an impact on these items.

Discussion topic

What role does lean construction have in the cost control of construction projects?

Cost control or controlling construction costs are methods that are used to increase added value in construction projects. In some cases this may be the development and application of proper and effective systems. In other situations it might mean reviewing some of the actual practices of organisation and management that are employed in both design and construction. For example, design and build is often championed as a least cost solution. This may be correct under certain conditions and especially where the design content is minimal and the process used for construction is relatively routine in nature.

Lean construction methods are, in many ways, like cost planning and value management. Each of these techniques or practices aim to reduce the unnecessary costs from construction projects. In today's world they will seek to do this within a context of whole life costing. Cost planning (see page 126) was introduced in the 1950s and whilst it sought to reduce expenditure it did this only against a principle of setting cost limits (see page 127). Both value management and lean construction set no cost limits and thus claim to offer more added value to the client.

Lean construction is a derivative of the lean manufacturing process. This concept has been popularised since the early 1980s in the manufacturing sector. The original thinking was developed from Japan, (Womack and Jones 1996) although its principles have since been adopted worldwide. It is concerned with the elimination of waste activities and processes that create no added value. It is about doing more for less. It fits neatly within the philosophy of John Ruskin (1819–1900), *It is not the cheaper things in life that we wish to possess, but expensive things that cost less.*

Lean production is the generic version of the Toyota Production System. Automobile manufacturing has seen spectacular advances in productivity, quality and cost reduction (Howell 1999). The construction industry by comparison, has not yet achieved these advances. Whilst it is possible to learn and adapt successful methodologies from other industries, it should be recognised that construction is a different activity.

The principles involved in lean construction are identified in Fig. 6.4. This suggests that lean construction is a combination of current management thinking combined with proven developments both in the UK and worldwide. It recognises that it is a process of continuous improvement, of doing more for less and that this will

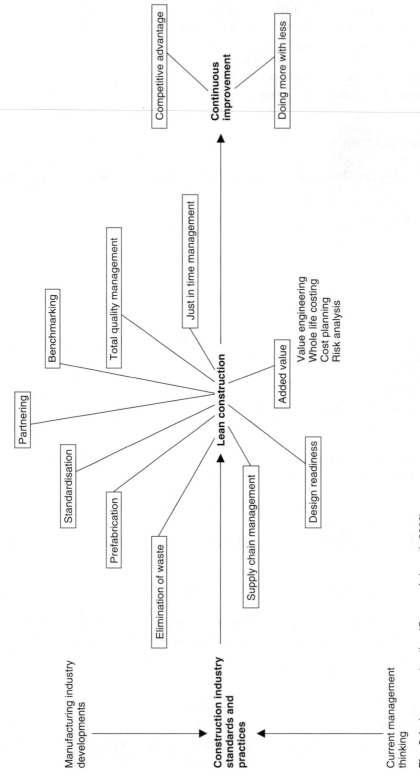

Fig. 6.4 Lean construction (*Source*: Ashworth 2006).

result in a competitive advantage for construction firms who adopt this process. The principles involved include:

- Elimination of waste, whether this is in design or construction practices
- Prefabrication of construction components off-site, where these can be manu-factured under factory conditions to a higher standard than site working is able to achieve
- Standardisation of building components to allow cost efficiencies to be made by using mass production techniques
- Partnering with a smaller number of subcontractors and suppliers and keeping successful teams together
- Benchmarking of practices against leading edge companies to maintain the most up-to-date practices
- Total quality management that will result in getting it right the first time and to a quality that cannot be matched
- Just-in-time management practices to keep investment and storage costs to a minimum
- Supply chain management practices where all firms involved are working to the same requirements
- Design readiness to allow the constructor the best opportunity of organising and carrying out the work efficiently and effectively. This also has the potential for reducing the amount of time contractors spend on site
- Adding value through the use of a range of techniques that focus on an holis-tic approach to the project.

Lean construction can be pursued through a number of different approaches. The lean principles have been identified as follows (Womack 1990):

- The elimination of all kinds of waste. This includes not just the waste of mate-rials on site, but all aspects, functions or activities that do not add value to the project
- Precisely specify value from the perspective of the ultimate customer
- Clearly identify the process that delivers what the customer values. This is sometimes referred to as the value stream
- Eliminate all non-added value steps or stages in the process
- Make the remaining added value steps flow without interruption through man-aging the interfaces between the different steps
- Let the customer pull; do not make anything until it is needed, then make it quickly. Adopt the philosophy of *just in time management* to reduce stock piles and storage costs
- Pursue perfection through continuous improvement.

Male (2002) has in addition recognised that lean thinking has more obvious appli-cations with large, regular procuring clients who can provide some continuity of work and are able to integrate the supply chain.

The use of lean production techniques must be placed within the context of the construction industry. The comparison between mass production in factories with generally bespoke buildings on a construction site must not be minimised. Furthermore, lean manufacturing appears to achieve the greatest improvements in efficiency and quality when design and manufacture occur in close proximity. With traditional procurement arrangements in the construction industry this is frequently not the case.

The vision of lean construction stretches across many traditional boundaries. It challenges our current practices. It has required changes in work practices and in understanding new roles and different responsibilities. It has required a change in attitudes and culture. It recognises that technology has an important role to play today. It expects that those involved in a construction project will share a common purpose. It is setting new standards that can be measured to indicate improvement. It is a vision for the future.

The response to lean construction from the construction industry is not unanimous (Green 1999). Like many new ideas, when put into practice, some of them do not always live up to the theoretical expectations. But lean construction is not just a theory since its techniques have been successfully applied in other industries. Also, there has to be the desire for success and a belief system is, therefore, very important. Lean construction is not a phenomenon that is likely to disappear, since there is a ground swell of opinion and worldwide interest in its principles.

References and bibliography

Ashworth A. *Cost Models, Their History. Development and Appraisal.* CIOB Technical Information Service No. 64, Chartered Institute of Building. 1986.

Ashworth A. Cost modelling. In: *Building in Value* (eds Best R. and De Valence G.) Edward Arnold. 1999.

Ashworth A. *Precontract Studies: Development Economics, Tendering and Estimating.* Blackwell Publishing. 2002.

Ashworth A. *Cost Studies of Buildings.* Pearson. 2004.

Ashworth A. *Contractual Procedures in the Construction Industry.* Pearson. 2006.

Ashworth A. and Hogg K.I. *Added Value in Design and Construction.* Pearson Education. 2000.

Ashworth A. and Skitmore M. *Accuracy in cost estimating.* Proceedings of the Ninth International Cost Engineering Congress, Oslo. 1986.

Barrett F.R. *Financial Reporting, Profit and Provisions.* CIOB Technical Information Service No. 12, Chartered Institute of Building. 1982.

Brandon P.S. (ed.) *Quantity Surveying Techniques: New Directions.* Blackwell Science. 1992.

Cooke B. and Jepson W. *Cost and Financial Control in Construction.* Macmillan. 1982.

DETR. *Best Value.* Department of Environment, Transport and the Regions. 2001.

Egan J. *Rethinking Construction.* Department of the Environment, Transport and the Regions. 1998.

Ferry D.J., Brandon P.S. and Ferry B. *Cost Planning of Buildings.* Blackwell Science. 1999.

Flanagan R. and Norman G. *Risk Management In Construction.* Blackwell Science. 1993.

Flanagan R. and Tate B. *Cost Control in Building Design.* Blackwell Science. 1997.

Fortune C. and Lees M. *The relative performance of new and traditional cost models in strategic advice to clients.* RICS. 1996.

Gray C. *Value for money: Helping the UK afford the buildings it likes.* Reading Construction Forum. 1996.

Green S.D. The dark side of lean construction: exploitation and ideology. In: *Proceedings of the Seventh Annual Conference of the International Group for Lean Construction*, IGLC-7, University of California, Berkeley. 1999.

Hackett M., Robinson I. and Statham G. *The Aqua Group Guide to Procurement Tendering and Contract Administration.* Blackwell Publishing. 2007.

Howell G.A. What is lean construction? In: *Proceedings of the Seventh Annual Conference of the International Group for Lean Construction*, IGLC-7, University of California, Berkeley. 1999.

Kelly J. and Male S. *Value management in design and construction.* E & F.N. Spon. 2000.

MacPherson J., Kelly J. and Male S. *The Briefing Process: A Review and Critique.* The Royal Institution of Chartered Surveyors. 1992.

Male S. Building the business value case. In: *Best Value in Construction* (edited by J. Kelly, R. Morledge and S. Wilkinson). Blackwell Publishing. 2002.

Nellis H.G. and Parker D. *The Essence of Business Taxation.* Prentice Hall. 1992.

Powell C. *The Challenge of Change.* Royal Institution of Chartered Surveyors. 1998.

Pullen L. What is best value in construction procurement? *Chartered Surveyor Monthly*, February. 2001.

Raftery J. *Principles of Building Economics.* Blackwell Science. 1991.

RICS. *Precontract Cost Control and Cost Planning.* The Royal Institution of Chartered Surveyors. 1982.

Saunders G. (ed.) *Tolley's Taxation Planning.* Tolley Publishing Company Ltd. 2000.

Somerville D.R. *Cash Flow and Financial Management Control.* CIOB Surveying Information Service No. 4, Chartered Institute of Building. 1981.

Womack J. *The Machine that Changed the World.* Simon and Schuster. 1990.

Womack J. and Jones D. *Lean Thinking.* Simon and Schuster. 1996.

Website

www.leanconstruction.org

7 Whole Life Costing

Introduction

Whole life costing is typically adopted by owners as part of a strategic reassessment of their facilities. It influences the procurement of new buildings and engineering structures and the choices about renewal, refurbishment and disposal. It is becoming much more important as long-term building owners start to demand evidence of their costs of ownership. PFI consortia must, as a matter of course, attempt to assess the financial risks of taking on long-term responsibility for building and engineering operation and maintenance.

It has long been recognised that to evaluate the costs of buildings and engineering structures on the basis of their initial costs alone is unsatisfactory. Some consideration must also be given to the costs-in-use which will accrue throughout the building or structure's life. The use of whole life costing for this purpose is an obvious idea, in that all costs arising from an investment are relevant to that decision. The image of the whole life of a building or structure is one of progression through a number of phases, with the pursuit of an analysis of the economic whole life cost as the central theme of the evaluation. The proper consideration of the whole life costs is likely to result in a project which offers the client better value for money. The earlier that whole life costing is considered, the greater will be the potential benefits.

There are a number of definitions and descriptions of whole life costing. The one that currently best describes it is, 'the systematic consideration of all relevant costs and revenues associated with the acquisition and ownership of an asset' (Construction Best Practice Programme 1998).

Anecdotal evidence suggests that for every unit of capital cost spent on the construction of a building, over a 30 year period there will be ten units spent on maintaining and 100 units on staffing the business activity. Improvements obtained through adopting a whole life cost approach can make a big difference, particularly when expressed in terms of the core business activity (Construction Best Practice Programme 1998).

Each building has a useful life. Planning costs in terms of that usage is as important as establishing costs at the inception. However, the prediction of whole life costs is problematic and often insufficiently used in the design process as an advisory tool.

There are a variety of factors that influence the whole life costs of building. These include the:

- Identification of costs incurred during a building or engineering structure's life and the inter-relationship with its use and maintenance
- Appreciation of forecasting techniques available and their use as a planning tool
- Consideration of the effects of time on the accuracy of cost advice with particular regard to technological advancement, government policy and fashion
- Understanding of the importance of the application of risk analysis techniques in the validation of cost advice
- Importance of long life, loose fit and low energy in managing the flexibility of designs during the life of buildings.

Brief history

Total building cost appraisal includes every aspect concerned with a built asset. Ashworth (2004) has identified how building cost appraisal in the middle of the twentieth century was restricted largely to a consideration of the initial building costs alone. During the latter part of the twentieth century a shift in practice was observed to focus more correctly on both initial and recurring costs, the latter relating to those costs expended during the project's lifetime. At the start of the twenty-first century this thinking has moved much further to include the costs of the activities that take place within buildings and as such impinge upon the client's entire business operations and how the built assets contribute towards these.

As Stone (1983) comments, the terminology used has gone under a variety of different names. This sometimes varies depending upon the country concerned or the context to which it is applied. The USA, for example, continues to use the term life cycle costs, whereas in engineering it was frequently referred to as terotechnology. Other terminology associated with this includes life cost, recurrent costs, operational cost, post-occupancy analysis, running costs and ultimate cost. The technique is related to cost–benefit analysis, cost effective analysis and threshold analysis. The discounted cash flow technique is an element in all of these forms of analysis.

During the 1960s it was often referred to as costs-in-use, although strictly speaking this term excluded anything to do with initial construction costs. In the 1970s, life cycle costing became the commonly accepted terminology but by the end of the century this had been replaced with whole life costing and this is the description under which it is now most commonly referred.

There can be few now in the construction industry who do not realise the potential of evaluating projects on the basis of whole life costs.

Government policy

The UK government has taken a decision to make all construction procurement choices on the basis of whole life costs. HM Treasury guidance stipulates this specifically. This general ruling has been crystallised in Private Finance Initiative (PFI) and Public Private Partnerships (PPP) contracting. The focus of PFI and PPP on long-term risk management and long-term maintenance and operation makes comprehensive whole life costing a necessity. Government procurement outside of PFI and PPP also has an emphasis on whole life costs at all levels, including local authority and housing association procurement.

Government procurement practices have had a huge influence on the construction industry as a whole, with major client organisations adopting similar policies. This has been reinforced by the move towards partnering that engenders longer-term strategies and risk management processes. These are naturally supported by whole life costing practices.

The Office of Government Commerce has produced *The Achieving Excellence* suite of procurement. One of these procurement guides is titled *Whole Life Costing and Cost Management* (2003). The new series reflects developments in construction procurement over recent years and builds on government clients' experience of implementing the *Achieving Excellence in Construction* initiative. This initiative has resulted in the publication of the following guides:

High level guides

Construction Projects
Construction Projects Pocketbook

Core guides

1. Initiatives into action
2. Project organisation
3. Project procurement lifecycle

Supporting guides

4. Risk and value management
5. The integrated project team
6. Procurement and contract strategies
7. Whole life costing and cost management
8. Improving performance
9. Design quality
10. Health and safety.

Whole life value

This is a term that has been developed by the Building Research Establishment (BRE) to describe the various aspects of sustainability in the design, construction, use, demolition and, where appropriate, the reuse of a built asset.

It involves achieving compromise and synergy between the following three sets of values:

- **Economic value.** This is focused on economically sound sustainability: growth-oriented and seeks to safeguard the opportunities of future generations.
- **Social value.** This is concerned with aspects of a societal nature and covers a wider scope of social, cultural, ethical and juridical impacts.
- **Environmental value.** This focuses on environmental aspects of development such as pollution, waste and CO_2 emissions. These issues involve the initial manufacture of construction materials, the construction of the project, its use and eventual replacement. In this context value is maximised when environmental pressures are minimised to the level of the carrying capacity of ecological systems while using natural resources effectively and safeguarding natural capital and its productivity.

The Whole Life Value (WLV) Framework established by BRE is a search facility that points users to a variety of design tools that deal with the issue of sustainability and are appropriate to the particular user group and stages in the procurement process and life of an intended facility. It enables designers and their clients to take account of the most significant aspects of sustainability and to predict the whole life value of their projects. The Framework has been populated with tools, guidance and procedures identified during a BRE review of material in the public domain, as well as being further populated with tools identified in the questionnaire and by the user group to ensure that the WLV Framework meets industry's requirements. The WLV Framework has also been populated with Embedded Knowledge, a term used to describe information that the designer is likely to find useful when trying to design a built asset that takes into consideration the Whole Life Value ethos. For further information refer to www.bre.org.uk.

Additional information can also be obtained from The Association for Environment Conscious Building (AECB). This was established in 1989 to increase awareness within the construction industry of the need to respect, protect, preserve and enhance the environment, principles that have now become known collectively as sustainability. The Association includes local authorities, housing associations, builders, architects, designers, consultants and manufacturers among its membership. Further information can be found at www.aecb.net.

Whole life costing applications

The following are some of the advantages of using whole life costing:

- It gives an emphasis on a whole or total cost approach undertaken during the acquisition of a capital cost project or asset, rather than merely concentrating on the initial capital costs alone

- It takes into account the initial capital costs, repairs, running and replacement costs and expresses these in comparable terms
- It allows for different solutions of the different variables involved and sets up hypotheses to test the confidence of the results achieved
- It is an asset management tool that allows the operating costs of premises to be evaluated at frequent intervals
- It allows, for example, changes in working practices, such as hours of operation, introduction of new plant or machinery, use of maintenance analysis, to be properly evaluated as tools of facilities management.

Whilst there has been an emphasis upon the use of whole life costing during the pre-contract period, its use can be extended throughout every phase of a building's life, as follows.

At inception   Whole life costing can be used as a component part of an investment appraisal. The technique is used to balance the associated costs of construction and maintenance with income or rental values.

During the design stage   It is used to evaluate the different design options in order to assess their economic impact throughout the project's life. It is frequently used alongside value management and other similar techniques. The technique focuses on those areas where economic benefits can be achieved.

During the construction stage   During this phase there are many different areas that can be considered for its application. It can be applied to the contractor's construction methods, which can have an influence upon the timing of cash flows and hence the time value of such payments. The contractor is able to apply the principles to the purchase, lease or hire of the construction plant and equipment. Construction managers and contractors' surveyors are able to offer an input to the scrutiny of the design, if involved sufficiently early in the project's life to be able to identify whole life cost implications of the design, manufacture and construction process.

During the project's use and occupation   It is a physical asset management tool. Costs-in-use do not remain uniform or static throughout a project's life, and therefore need to be reviewed at frequent intervals to assess their implications. Taxation rates and allowances will change and have an influence upon the facilities management policies being used.

At procurement   The concept of the lowest tender bid price should be modified in the context of whole life costing. Under the present contractual and procurement arrangements, manufacturers and suppliers are encouraged to supply goods, materials and components which ensure their lowest initial cost, often irrespective of their future costs-in-use. It is now accepted by many clients that a greater emphasis should be placed upon the overall economic performance of the different components.

In energy conservation   Whole life costing is an appropriate technique to be used in the energy audit of premises. The energy audit requires a detailed study and investigation of the premises, recording of outputs and other data, tariff documentation and appropriate monitoring systems.

Whole life costs

In addition to initial capital construction costs, other costs accrue as changes occur throughout the project's life (Fig. 7.1). The costs associated with the major refurbishment of an existing project are sometimes classified as initial costs, particularly where an existing project is undergoing extensive works or a change in building use.

In addition to the items shown in Fig. 7.1, there are many other costs-in-use that must also be considered in connection with the whole life costing of a project. These include:

- Fuel for heating, lighting and communications
- Facilities management
- Business rates and insurances
- Redecoration (much of this will coincide with an aspect from Fig. 7.1)
- Cleaning

Terminology description	
Maintenance	Regular ongoing work to ensure that the fabric and engineering services are retained to minimum standards. It is frequently of a minor nature.
Repairs	Associated with the rectification of building components that have failed or become damaged through use and misuse.
Renewal	The upgrading of a building to meet modern standards.
Adaptation	Frequently includes building work associated with conversions, such as a change in function and typically includes alterations and extensions.
Renovation	Repairing and rebuilding.
Retrofitting	The replacement of building components with new components that were not available at the time of the original construction.
Conservation	Building works carried out to retain the original features or restore a building to its original concept.

Fig. 7.1 Building renewal.

Main factors to consider

Building life

Over time existing buildings decay and become obsolete and require maintenance, repair, adaptation and modernisation. There also lies a varied pattern of existence, where buildings are subject to periods of occupancy, vacancy, modification and extension. Figure 7.2 identifies some types of building life.

The useful life of any building is governed by a number of factors, such as the methods of construction envisaged at the initial design and the way that the building is cared for whilst in use, e.g. the amount of wear and tear and the levels of maintenance that are applied. The life of buildings can be assessed in different ways and these are briefly considered later in this chapter.

The design must also recognise the difference between those parts of the building with long, stable life and those parts where constant change, wide variation in aesthetic character and short life are the principal characteristics. There seems to be

Condition	Definition	Examples
Deterioration		
Physical	Deterioration beyond normal repair	Structural decay of building components
Obsolescence		
Technological	Advances in sciences and engineering result in outdated building	Office buildings unable to accommodate modern communications technology
Functional	Original designed use of the building is no longer required	Cotton mills converted into shopping units, chapels converted into warehouses
Economic	Cost objectives are able to be achieved in a better way	Site value is worth more than the value of the activities current on the site
Social	Changes in the needs of society result in the lack of use for certain types of buildings	Multi-storey flats unsuitable for family accommodation in Britain
Legal	Legislation resulting in the prohibitive use of buildings unless major changes are introduced	Asbestos materials, Fire Regulations
Aesthetic	Style of architecture is no longer fashionable	Office building designs of the 1960s

Fig. 7.2   Building life and obsolescence (adapted from RICS 1986).

little merit in including building components with long life in situations where rapid change and modernisation are to be expected.

Component life

The life span of the individual materials and components has a contributory effect upon the life span of the building. However, data from practice suggest widely varying life expectancies, even for common building components (RICS 1992). Whole life costing is concerned not so much with how long a component will last, but for how long a component will be retained. The particular circumstances of each case will have a significant influence upon component longevity. These will include:

- Correct choice of component specification
- Use of appropriate design details
- Installation in accordance with the manufacturer's directions, relevant Codes of Practice and British Standards
- Compliance with the conditions of any relevant third party assurance certificates
- Appropriate use by owners, users and third parties
- Frequency and standards of maintenance.

The management policies used by owners or occupiers are perhaps the most crucial factors in determining the length of component life. There is a general absence of such characteristics in retrieved maintenance data (Ashworth 1996b).

Discount rate

The selection of an appropriate discount rate to be used in whole life costing calculations depends upon a wide range of different factors. The discount rate to be chosen will depend to some extent upon the financial status of the client. For example, public sector clients are generally able to obtain preferential rates of interest. Consideration should also be given to whether the finance is borrowed or obtained from retained profits. The choice of a discount rate is frequently inferred to mean the opportunity cost of capital. The choice of a discount rate is one of the more critical variables in the analysis.

The choice of an appropriate discount rate is a difficult task dependent upon a combination of interest rates and inflation within an economy. The rate selected will vary over time and in different parts of the world.

Taxation

Since whole life costing is an attempt to aggregate the initial and recurring costs of building, taxation must also be considered. The effects of taxation may result in building projects becoming more economically viable. The application of taxation to initial and future costs in not uniform and is subject to frequent change. This must be considered when calculating the whole life costs of buildings.

Spending category	Commercial %	Recreation %	Educational %	Residential %
Utilities/energy	35	22	32	22
Overheads	24	19	23	13
Administration	12	29	14	20
Cleaning	12	13	16	25
Fabric maintenance	9	6	5	6
Services maintenance	5	7	7	6
Decorations	3	4	3	8

Fig. 7.3   Breakdown of operational costs (*Source*: Construction Best Practice Programme 1998).

Targeting the major elements of costs-in-use

Figure 7.3 provides a breakdown of suggested average operational costs for different categories of buildings. This indicates where the major areas of possible savings in ongoing expenditure might be made. However, historic cost data analyses hide significant variations between the best and worse performers in each category. For example, the most efficient examples may reduce the cleaning cost to about half those of the least efficient. Large savings in energy costs can be achieved, both initially and over the building's life, if air conditioning can be avoided.

Depreciation and obsolescence in buildings

A distinction needs to be made between obsolescence and the deterioration of buildings (Fig. 7.4). The physical deterioration of buildings is largely a function of time and use. Whilst it can be controlled to some extent by selecting the appropriate materials and components at the design stage and through correct maintenance whilst in use, deterioration is inevitable as an ageing process. Obsolescence is much more difficult to control since it is concerned with uncertain events such as the prediction of changes in fashion, technological development and innovation in the design and use of buildings. Deterioration eventually results in an absolute loss of use of a facility, whereas with buildings that become obsolete it is accepted that better facilities are available elsewhere. Whilst deterioration in buildings can be remedied at a price, obsolescence is much less easy to resolve. Obsolescence can be defined as value decline that is not caused directly by use or passage of time (Ashworth 1997).

Due to the large investment that is required in a building, demolition due to obsolescence is a last resort and will only take place where the building is either not capable of renewal or is in a wrong or decaying environment. In some cases the site on which the building stands may be required for other purposes.

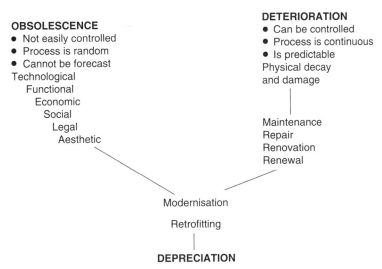

Fig. 7.4   Obsolescence, deterioration and depreciation (*Source*: adapted from Flanagan *et al.* 1989).

Long life, loose fit and low energy

Physical deterioration occurs more slowly than the various forms of functional and other types of obsolescence. The blame for a great deal of obsolescence is due to inflexible planning and designing buildings that were unsuitable for adaptation should their original function cease. However, a majority of building clients commissioning buildings require bespoke design solutions to meet their individual needs. Some of these solutions, perhaps decades later when the property is no longer required, are difficult to adapt to changing circumstances. Obsolescence is also to some extent to be coupled with population movement which may make even the most adaptable structure obsolete if it is located in an area of declining desirability or usefulness. The architect Alex Gordon (1977), over three decades ago coined the phrase 'Long life, loose fit and low energy', describing these as the '3 Ls principle'. He recognised that obsolescence rather than physical decay was the major reason for buildings falling into disuse. He emphasised the importance of designing flexible solutions to clients' problems so that these could easily be adapted to suit the changing needs of work, leisure and living spaces. Since 1977 changes throughout society have been at an accelerating rate, emphasising the responsibility to ensure that the scarce resources are used with as much longevity in mind as possible.

Calculations

One of the apparent difficulties of using whole life costing in practice is the mathematics associated with the evaluation. An understanding of the principles involved in discounting the value of future receipts and payments is an essential feature of such an evaluation. Although the arithmetic associated with discounting may appear complicated, the concept is simple. This is that the capital in the hand today is worth more than the capital at some time in the future. Even ignoring inflation, it would be more beneficial to choose to receive an amount of money today than the same amount next year.

If the current rate of interest is 5% per annum, then £100 invested today will yield £105 in twelve months' time. Conversely, £100 to be received next year has what is known as a present value (PV) of:

$$£100 \times \frac{100}{105} = £95.24 \text{ today}$$

This example is the principle associated with discounting or discounted cash flow calculations. Whilst the amounts can be calculated through the use of valuation formulae or equations, it is still common to use standard valuation tables to allow comparisons between money spent or received at different times. It is necessary to convert these sums to a common timescale and valuation tables are a means of making this conversion.

Amounts are generally expressed in the following ways:

- Actual values, e.g. capital costs
- Present values of a lump sum to be paid at some time in the future. The amount is calculated using the 'Present value of £1 table'. (See replacement cost below)
- Present value of a regular annual amount. The amount is calculated using the 'Present value of £1 per annum table'. (See cleaning costs below).

Each of the above can also be represented as the 'annual equivalent' amount by converting present values to annual sums by using the 'Present value of £1 per annum table'. The technique of discounting is best explained by the use of the following example.

A restaurant proprietor is seeking advice on floor finishes for a new restaurant that is being constructed. The data on the floor finishes are shown below.

Floor finish	Initial cost	Annual maintenance	Annual cleaning	Expected life
A	£20000	£750	£1000	15 years
B	£30000	£500	£500	30 years

The situation implies a time horizon of 30 years (or multiple of 30 years for comparison purposes). A discount rate of 5% has been selected. (See Fig. 7.5.)

The calculation in Fig. 7.5 suggests that floor finish B is the preferred choice based upon the data provided. However, this is not done on the comparison of the cost calculation alone. Other subjective factors must also be considered. In this example these might include the following:

	Discounted amount	Notes
Floor finish A		
Initial cost	20 000	
Maintenance 750 × 15.3725	11 529	Use PV of £1 per annum table
Cleaning 1000 × 15.3725	15 373	Use PV of £1 per annum table
Replacement (year 15)		
20 000 × 0.481017	9 620	Use PV of £1 table
Net Present Value (NPV)	56 522	
Floor finish B		
Initial cost	30 000	
Maintenance 500 × 15.3725	7 686	Use PV of £1 per annum table
Cleaning 500 × 15.3725	7 686	Use PV of £1 per annum table
Replacement life = 30 years	0	
Net Present Value (NPV)	45 372	

Fig. 7.5 Whole life costs of floor finishes.

- As well as being more durable, expensive construction is generally more pleasant in appearance
- Future replacement or repairs may be inconvenient
- Replacement or repairs may be difficult and therefore expensive (costs and revenue)
- The saving of money on a specific item may involve costs out of all proportion to the possible savings
- Obsolescence may not be a factor to be considered
- Prestige associated with more expensive construction
- A client may be more concerned with higher initial costs than the reduction of future costs in use.

Forecasting the future

The fundamental problem associated with the application of whole life costing in practice is the requirement to be able to forecast a long way into the future. Whilst this is not done in absolute terms, it must be done with a reliability to allow the confident selection of project options that offer the lowest whole life economic solutions (Ashworth 1996a).

All long term forecasting is fraught with some sort of confidence credibility, but this does not mean that such forecasting should not be attempted. Weather forecasting, for example, has access to huge amounts of physical science data which have been systematically researched and collected over a long period. Even so, reliable forecasts of weather patterns for even a few days ahead, are not always guaranteed. Whole life costing by comparison is a social science skill, where human perceptions and experiences including their vagaries are manifest, and where aspirations, objectives and desires fluctuate and evolve. It is in this context that whole life costing is to be applied, and predictions, if they are to be useful, need to be reliable for at least the next few years.

Whilst some of the future events can at least be considered, analysed and evaluated, there are others that cannot even be imagined today. These remain therefore largely outside of the scope of prediction and probability, and are thus unable to be even considered, let alone assessed, in the analysis.

There are a number of major issues that influence the application of whole life costing to construction projects. Also, forecasting can only be carried out in the context of present day knowledge and information.

Life expectancy

Traditionally many commentators on whole life costing have based their calculations on a 60 year life span of buildings. This originated from a development surveyors' perspective associated with rentals and yields, and from the technologists' view regarding construction longevity and obsolescence. It does not now really seem sensible to even attempt to forecast costs up to the year 2060 and beyond. It is now generally accepted that the whole life time horizon should be more related to current use expectations associated with the building's or structure's own cycle or related to the cyclical effect of population movements associated with the project. Different buildings have different life expectancies as indicated in Fig. 7.6. Whole life costs should attempt to use time horizons that are realistic for the particular buildings concerned. Even then calculations in a rapidly changing world may be difficult to substantiate.

Data difficulties

The reason why this technique has not been more widely used is frequently said to be the lack of appropriate, relevant and reliable historical cost information and data (Ashworth 1996c). Maintenance cost data have in the past largely been collected solely for accounting purposes and to satisfy the reconciliation of budgets. These

Project	Renewal time (years)
International fast food outlet	3
Car factory	10
Office development	10
Local authority school	20
Prison	60
Cathedral	100

Fig. 7.6   Indicative renewal time intervals (*Source*: Hoar and Norman 1992).

accounting headings are often unsuitable for use in whole life costing applications. Where data has been found to be available, it is often so contradictory in its nature that its satisfactory reuse becomes almost impossible. Further studies have emphasised the mismatch between the cost headings used by maintenance departments and those which are required for use in the whole life costing of future projects.

The data used for estimating by contractors for tendering purposes also have wide variations in outputs but these are relatively insignificant when compared with the inconsistency of maintenance cost records. Furthermore, little indication is provided of a qualitative nature, which is required if the costs-in-use data is to be properly interpreted. The broad categories of accounting headings such as general maintenance or repairs disclose too little information to prospective data users for reliable or sound judgement to be made.

The application of historic cost records to initial building costs requires a good deal of judgement if the desired results are to be achieved. Maintenance cost data is of a more critical nature than this, due to the other influences which affect its quality and reliability. The inherent characteristics of such data must be known to the user. For example, it is necessary to have some insight into the causes of component failure or deterioration. Was the repair carried out due to a failure of other items of work, vandalism, misuse or simply normal wear and tear? Furthermore the life of components or materials may have become shortened due to the time-lag delay occurring between reporting, remedy, invoicing and payment stages. The historic information can therefore easily be misleading in terms of both component life and its attributed costs.

Technological change

It is difficult to forecast the possible changes in technology, materials and construction methods that may occur even for the next few years. The construction industry, its process and its product are under a purposeful change and evolution. There is a constant striving to develop excellence in both design and manufacture and to introduce new materials having the desired characteristics of quality and reliability in use. The changes in technology can often be sudden and unexpected. The introduction of new technology and workable solutions to existing problems can have a major impact upon the whole life cost forecasts.

Fashion changes

A further difficulty facing the application of whole life costing in practice is the changes in fashion. These changes are less gradual and more unpredictable than changes in technology. Themes within the construction industry have been developed in different eras, such as built to last, inexpensive initial cost, industrialisation, long life, loose fit, low energy, and the current attitudes of conservation and refurbishment. Changes in the way that buildings might be used in the future are already being predicted. Some of these are hopelessly fanciful. Others reflect a more reasoned attitude towards work and leisure, changes in personal expectations, demographic trends and developments generally in society. Whole life costing must attempt to anticipate future trends and their effect upon an overall economic solution.

Cost and value changes

The erratic pattern of inflation throughout the past 25 years could not have been predicted even a decade earlier. Slumps in the construction industry follow booms and vice versa, but even so, these are currently beyond the scope of present indicators and economic forecasters. Costs and values do not move in tandem, and the respective indices for the different materials, products or components do not follow similar patterns but are subject to wide degrees of cost fluctuations.

Policy and decision-making changes

One of the most important whole life costing variables is the owner's future use and maintenance policy. This factor is absent in historic cost data sources. Also maintenance work is not needs oriented but budget led. Maintenance work is thus largely determined by the amount of funds which are available for this purpose. The policy adopted by the owner and adapted by the occupants is at least as important as the design and construction criteria in the determination of maintenance cycles and costs.

Accuracy

One criterion in any estimate is its reliability or accuracy. By definition an estimate will never be spot-on. The accuracy of capital cost estimating in the forecasting of contractors' tender sums has been measured to be about 13%. Contractors' estimating of their own costs is marginally better. The processes used for both of these types of forecast have been refined through many years of use, experience and practice. Whole life costing is realatively new with limited experience of practice and a quality of data which is very subjective and inferior to that used for capital cost estimating. The reliability of the results achieved will be subject therefore to much larger variations and possible errors than those indicated for capital cost estimating (Ashworth and Skitmore, 1982).

A key criterion of whole life costing is that it enables the correct economic choice to be made between competing proposals. However, this choice must be made in the knowledge that there could be large estimating inaccuracies.

Whole life cost forum (WLCF)

The WLCF was formed in 1999 by a cross-section of construction industry representatives. The Forum organisation that is financially supported by its members includes property clients, property owners and operators, construction consultants, contractors, PFI contractors, suppliers and material manufacturers. Whilst government do not offer any direct financial support, several government departments are members.

The remit of the WLCF was to develop a set of definitions and methodologies that could be accepted by the construction industry as a whole that would provide a degree of certainty when talking about WLC or providing quotations on a WLC basis.

The WLCF has developed a web-based WLC comparator tool which uses the WLCF methodologies and provides a secure process that does not allow users to access the calculations. Therefore, any output or results from the WLCF comparator tool can be considered as authenticated as there is no possibility of manipulation of the results. Further information can be obtained from www.wlcf.org.uk.

Conclusions

The importance of attempting to account for future costs-in-use in an economic appraisal of any construction project has already been established in theory. The problem of its application in practice is twofold: the known or predictable and the unknown or uncertain.

The first real difficulty, and this should not be underestimated, is the application of statistical analysis to life expectancies, discount factors and other data to be used in the calculations. This is the assessment of risk, where the use of techniques such as probability and sensitivity analysis can be used to interpret the results and provide confidence limits to such assessments.

The second problem is much more difficult to deal with. This concerns the possibility of events not being imagined at the inception stage of a project that may have a total life of 100 years. Such events are in the realms of uncertainty and fantasy and cannot be measured or evaluated. During the past 50 years society has been under rapid change. This change is expected to accelerate in the future. It is therefore difficult to predict what influence such changes will have on the validity of whole life costing, even for a few years ahead.

Whole life costing does however offer potential. Its philosophy of whole cost appraisal is preferable to the somewhat narrow initial cost estimating approach. The widespread efforts so far expended in its research and development are a positive move. Whole life costing is at best a snapshot in time in the light of present day knowledge and practice and anticipated future applications.

Discussion topic

How can the large amount of uncertainty associated with the use of whole life costing be dealt with in practice?

Whole life costing is intrinsically linked with events in the future and, since the future is unknown, predictions can never be made with certainty. Ashworth, as long ago as 1987, posed the question, *Can it really work in practice?* In this he presented a number of different scenarios, that are still valid today, suggesting that our lack of knowledge about the future can easily allow us to make incorrect predictions based upon our current understanding. The future includes those events that are both unknown and in some cases even unimaginable. Undoubtedly, and particularly as practice wants to make use of whole life costing, there is a feeling of a greater sense of confidence and security. Kishk *et al.* (2003) noted that uncertainty is endemic to whole life costing. They also stated that there is a need to be able to forecast a long way ahead in time with factors such as life cycles, future operating and maintenance costs, and discount and inflation rates. This difficulty is also exacerbated not only by a lack of adequate data but that when such data is available it often produces conflicting results. This means that the treatment of uncertainty in information and data remains crucial to a successful implementation of whole life costing (Fig. 7.7). A number of risk assessment techniques can be used but these alone will never eliminate the uncertainty that should be anticipated.

Sensitivity analysis

This is a modelling technique that is used to identify the impact of change in variables such as building life, component life, interest rates, etc. Whole life costing is a technique that is able to assist users in the best or most economic use of the alternative assumptions that are available. Different values are therefore attributed to these different variables and comparative models developed to test out the different assumptions. It needs to be noted that only assumptions can be tested but these might not provide an optimum solution, since other, as yet, unknown factors cannot be included in these models. As the technique of whole life costing is developed then smarter solutions will undoubtedly be made possible. Kirk and Dell'Isola (1995) suggest that the objective is usually to determine the break-even point defined as the value of the input-data element that causes the least-cost alternative to be better than that of the next-lowest-cost alternative. A major advantage of this

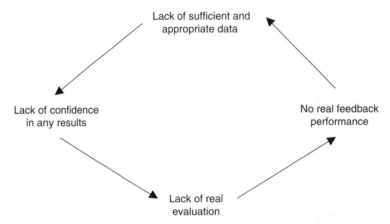

Fig. 7.7   The vicious circle of whole life cost implementation (*Source:* Al-Hajj 1991).

technique is that it explicitly shows the robustness of the ranking of alternative solutions (Flanagan and Norman 1983).

However, sensitivity analysis can only really be interpreted by changing the value of one variable at a time. Also, it does not aim to quantify risk but rather to identify factors that are risk sensitive. Thus, it does not provide a definitive method of making the decision.

Probability techniques

In the probabilistic approach to risk analysis, all uncertainties are assumed to follow the characteristics of random uncertainty. A random process is one in which the outcomes of any particular realisation of the process are strictly a matter of chance. The following are two suitable probability-based techniques.

Confidence index approach

The confidence index technique (Kirk and Dell'Isola 1995) is a simplified probabilistic approach. It is based on two assumptions:

- Uncertainties in all cost data are normally distributed
- High and low 90% estimates for each cost do in fact correspond to the true 90% points of the normal probability distribution for that cost.

Monte Carlo simulation

This is a means of examining problems for which unique solutions cannot be obtained. It has been used in whole life cost modelling by many authors such as Flanagan *et al.* (1987). In a typical simulation, uncertain variables are treated as random variables, usually, but not necessarily, uniformly distributed. In this probabilistic framework, the whole life costing variables, usually the net present values, also become random variables. In the last phase of evaluation, various alternatives are ranked in order of ascendant magnitude and the best alternative is selected such that it has the highest probability of being first. The decision-maker must weigh the implied trade-off between the lower expected cost of one alternative and the higher risk that this cost will be exceeded by an amount sufficient to justify choice of an alternative. Although the technique provides the decision-maker with a wider view in the final choice between alternatives, this will not remove the need for the decision-maker to apply judgement and there will be, inevitably, a degree of subjectivity in this judgement.

Fuzzy set theory

The above techniques recognise shortcomings suggesting that other alternative approaches might be more appropriate. There has been a growing interest in many science domains in the idea of using the fuzzy set theory (FST) to model uncertainty. Fuzzy set theory is an appropriate technique where human reasoning, human perception, or human decision-making are inextricably involved. In addition, it is easier to define fuzzy variables than random variables when no or limited information is available. Mathematical concepts and operations within the framework of FST are much simpler than those within probability theory especially when dealing with several variables.

Byrne (1997) carried out a critical assessment of the fuzzy methodology as a potentially useful tool in discounted cash flow modelling. However, this work was mainly to investigate the fuzzy approach as a potential substitute for probabilistic simulation models. Kaufmann and Gupta (1988) described how to manipulate fuzzy numbers in the discounting problem. They introduced an approximate method to simplify the mathematical calculations with fuzzy numbers.

Integrated approach

Kishk and Al-Hajj (1999) proposed an integrated framework to handle uncertainty in whole life costing. It is based on the simple idea that a complex problem may be deconstructed into simpler tasks. The appropriate tools are assigned a subset of tasks that match their capabilities.

Conclusions

Although the sensitivity analysis approach is simple, it is effective only when the uncertainty in one input-data element is predominant. Furthermore, it does not provide a definitive method of making the decision. The confidence index method is a simplified probabilistic method that has been found to lack the generality of application. Simulation techniques are more powerful but they have been criticised for their complexity and expense in terms of the time and expertise required. Probability theory can deal only with random uncertainty. The fuzzy algorithms are superior due to their ability to deal with judgmental assessments of all stated variables. The integrated approach can handle both statistically significant data and expert assessments within the same whole life cost model.

References and bibliography

Al-Hajj A. *Simple cost-significant models for total lifecycle costing in buildings*. PhD Thesis, Department of Civil Engineering, University of Dundee. 1991.

Ashworth A. Life cycle costing – can it really work in practice? In: *Building Cost Modelling and Computers* (edited by P.S. Brandon). E & F N Spon. 1987.

Ashworth A. Life cycle costing: Predicting the unknown. *Journal of the Association of Building Engineers*, April. 1996a.

Ashworth A. Estimating the life expectancies of building components in life cycle costing calculation. *Journal of Structural Survey*, **14** (2). 1996b.

Ashworth A. Data difficulties of building components for use in whole life costing. *The Surveyor. Journal of the Institution of Surveyors Malaysia*, fourth quarter. 1996c.

Ashworth A. *Obsolescence in buildings: Data for life cycle costing.* Chartered Institute of Building Technical Information Service. 1997.

Ashworth A. *Cost Studies of Buildings.* Pearson. 2004.

Ashworth A. and Skitmore M. *Accuracy in Estimating.* Chartered Institute of Building. 1982.

Boussabaine A. and Kirkham R. *Whole Life-Cycle Costing.* Blackwell Publishing. 2004.

Byrne P. *Fuzzy D C F: a contradiction in terms, or a way to better investment appraisal? In: Proceedings of RICS Cutting Edge Conference.* 1997.

Construction Best Practice Programme. *Whole life costing.* Department of the Environment, Transport and the Regions. 1998.

Flanagan R. and Norman G. *Life Cycle Costing for Construction.* RICS. 1983.

Flanagan R., Kendell A., Norman G. and Robinson G. Life cycle costing and risk management. *Construction Management and Economics,* 5. 1987.

Flanagan R., Norman G., Meadows J. and Robinson G. *Life Cycle Costing: Theory and Practice.* Blackwell Science. 1989.

Gordon A. The three 'Ls' principle: Long life, loose fit, low energy. *Chartered Surveyor Building and Quantity Surveying,* Quarterly. 1977.

Hoar D. and Norman G. Life cycle cost management. In *Building Cost Techniques New Directions* (ed. P.S. Brandon). Blackwell Science. 1992.

Kaufmann A. and Gupta M.M. *Fuzzy Mathematical Models in Engineering and Management Science.* Elsevier Science Publishers. 1988.

Kirk S.J. and Dell'Isola A.J. *Life Cycle Costing for Design Professionals.* McGraw-Hill Book Company. 1995.

Kishk M. and Al-Hajj A. *An integrated framework for life cycle costing in buildings.* Proceedings of RICS COBRA. 1999.

Kishk M., Al-Hajj A., Pollock R., Aouad G., Bakis N. and Sun M. Whole life costing in construction – a state of the art review. *RICS Research Papers,* 16. 2003.

Office of Government Commerce. Procurement Guide 7: *Whole Life Costing and Cost Management.* HMSO. 2003.

RICS *A Guide To Whole life Costing for Construction.* Surveyors Publications. 1986.

Royal Institution of Chartered Surveyors and the Building Research Establishment. *Life expectancies of building components. Preliminary results from a survey of building surveyor's views.* Royal Institution of Chartered Surveyors Research Paper No. 11. 1992.

Stone P.A. *Building Economy.* Pergamon Press. 1983.

8 Value Management

Introduction

The opportunity that value management (VM) affords the practising surveyor to improve value to the client has now been well demonstrated for several years in the UK construction sector. Although this may be regarded as a specialist area, many quantity surveying practices provide this as one of the growing areas of professional service. Having said this, the development of value management within the profession, despite its recognised benefits, has been slower than ideal. In this chapter, value methodology is outlined and supported with examples that will aid the reader in understanding how the service may be performed and how value can be enhanced by its application. The execution of value management requires an understanding of the processes involved and an understanding of how to determine an appropriate VM approach. Whilst this knowledge is essential to the practice of value management, the principles and philosophy provide surveyors with additional tools and techniques, and possibly new ways of thinking, all of which may be used in other areas of professional activity. Knowledge of value methodology is important to the surveyor, either from the perspective of actual service provision, or from an appreciation of its benefits and application when advising clients.

The list of clients using value management in the UK contains many large organisations, and the high profile endorsement of the practice, at professional institution and government level, was fairly continuous during the 1990s. Value management has been recognised as an important component that is important to the success of projects in providing the foundation for improving value for money in construction. It therefore provides practitioners with an excellent opportunity of contributing further to the value added service they provide. This may occur at several points in the life of a project, for example, to assist in the development of a project brief or in response to a problem at any stage during design or construction. Value methodology may also provide the surveyor with opportunity beyond the usual boundaries of the construction industry, for example, prior to the decision to build, to examine a strategic problem that a company may face. In such areas of service provision, competition may be more apparent from management consultancies than construction professionals. Value methodology does not belong to the construction industry; in fact its origins lie elsewhere.

Background

Value management developed from the demands of the manufacturing industry in the USA during World War II. Lawrence Miles (1972), an electrical engineer with the General Electric Company, who adopted a functional approach to the purchasing requirements of his company, developed the value analysis concept. This involved the functional analysis of a component in terms of what it did and invited a search for an alternative solution to the provision of that functional requirement at a lower cost. The use of the concept further developed during the 1940s and 1950s and grew within the USA, becoming a procedure that could be used during the design or engineering stages. The term value engineering was initiated in 1954 by the US military, an organisation that has a long history of involvement with value techniques. Value engineering spread to the UK manufacturing industry during the 1960s and, at around the same time, was introduced to the US construction industry.

The value management concept was first used within the UK construction industry in the 1980s. Although manufacturers around the world today use value techniques, their use within the construction sector is largely restricted to the USA, UK and Australia; however, interest in other parts of the world is growing, for example, in Far East countries such as Hong Kong and Malaysia. The status of value management is now recognised by legislation in the USA and New South Wales, Australia. In the UK, several bodies have issued documents providing recommendations and guidelines; however, the government has stopped short of any obligatory requirement. Several countries now have representative organisations that serve to promote the use of value methodology and control standards of service provided. A discussion on training opportunity and membership of relevant professional societies is provided later in this chapter.

Despite the positive trend in the UK during the 1990s, it appears that the wave of development in value management may have lost its momentum. It is now reasonably common for organisations to restrict their level of VM involvement in response to a desire for a minimalist, lower-cost approach resulting in a reduction in the scope of VM studies.

Terminology

The terminology used in value management may be confusing to those being introduced to the subject for the first time. For example, the terms value management, value engineering and value analysis are frequently used synonymously, particularly value management and value engineering, to mean the entire concept. Although the semantics are considered unimportant, it is vital that the meaning of each term is understood in the context in which it is read or discussed. Therefore, the following definitions are provided as those used throughout this text:

Value management This is the overarching term used to describe the total philosophy and extent of the practice and techniques. Value planning, value engineering and value analysis, shown below, together form a subset of value management.

Value planning   Value planning is carried out in the early part of a project before the decision to build or at briefing or outline design stage. Value techniques assist in arriving at a group decision in terms of the available criteria. It is a common misconception that the use of value management techniques is intended solely for the resolution of a problem, for example, for the reduction of planned expenditure in cases of a budget overrun following the production of a scheme design.

Value engineering   The use of value techniques when completed designs or elements of the design will be available for study during the detailed design and construction stages.

Value analysis   The use of value techniques that are carried out following the completion of a building.

When should surveyors use value management?

To examine this aspect, it is first important to define what value management is and what it is not, which may be considered subjective. Although value management does not provide a strict discipline or science to which every practitioner must rigidly adhere, it is considered to have some essential components, as follows.

- *The use of a structured job plan*   Central to the value management process is the structured workshop, or series of structured workshops, at which group decisions are made. An important aspect of the value management workshop is its structure that usually follows a five-phase process known as the job plan. This is described in detail below.
- *The involvement of a multi-disciplined team*   Value management is a *multi-discipline* exercise in that it ideally requires the participation of consultants from all relevant design disciplines and client representatives who share a common interest (and thus are *stakeholders*) in the success of a project. To be effective, and it is regarded as critical to the success of value management, the team should have an appropriate mix of experience, knowledge and skills and, dependent upon workshop objectives, a range of stakeholder perspectives. An important consideration when selecting a value management team is whether to use the existing design team members or an independent workshop team – although the lack of use of the latter in the UK suggests that this may be rather academic. The participation of independent design consultants would appear to be unusual other than in the US public sector. Use of the existing design team brings several advantages to the proceedings. Where existing project members are used, there is less likelihood of difficulties with the implementation of 'outsiders' ideas, costs are curtailed, there is a saving in time due to the existing knowledge state of the project and it could prove to be a useful team building exercise. However, a potential major disadvantage could be the dominance of original design con-

cepts (if these have been established), which may be strongly defended by designers. The prospect of this occurring is supported by research into design practice and it is something that facilitators are and need to be aware of.

- *Maintenance of the basic project function*   An often-made comment by quantity surveying practitioners is that they carry out value management as part of their traditional duties. This belief is held due to the contribution the quantity surveyor frequently makes relating to the establishment of cost reductions. This reveals a fundamental misunderstanding of value management. The task of reducing costs is usually carried out by the quantity surveyor, irrespective of function. Value management attends to the elimination of unnecessary costs, costs that relate to elements of a design that provide no function.
- *The management of a competent facilitator*   The facilitator is central to the success of the value management process. The role of the facilitator includes advising upon the selection of the VM team, co-ordinating pre-workshop activities, deciding upon the most appropriate timing and duration of workshops, workshop management and preparing reports. Workshop management can be a difficult challenge requiring a variety of skills. These include:
  - The ability to determine and adhere to an appropriate agenda
  - Identifying and gainfully using the characteristics of team members
  - Promoting the positive interaction of participants
  - Motivating and directing workshop activity
  - Overseeing the functional analysis process
  - Encouraging an atmosphere conducive to creativity whilst at the same time maintaining a disciplined structure.

Therefore, although value techniques may be applicable in many situations, the costs and time associated with a formal value management study incorporating the elements outlined above may restrict its use. In practice, value management tends to be utilised in high cost and more complex projects. However, since the savings achieved in a VM study are likely to be significant (say 5% to 15%), and external VM costs could be less than £5000 to £6000, VM is a feasible option on relatively small projects.

Examples of how value management may be used to the benefit of clients include the following.

To obtain a reduction in costs

Consider the value management approach compared to the normal quantity surveying approach:

At tender stage, a proposed library building is over budget by, say £250000.

Traditional cost reduction exercise by the quantity surveyor

The quantity surveyor considers that the internal finishes element appears to offer scope for savings, and proposes for consideration a schedule of cost reductions.

This includes a reduction in the specification of floor tiles (the substitute having a reduced life span, being more difficult to maintain and having a less attractive appearance) and the omission of access flooring from two floors. Clearly, the required budget can be achieved by this action, but with a loss of function that the client would have preferred to retain.

Value management approach

A facilitator co-ordinates a value engineering workshop with the objective of achieving savings of £250 000. The workshop team, using value techniques, ascertains that the function of 'accommodate staff' has been provided by the inclusion of personal office space for all members of staff based upon previous library practice. The workshop team find that the cost of this function (£750 000) can be halved, without loss of function, since, with the increased use of IT and 'hot desking', private staff accommodation can be provided on a staff ratio of 33%. It may be observed that if value management had been used early in the design process, this element of poor value would probably not have occurred. Since it is likely that the retrospective improvement to design is less efficient than *getting it right first time*, the beneficial use of value management early in a project's life is emphasised.

The difference between cutting costs and cutting unnecessary costs is fundamental to the value management approach (Ashworth and Hogg 2000).

The improvement of concept briefing

As referred to previously, value management is not restricted to the resolution of a problem that has occurred during design development but may be used to assist in the concept briefing stage.

Construction design is a complex task and in order to simplify the process, designers tend to adopt a strategy that results in the early production of sketch proposals. These early designs, often based upon relatively little information, are dominant in determining the final outcome of the eventual design. On the understanding that 80% of costs are committed at concept design, the opportunity to add value at this stage of a project is significant.

Value planning, which is carried out in the early stages of a project, before the decision to build or at briefing or outline design stage, provides good opportunity for a multi-discipline team (including client *stakeholders*). With the use of value techniques, the VM team will be supported in reaching a group decision in terms of the criteria for a proposed design. The production of a functional analysis diagram may be one of the methods used and the example in Fig. 8.4 indicates the use of this technique at a strategic level. This diagram, which can be weighted to indicate design priorities, will assist in directing members of the design team toward the project objectives and may be used to monitor design output. Value management can therefore be used to mitigate the well-known inadequacies of traditional concept briefing (Ashworth and Hogg 2000).

It is relevant to note that the importance of construction related design has come to the fore in recent years and several significant initiatives have emerged. That the variability and impact of design quality can be seen everywhere is quite clear. Lord Rea, speaking in the House of Lords in January 2003, put this very well:

'Good design may initially cost a little more in time and thought, although not necessarily in money. But the end result is more pleasing to the eye and more efficient, costs less to maintain and is kinder to the environment.'

In response to the drive for improvement in design quality, a tool has now been developed by the Construction Industry Council (CIC) with support from CABE (Commission for Architecture and the Built Environment), the DTI, Constructing Excellence and the Office of Government Commerce. This process, known as the Design Quality Indicator (DQI), is used to review design quality and practitioners are referred to the associated website and on-line toolkit for further information.

The application of value management

When examining the practice of value management, there is a tendency to focus upon the activities of the workshop and pay too little attention to the preparatory and completion stages. Since the essence of the value management philosophy is held in the workshop activities, this may be understandable. However it is important to stress the importance of both the pre- and post-workshop activities of a value management study. For example, at the pre-workshop stage there is a need to identify the nature of workshop events, and during the post-workshop phase, action is essential if the client is to benefit from the achievements of the study.

Pre-workshop stage

The pre-workshop stage is necessary to:

- Establish why the client wishes to undertake a value management study, and the expected outcomes. Learning about the problem before action is taken is an important step in the value management process. Although a client may have identified a problem at the onset, a preliminary investigation may reveal a different problem to that outlined. For example, the reduction in building costs may be seen as the solution to a budget shortfall. However, the real problem may lie in the need to enhance project revenue and thus establish a higher project budget. Although this example is fundamental, the suggestion is also quite radical in terms of our traditional approach.
- Decide upon the nature of the workshop including duration, timing, core activities and location. Although the benefits of value management may be extended to any problem or decision-making situation, specific project needs and resources have unique demands. For example, the detail relating to a study at the concept stage for a new rail tunnel between Portsmouth and the Isle of

Wight, will clearly be different from that required in connection with the need to attain savings, one week into an inner city refurbishment project. Factors such as project complexity, size, public profile, available resources and programme may all have a significant bearing.

- Determine the composition of the value management team. As stated, stakeholders' participation is a fundamental requirement of value management since value needs to be improved in terms of their perspectives, not those of consultants or a single client agent. The term stakeholder can be considered to incorporate any person with an interest in the proposed project. The composition of the team will vary with each project, not only in terms of personnel, but also the number and experience of participants. For example, in the above mentioned comparison between the Portsmouth/Isle of Wight Tunnel and inner city refurbishment project, the tunnel project would likely attract a greater number of stakeholders, many with little relevant experience. In such situations, briefing members of the value management team prior to the workshop with regard to the project, study objectives and the value management process will be an important factor. In any value management study, briefing the participants of the study objectives will be beneficial.
- Collect information and circulate to the selected workshop participants. This may necessitate substantial research and the support of client representatives and members of the design team. Throughout the value management process, the support of senior management is seen as crucial to a successful outcome.

An example of the type of pre-workshop activity that may occur is shown in the scenario below.

Scenario

A cinema organisation proposes to construct a new multiplex cinema in a major city location. They have been advised to undertake a value management study at feasibility stage as a response to a lower than anticipated project return. A facilitator has been appointed. This term is commonly used in lieu of value manager and has been used throughout this example.

Following appointment, the facilitator meets with the client and ascertains the following information:

- A site has been obtained and outline planning permission has been given
- The client's architect has been appointed to the proposed scheme and has prepared an outline sketch proposal. The quantity surveyors have prepared a feasibility report.

It is agreed that a first value management workshop should be held two weeks hence. Following this initial briefing, the facilitator also recommends a two-day event that will be held in the business suite of a hotel located close to the proposed development.

Study team

In cognisance of the information obtained from the client, the facilitator recommends a value management team comprising:

- Project consultants: architect, quantity surveyor, structural engineer, services engineer. Each of these consultants has been involved in other cinema developments for the client
- Client stakeholders: commercial director, sales and marketing director, regional operations director.

An important aspect of the team composition in this study is seen as the presence of decision takers. Absence of the need to refer back or be concerned about the approval of a senior authority before giving advice or taking decisions will contribute to the successful outcome of the study.

Information gathering and briefing

The facilitator arranges a meeting with members of the design team and the client organisation to discuss details of the project. At this meeting, information required for the workshop is identified and key information is collected for distribution to all study participants. The design team are also briefed with regard to the study and its objectives. Their contribution to the workshop is outlined, namely brief presentations from each member at commencement of the study. The quantity surveyor is advised that cost advice will be required at key points during the workshop and is requested to review, in advance, the budget estimate and to provide an elemental cost comparison (see Chapter 6, Cost Control) with similar developments previously executed by the client. A spatial cost model showing a breakdown of costs for key areas in the cinema is also requested. The zones to be considered include entrance lobby, ticket sales, pre-entertainment areas, refreshment/merchandise areas, theatres, staff and management areas, ancillary space provision and general circulation space. It is agreed that an additional quantity surveying representative will also attend the workshop to assist with cost advice needed during the workshop.

In addition to briefing the participants as to the purpose and format of the workshop, the facilitator is able to obtain feedback with regard to particular views and perspectives of each stakeholder. This is important to the success of the study, allowing some forethought as to workshop management and identifying some items of focus.

It is of some importance to note that the arrangements for the workshop that have been outlined above can and should vary dependent upon individual situations. Some practitioners believe that a minimum study duration should be three days. Whilst it is possible that in practice many studies would benefit from more time, value can be gained from workshops of shorter duration. Clearly, the required scope and depth of the study will be major determinants.

Workshop stage

The structured job plan approach was developed by Lawrence Miles and, whilst academics and practitioners have refined and contextualised a variety of differing methods, it is normally adhered to in all approaches to value management in some form. Whilst this suggests a rigid approach, it should be regarded as an outline; situations and projects will differ and make their own demands on workshop approach and activity. For example, a workshop is perhaps more efficient when limited to the information, creative and evaluation phases (described below) where time is restricted or resources are not available, with provision for development and presentation in a follow up meeting. Likewise, the job plan structure indicates a logical and sequential path, which in practice may vary and necessitate iterative action. The stages of the job plan are discussed below.

The information phase

The workshop begins with an information phase in which particulars of the problem or project are presented to the value management team. If the value management study relates to a proposed building, this can include a contribution from the client's representatives, the architect, structural engineer, quantity surveyor and other members of the design team which will offer details of project background, aims (distinguishing between a client's needs and wants) and constraints (e.g. site, budget, time). Although the primary aim of the information phase is to provide all team members with sufficient detail to allow a good understanding of the project, it also serves as a team building opportunity that, if well managed, is useful in preparing a good base for the remainder of the workshop. A feature of this phase of the workshop is some form of functional analysis that is frequently carried out via the production of a function logic diagram (see the functional analysis section later in this chapter). This results in an enhanced project understanding and allows unnecessary costs to be identified in terms of cost worth (see functional analysis section) and forms a focus of further study. An example of a FAST diagram relating to the cinema scenario outlined above is also considered in the functional analysis section.

Creative phase

Once the VM team have a good insight into the project, including an understanding of its functional needs, the participants are requested, in the creative phase, to engender alternative solutions and ideas. This part of the proceedings is usually performed with the aid of brainstorming and other creative thinking methods to stimulate members to generate ideas that will improve value. With adherence to the key brainstorming rules – that as many ideas are produced as possible and that participants are reserved in their judgement of the suggestions until the creative phase is complete – a large number of suggestions will be generated for future evaluation.

The evaluation phase

There is a range of methods used during the evaluation phase to evaluate the merits of the proposals made during the creative phase. How best to obtain the agreement of workshop participants as to the selection of ideas for further development is a matter that needs to be resolved. Since the work carried out in the development phase is likely to be very detailed and time consuming, only those ideas able to demonstrate good value improvement should be selected. The method used to evaluate the ideas generated in the creative phase will be situation dependent (e.g. influenced by available time, workshop timing, workshop team, project complexity) and may rely on a democratic procedure or the facilitator's ability to get open accord. One technique that is occasionally used is 'championing' which depends upon team members volunteering to 'champion' a particular idea (i.e. accepting responsibility for its development). Therefore, ideas without champions are rejected and thus the best ideas retained. The outcome of the process, irrespective of the method used, is to carry forward the most beneficial ideas to the development stage.

One approach, which may be used to evaluate brainstormed suggestions, is to provide qualitative estimates as to the likely impact in terms of time, cost and quality, and ease of implementation for each proposal. This procedure will allow the identification of those items that offer high value gain and are relatively easy to incorporate within the design. An example of a value proposal evaluation form is shown in Fig. 8.1.

VMA VALUE MANAGEMENT ASSOCIATES							
Project:				Sheet:			Date:

Nr	Description	Impact			Opp.	Action	Comment
		T	£	Q			
1	Redesign entrance/ticket dispensing area	2	5	3	5	D	
2	Reduce storey height	2	3	1	1	R	
3							
4							
5							
6							
7							
8							
9							
10							
11							
...							

Notes:
Impact (T = time, £ = cost, Q = quality): 1 = low; 5 = high
Opp. (i.e. Opportunity) – ease of implementation: 1 = impossible; 5 = easy
Action: Accept, Develop, Reject

Fig. 8.1   Value proposal: evaluation form.

In this example, the suggestion 'Redesign entrance/ticket dispensing area' is considered likely to have a very high impact upon cost and could be incorporated within the design easily. Alternatively, the suggestion 'Reduce storey height' is considered to have a much lower cost impact and will be impossible to achieve. Clearly, suggestion one will be worthy of development and suggestion two rejection.

The development phase

The evaluation of ideas generated in the creative phase has probably been based upon no more than an outline perception by this stage of the workshop. The development phase accommodates the further work that is necessary to establish whether an idea should become a firm proposal or, if the workshop is at briefing stage, to consider the incorporation of ideas within a revised brief. This detailed work is time consuming and will probably involve much technical input. It is nevertheless essential to execute this developmental work before the presentation of formal proposals. Because of the time involved during this development phase, it is often advantageous to complete the related work beyond the confines of the workshop and present it at a subsequent meeting. The work performed during this phase will include the preparation of alternative designs and cost exercises in order to justify the merits and feasibility of the new proposals. The services of non-consultant stakeholders are, therefore, unlikely to be required at this stage.

Whilst details of whole life costing techniques are outlined elsewhere in this book, it is important to further enforce their importance in the preparation of adequate value management proposals. In the life of a building, occupancy costs will be more significant than initial costs. Therefore, when considering alternative design proposals within the value management process, whole life costs must be considered to assist the decision making process.

It is usual to submit proposals on a pro-forma such as the one shown in Fig. 8.2. The form serves to illustrate the level and type of information that will be prepared in the development stage. In addition to the information contained on the form, additional detail in the form of drawings, calculations, etc. may also be relevant and necessary for the client to give full consideration to the design change proposed.

The presentation phase

The objective of the presentation phase is to present the team's proposals to the client representatives. The presentation of proposals, which will probably include adjustments to original design proposals, is something that may be very sensitive to consultants and possibly the client.

The presentation normally occurs at the end of the value management workshop process and is intended to communicate the proposals to the client representatives. Decisions relating to the proposals will probably be deferred until after workshop closure. Although the proposals may have been prepared in detail and with as much accuracy as possible, some review is likely to be necessary and further investigation and consultation may be required.

VMA VALUE MANAGEMENT ASSOCIATES			
Project	New City Cinema	**Date:** 2/3/07	**Impact**
Proposal:	Revise reception/entrance area	**Ref:** 21 A	**considerations**

Detail:

The original entrance/reception area accommodates a ticket counter and lobby which is designed to handle all customers. With reference to the FAST diagram, the system to receive viewers/dispense tickets/inspect tickets is high cost in terms of both designed accommodation and operation. There is an increasing trend toward auto-dispensers and inspection systems which is changing the space requirements and staffing location and levels.

The new design proposal shows a revised entrance layout which accommodates IT displays/ticket dispensers at 12 customer stations. Anticipated reduction in staffing – at reception, 1 at entrance.

Time:
No revision to exterior. Interior redesign: project at concept – minimal time implications

Resources:
No significant resource implications to implement

Rationale/advantages/disadvantages:

Entrance to the cinema is more rapid and comfortable and requires less space that can be dedicated to ancillary entertainment and sales. This will improve revenue and reduce running costs. As a result of reduced staffing levels, customer care may be seen to reduce and security problems may increase.

Action:

❏ **Reject**
❏ **Develop**
❏ **Accept**

Cost summary	Drawing refs:		Design status: Outline/ working		Implementation details:
	Initial cost	NPV running	NPV maintenance	Total cost	
Original design					
This proposal					
Saving/add					

Notes:

1. The accuracy of costs provided will depend upon the level of information and time available for the calculations.
2. The net present value of running costs and net present value of maintenance costs must relate to a realistic business life projection.
3. The proposal should be prepared in such a way as to allow various members of the design team and client organisation to understand the rationale.
4. In addition to providing details of the proposal, the form acts as a checklist and prompter of action and implementation.
5. When considering the proposal, consideration should be given to the impact of accepting the change in terms of time or resource implications, e.g. time to redesign and associated fees, planning approvals, delivery dates…

Fig. 8.2 Value management proposal form.

Post-workshop stage

The completion and monitoring of an implementation plan to ensure post-study action is taken as an essential component of successful value management. Although the benefits of value management include team building and 'buy in' to a project, project value improvement necessitates action beyond words. Despite this, in some situations, the post-workshop stage may fail to fulfil the findings and undertakings of the value management study. These will be included in a detailed report, but reports do not guarantee action.

The post-workshop phase is improved if action determined in the value management study is fully accepted by participants (rather than imposed) and is carefully monitored. Personnel need to be identified to follow up the action outlined in the report and a post-workshop meeting, at which the outcome of the action phase is reviewed, will promote success.

The activities included in the post-workshop stage include the following.

Report

Following completion of the value management study, it is necessary to prepare and submit a detailed report of the activities of the workshop. This will include a review of the value management process carried out, indicating key aspects such as project background, study objectives, value management team membership, information base, timing and duration, workshop activities, function logic diagrams and other supporting models and details of the workshop proposals.

Implementation

Irrespective of the level of success achieved during the workshop stage, the real measure of success in a value management study lies in the extent of implementation of the proposals. This will be dependent upon several factors including the level of client support and commitment, the attitude of members of the design team and the time available. The implementation of 50% of value management proposals may indicate a successful value management study (Norton and McElligott 1995).

Implementation will occur via a detailed response to the facilitator's report from the client body. This will indicate those proposals outlined in the report which are accepted, rejected or requiring further development before a decision can be made. Following this, it is good practice to hold a further meeting at which all outstanding matters can be clarified and confirmed.

Functional analysis

There is a range of opinion regarding the importance of functional analysis within construction related value management. Its benefits are well recognised and considered by some to be vital to the process. This is not a universal opinion however and some practitioners consider it of little value.

At this point, it is useful to reflect that functional analysis is a method of examining a product or process in terms of what it does rather than simply what it is. It is generally recognised as a distinguishing feature of value management, dating back to Larry Miles and value analysis. However, the analysis of function should not be seen purely as an integral part of the value management process but should be recognised as a useful methodology that may be used in the absence of the value management framework. For example, there is nothing to prevent a site team (or a single individual) examining a process or aspect of the design with the use of an internally executed functional analysis exercise. This may be relatively quick to perform and may be helpful in searching for value improvement or the mitigation of a problem. This is not to suggest that this is a substitute for a formally executed value management study. This informal approach to functional analysis will clearly have limitations; an individual or small and unrepresentative team will have a restricted view of function and in any event, it is only one part of the value management package which, as stated above, some practitioners now downplay in terms of construction related service.

The main reason for the use of the simple and effective technique of functional analysis is clear. When applying the method to a building component or element, the question 'What does it do?' as opposed to 'What is it?' is asked. Therefore, when looking for alternatives, we search for something that will provide the required function rather than try to find a substitute for the previous solution. Establishing the true functions of a product and considering the costs of each function identified can in itself be an illuminating exercise. The potential impact of the approach can be seen when considering the wristwatch. Most people would state that its function was to indicate the time. If this were simply the case, why do people pay a wide range of prices for such a function? If you can obtain this function for £3.50, why spend £2000. The answer lies in the additional functions required by some people; improving image (esteem function), extending life/providing date (additional use functions) may well justify, to some, an additional expenditure of £1996.50.

Figure 8.5 shows the range of application of functional analysis, from the strategic level to that of the individual component. The applicability of functional analysis within the construction industry is determined by level of operation. Although a one week value management study examining, in detail, the design of a door closer may provide value gain to a manufacturer (in a production run of 100 000), there can be no place for such examination within the context of a new building. The relative costs involved are insignificant and in any event, suggested design changes are likely to lead to non-standard components resulting in cost increase rather than decrease. However, the application of functional analysis at a higher level of abstraction, for example a high cost element or entire building, is valid.

The matrix shown in Fig. 8.3 (Ashworth and Hogg 2000) further explains some of the principles of functional analysis. It shows a hypothetical analysis of the costs of a softwood window relative to functional requirements. Note that the values and function allocation indicated are entirely notional and are there purely to serve the explanation. The logic shown in the table is related to cost/worth and is based upon the convenient assumption that all costs can be allocated to some particular

Component	Function											
	Permit ventilation	Control ventilation	Exclude moisture	Retain heat	Transmit light	Improve security	Reduce sound	Reduce glare	Extend life	Assist cleaning	Enhance appearance	Component cost £
Lintel	15	–	–	–	15	–	–	–	–	–	–	30
Opening	10	–	–	–	10	–	–	–	–	–	–	20
Frame	–	5	5	5	5	–	–	–	10	–	5	35
Casement	–	15	15	15	15	–	–	–	30	–	35	125
Ironmongery	–	10	–	–	–	5	–	–	–	20	5	40
Glass	–	5	5	5	–	5	10	5	–	–	10	45
Paint	–	–	–	–	–	–	–	–	5	–	10	15
Function cost	25	35	25	25	45	10	10	5	45	20	65	310

Fig. 8.3   Functional matrix 'softwood window' (*Source:* Ashworth and Hogg 2000).

function. Worth is defined as the least cost necessary to provide the function. Thus if we focus upon the functional costs of the 'casement' we can observe the following:

- The minimum cost of a casement to serve the basic functions of 'control ventilation', 'exclude moisture' 'retain heat' and 'transmit light' is £60 (assumed to be present in all windows). This amount has been allocated in equal amounts to each of the functions.
- An additional cost of £30 is attributed to the increased specification of softwood – the function of which is to 'extend life' (and reduce the maintenance) of the window.
- An additional cost of £35 is attributed to the window style, say of Georgian appearance in small panes incorporating moulded sections, the function of which is to 'enhance appearance'.
- Function is not always as it first may seem. The function of 'permit ventilation' is achieved by forming an opening; with regard to ventilation, the purpose of the casement is to 'control ventilation'.
- The functions can be divided into: basic – those which are essential (shaded) in Fig. 8.3 and secondary – those which are not essential (but possibly unavoidable or necessary to sell the product), often provided in response to the design solution (e.g. reduce glare). The choice of what is basic and secondary may be subjective and dependent upon individual perception (hence the need for stakeholder participation). The separation of basic and secondary functions assists with the understanding of a project and may identify areas to target for value improvement.
- In terms of cost/worth, if we only require the basic functions as shown (as stated above this is a subjective view), the window is only worth £165 (the sum of the

shaded function totals). Thus, if we have expended £310 and do not require 'reduce sound', 'reduce glare', 'extend life', 'assist cleaning', and 'enhance appearance' (e.g. we live in a bungalow, in a quiet rural hamlet, surrounded by trees and have simple aesthetic tastes) we have not achieved good value.

This shows the way in which functional analysis may be used to identify unnecessary costs or to highlight a disparity of expenditure, possibly leading to substitution with a design alternative. The benefits of functional analysis in practice on a component such as a softwood window will probably be minor and therefore not worthwhile; however, the benefits of the technique can be seen (Ashworth and Hogg 2000).

Functional analysis diagramming techniques

A common method of performing functional analysis is with the aid of functional analysis diagramming techniques such as FAST (functional analysis systems technique) or a value hierarchy. Please note that these diagrams are a means of carrying out and expanding the usefulness of a function analysis; functional analysis may be performed, as shown above, without such techniques.

Functional analysis diagramming techniques can be used across the varying levels of a project development, for example at a strategic stage to possibly identify the building need, at built solution level to assist in the briefing process, and at elemental level to identify unnecessary costs. In practice however, within the construction domain, the use of the technique at technical level is doubtful. Examples of these, and how they may inter-relate, are shown in Fig. 8.5.

Cinema scenario

As part of the information phase of the value management study, a FAST diagram is constructed. A simplified example is provided in Fig. 8.4. The following notes describe the main features:

- The diagram is constructed by using the identified functions and applying 'How/Why' intuitive logic to develop a structured model. Functions are identified by the value management team and conveyed in the form of verb–noun descriptions. In practice, the use of Post-it Notes solves the problems associated with continual development and amendment of the diagram.
- Applying the questions 'How?' and 'Why?' to the relationship between identified functions tests the logic of the diagram. For example, the function 'show movies' is served jointly by 'project image' and 'accommodate people'; likewise, the reason for projecting image and accommodating people is to show movies. A further question which may also be used when constructing the diagrams is 'When?'. In the cinema example, 'When' accommodating viewers, there is also the need to receive viewers and dispense tickets. This approach assists in the practical construction of the diagram and also in the maintenance of the applied logic.

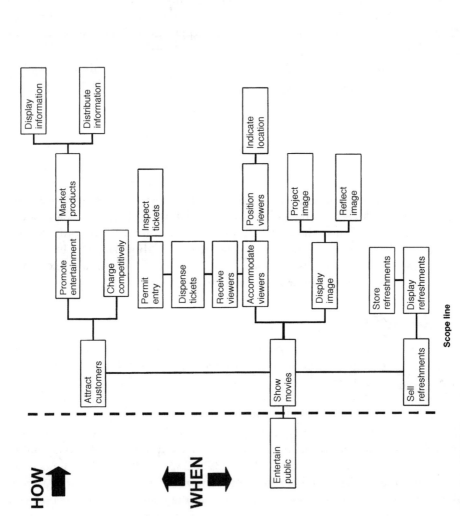

Fig. 8.4   FAST diagram.

- The left scope line has been established with 'Show movies' as the basic function. This is effectively the limit of the study. Although a higher level of abstraction would provide the opportunity to analyse other alternatives to the function 'entertain public', or even beyond, e.g. 'improve profit', the core business activity of the client in this study is to 'show movies'.
- The scope line to the right of the diagram provides the limit at which further dilution to the question 'How' would serve no practical purpose. Design solutions generally appear to the right of this line.

The construction of FAST diagrams can be a difficult task and the notes above are intended to provide an outline explanation only. In practice, the use of FAST diagramming techniques will require training. Simpler forms of function/logic diagrams may be more easily used and are more common. These follow the principles of How/Why logic in the form of a tree diagram, but are generally looser in terms of adherence to convention. The outline examples shown in Fig. 8.5 are indicative examples.

The use of function logic diagrams helps to describe the problem environment and allows participants in a value management workshop to understand the relationship between project objectives and solutions. The application of costs to a function logic diagram will also assist in the identification of high cost functions, which may help in directing further value management activity.

Figure 8.5 shows the inter-relationship between the various levels of use of function logic diagrams.

Supporting the case for value management

General claims of the success of value management in financial terms suggest that for a value management fee of 1%, a 10% to 15% cost saving can be achieved. It is hard to imagine that practitioners and clients would not use value management, in the belief that this level of result could be attained. However, general and vague statements such as the one above are difficult to prove, particularly perhaps to quantity surveying practitioners who frequently achieve the same order of cost savings for clients in their traditional role.

Despite the acclaim given to value management, although its application within the domain of the large quantity surveying practice is increasing, the extent of its growth and application is inconsistent. Also, the application of value management within the domain of the smaller quantity surveying practice appears to be at a very low level. This may be due to several reasons including:

- There is insufficient time to carry it out
- Clients are unwilling to pay for the service
- Clients do not request the service
- The quantity surveyor provides the service already
- Value management skills are unavailable
- There is resistance from design consultants.

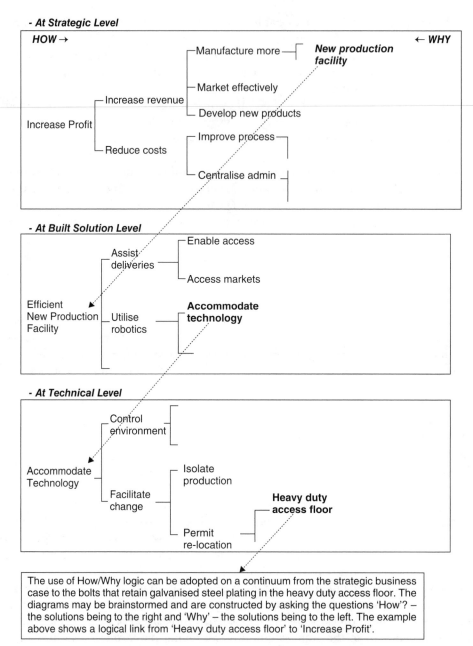

- At Strategic Level

HOW →   ← WHY

Manufacture more —⌐ **New production facility**

Market effectively

Increase revenue

Develop new products

Increase Profit

Improve process

Reduce costs

Centralise admin

- At Built Solution Level

Enable access

Assist deliveries

Access markets

Efficient New Production Facility

Utilise robotics

Accommodate technology

- At Technical Level

Control environment

Accommodate Technology

Isolate production

Facilitate change

Heavy duty access floor

Permit re-location

The use of How/Why logic can be adopted on a continuum from the strategic business case to the bolts that retain galvanised steel plating in the heavy duty access floor. The diagrams may be brainstormed and are constructed by asking the questions 'How'? – the solutions being to the right and 'Why' – the solutions being to the left. The example above shows a logical link from 'Heavy duty access floor' to 'Increase Profit'.

Fig. 8.5 The use of HOW/WHY logic and inter-relationships at varying levels of indenture (*Source:* Ashworth and Hogg 2000).

Research into the factors inhibiting the use of value management within the quantity surveying domain (Hogg 2000) indicates that practitioners who have had no previous involvement with value management consider each of the stated possible barriers to its use more highly than practitioners who have had direct value management experience. This may be the result of exposure to value management;

however it could also indicate an existent negative view, prior to direct experience, that inhibits its application in practice.

If a major reason for the low use of value management is that clients do not request it, the reasons for this could include:

- It is likely that many clients may be unaware of value management, particularly those that are smaller and/or more occasional. It appears that many practitioners, both with and without value management experience, believe that the quantity surveyor already provides the value management function as part of the quantity surveying role. If quantity surveyors believe this, they may communicate it to clients, not by direct reference but by the exclusion of value management from services offered.
- Clients may not think the service warrants an extra fee. If this is the case then surely clients are failing to fully understand the cost–benefit trade-off – which can be expected to show a high return for any value management costs. It could also be that some clients expect that such a high value service should be part of the standard design team remit.

Within the design team, it is reasonable to believe that the quantity surveyor is in a position of great influence with regard to the implementation of value management. This is in some way demonstrated by the inconsistent growth and application of value management within the quantity surveying domain, which suggests that its use is influenced by advice from specific quantity surveying consultants.

Value management can add great value to a project and the quantity surveyor should appreciate the full nature of the process and provide opportunity for its application in practice.

Professional development and accreditation

There are several value management societies in various parts of the world that aim to assist value managers by promoting, developing and controlling standards of practice and qualifications. The Institute of Value Management in the UK has introduced a programme of formal training which provides practising and prospective value managers with the opportunity to obtain a recognised education and comprehensive training, resulting in a UK qualification.

By far, the most established international VM organisation is the Society of American Value Engineers. Of particular interest to students should be their rapidly developing website. The information available through the section 'Publications and Product Catalogue' where the annual SAVE International conference proceedings are published is a good resource, although students will need to be selective in their use of material. Also, a recent development is the 'Knowledge Bank' which has its own search tool. This being said, a large number of publications are very much focused on American practice issues and may seem meaningless to the UK practitioner.

Discussion topic

A client has enquired about value management and has asked you to: (a) outline its key features and benefits; (b) explain how you would go about providing a VM service. Prepare a suitable response.

(a) Key features and benefits of value management

Value management has its origins in manufacturing, however, the processes involved allow the approach to be applied equally to construction. It is seen as distinct from other management approaches bringing together a unique set of tools and techniques.

In construction, it involves a multi-discipline team which looks at a project in terms of what it needs to do for the greater client body. This functional perspective can achieve large gains in value and is different to the QS service normally provided. (Therefore, why haven't we suggested this to the client?)

Dependent upon project stage, we can adopt either a 'hard' approach whereby we attempt to achieve better value from an existent design, or a 'soft' approach whereby VM supports the decision-making process at briefing stage. The greatest benefits, although difficult to authenticate, are seen to come from a soft approach which allows a multi-discipline team, including stakeholders, to contribute in the establishment of a design brief. However, the early days of VM were more focused upon a hard, value engineering approach and this can also help clients to achieve better value from their projects. For example, consider a project for which we are having difficulty in achieving the required budget. Typically, as QS practitioners, we may suggest a package of savings options for the client and design team to consider. Primarily, these would normally focus on level of finishes or specification, and if adopted would result in a decrease in overall building quality. Alternatively, with value management, we review the design in terms of the functions required and attempt to eliminate items which make no contribution to these.

The value management process involves a multi-discipline workshop which is led by a facilitator and is structured around an established set of activities, known as the job plan. It also requires substantial preparation and post-workshop activity, the process in total taking several weeks, although possibly shortened to say 1–2 weeks where needs arise. The atmosphere of a VM workshop is intended to be one that encourages creativity and allows the team to focus on client requirements. In addition to the benefits of improved value, it is excellent as a team-building exercise in itself helping the team to find a common solution, and allowing a better shared understanding of a project. In some situations, for example large and complex organisations, it allows an esprit de corps to emerge with the benefits of team ownership.

The most apparent benefit coming from VM application is added value through an improvement in decision-making and, importantly, VM clients can see what is actually happening. Normally, the outcomes include improved project performance and/or better services, often at a reduced cost.

(b) How would you provide a VM service?

In terms of 'how?', this is very much dependent upon situation and resources, however, if we assume a client is wishing to develop a large industrial complex in the West Midlands, the following would be a reasonable approach.

To begin with, planning and preparation before a formal workshop event are a vital stage of the VM process. Following first contact by the client, a meeting to discuss the project and VM process needs to be arranged. At this meeting, which should be attended by client stakeholders and members of the design team, the workshop process is outlined, together with examples of anticipated outcomes. In addition, a schedule of required/available information is often agreed, for example, the provision of briefing guides, correspondence files, newspaper articles, applications for funding, feasibility studies, post-occupancy evaluations. It is also beneficial to view the project site, where possible.

There is a range of tools and techniques that you may find helpful at this stage and one of the most helpful is the use of checklisting. In simple terms, this involves the use of a 'recognised' checklist to review as much relevant matter as possible before the project workshop. Table 8.1 provides a good outline of a checklist (based on Male *et al.* 1998) which could be used for the given scenario. Most of these items may seem 'obvious' considerations to make; however, the important point is the routine adherence to the schedule.

However, the problem with checklists is that you may not look beyond them for other information, although the above are generic and therefore inclusive. Each project, project type or client type will have unique characteristics which could be used to add to the list.

At this pre-workshop stage, a date, time and venue for a workshop are agreed. In this case, say two weeks ahead, with a one-day workshop at a hotel adjacent to the project site.

At commencement of the workshop it is usual for facilitators to attempt some team-building exercises. There are some artificial approaches, which depending upon audience, can be useful, however, it may be better to generate activity around the theme of the project in hand. For example, asking participants to explain what they consider to be the most important project objective or constraint. At this point in a workshop, an experienced facilitator may be able to recognise particular personalities within the group and deal with this perception in the most effective way. In most groups of say 6–12 people, there is a high chance that personalities will clash in some way.

Following this, the job plan will be adhered to, although, due to limited time available, possibly restricted to information, speculation and evaluation stages, allowing development activity to occur post-workshop. It is common to carry out some form of functional analysis during the early part of the workshop.

Creativity is central to the success of value management and techniques to improve it are therefore of great importance. The main objective in the speculation phase is to generate as many ideas as possible and at this time there should be resistance to judgemental activity which should be reserved for the separate evaluation stage.

Table 8.1 Checklist

Checklist item	Considerations	Examples
Project environment	The nature of the environment (using the word in its widest sense) in which the project will be developed	An industrial building at a time of rapid changes in technology
Community	Community interests have come to the fore in recent years – in the past, the voice of the people has been less audible	The construction of a new chemical processing factory bringing severe objection from local residents
Politics	Consider a change from local to national government or the implications of Europe, or a change of personnel within a political party? If there is a change during the life of the project, what effect will this have?	Incentives in support of designated development areas, say tax breaks or grants. How long will they survive in the aftermath of an election reversal?
Finance	*He who pays the piper calls the tune* . . . and how a project is financed may well impact upon a project development	A funding body, say a regional development board, may require a *Post Occupancy Evaluation* on previous projects . . . feeding into the new project design
Organisation	Most organisations are complex and determining who the key *stakeholders* are is sometimes difficult. The decision takers and structure of the internal project management team need to be known	Who exactly is the client? Who do you need to satisfy? Directors? Staff? Customers?
Schedule/time	How long have you got to finish design? When does it need to be finished? These are fundamental considerations but, surprisingly, are sometimes 'ignored' in VM deliberations	With the advance of 'tick boxing', VM studies may be severely constrained because of first involvement at an advanced project stage, e.g. the contractors start on site next week!
People, skills and expertise	Organisations are managed by personalities with varying degrees of power and egos, and perhaps with different personal agendas. Likewise, their level of participation in a project may be an important factor	Consider the potential difference of working for a highly educated and career minded regional manager, as opposed to a hard hitting, 'off the tools' self-made millionaire
Project concept	Will the project concept change? With the pace of change that now exists, advancing technology or changes in society can change the basis of project design and are difficult to predict with any certainty	How fixed is the concept of a twenty-first century business park?

Once a list of project improvements has been identified, they will then be evaluated in terms of opportunity and cost benefit and those ideas identified as 'high quality', noted for further development.

At this point, it is fairly common to close the workshop and request identified champions to take away the ideas for further development. This may include new drawings, costings, whole life considerations and so forth; say two weeks later, the outcome of this work will be presented to the client and decisions will be finalised.

Typically, a one-day workshop would accommodate 2–3 hours information, 2–3 hours speculation and 2–3 hours evaluation.

References and bibliography

Ashworth A. and Hogg K.I. *Added Value in Design and Construction*. Pearson Education. 2000.

Dallas M.F. *Value and Risk Management – a guide to best practice*. Blackwell Publishing. 2006.

HM Treasury. *Central Unit on Procurement Guidance Note No. 54 Value Management*. The Stationery Office. 1996.

HM Treasury. *Procurement Guidance No. 2: Value for Money in Construction Procurement*. The Stationery Office. 1996.

Hogg K.I. *Value management; a failing opportunity?* RICS COBRA Conference Proceedings. 1999.

Hogg K.I. *Factors inhibiting the expansion of value methodology in the UK construction sector*. Society of American Value Engineers. 2000.

Hunter K. and Kelly J. *Is One Day Enough? The argument for shorter VM/VE studies*. Society of American Value Engineers. 2004.

Kelly J., Male S. and Graham D. *Value Management of Construction Projects*. Blackwell Publishing. 2004.

Latham Sir M. *Constructing the Team*. The Stationery Office. 1994.

Male S.P., Kelly J.R., Fernie S., Grönqvist M. and Bowles G. *Value Management – The value management benchmark: A good practice framework for clients and practitioners*. Thomas Telford. 1998.

Miles L.D. *Techniques for Value Analysis and Engineering*. McGraw-Hill. 1972.

Norton B.R. and McElligott W.C. *Value Management in Construction – A Practical Guide*. Macmillan. 1995.

Office of Government Commerce. Procurement Guide 9. *Design Quality*. Achieving Excellence in Construction. 2004.

9 Risk Management

Introduction

Risk management is a practice that many of us use on a regular basis. As an example, consider briefly our possible concerns and preparations relating to the purchase of a secondhand car, seen by many to be a risky investment. The level of risk involved will be relative to our incomes, the cost of the car, the characteristics of the seller of the car and so forth.

The major concerns and actions we take may include:

- Risk of false ownership? We may request a formal search to verify ownership.
- Risk of latent mechanical failure? In the first instance, we will probably examine the car thoroughly, perhaps with the assistance of an expert. We may also obtain some form of warranty against such failure.
- Risk of tampering with the stated mileage? We may invite a friend with a good knowledge of car mechanics or a professional advisor to inspect for signs of inconsistent wear and tear.

This example demonstrates the principles that we apply instinctively at a personal level. Whilst this level of risk consideration may be appropriate at an individual level, the importance of active risk management in construction demands greater attention. The complexity and scale of most building projects is such that good risk management in the construction industry requires more than purely common sense and instinct.

Construction projects are full of risks and include those that may relate to external commercial factors, design, construction and operation. The principles outlined in this chapter can, in the main, be transferred to any stage of a project development, although in this text we have largely focused upon risk management in the context of the construction phase.

An increasing number of companies, professional organisations, academics and risk management practitioners advocate the benefits of risk management, and the promotion of the practice at a high level continues. The success of risk management is supported by the growing list of clients using it within the UK, a list that contains many major organisations. Risk management is understood to be an important factor that is critical to the success of projects in providing a method with which to improve value for money in construction. It is therefore of great interest to the quantity surveying practitioner, in private practice or in contracting, in that it provides an opportunity of contributing further to the 'added value' service they provide.

When should surveyors use risk management?

Risk management is seen as a major element of project management and therefore it could be argued that this chapter belongs within that domain. However, most experienced surveyors will recognise that in many projects, perhaps most, the responsibility for project management is rather unclear. In some situations, a dedicated project manager will be appointed (frequently with a quantity surveying background), whilst in most projects the role is performed by a member of the design team, typically the project architect, or is informally shared by several of the client's consultants. Irrespective of the prescribed roles and responsibilities, the leadership and management of projects is often a natural consequence of the abilities, experience and personalities of the individual project team members. The quantity surveyor is well placed and possibly the most suited and motivated toward the management of risk amongst all client advisors. It should therefore be a service that can be provided at varying levels. The absence of a dedicated project manager should not mean the absence of risk management, although in practice this seems to be the case.

The scale of risk management applied to projects will vary in accordance with needs. In most situations, complex risk analysis is likely to be both unnecessary and impractical. A great deal can be achieved without major fee implications and therefore quantity surveyors should give careful consideration to the provision of a risk management service. Clients will benefit and professional reputations will escalate. In the most basic form, and dependent upon project size and complexity, great benefit will come from a half day workshop during which risks can be identified and analysed and management response determined. The technical skills required at this level are not great; the most important aspect is ability to communicate and understand the context and demands of the project.

The construction industry has a poor reputation that is due, in the main, to its perceived inability to meet the needs of clients in achieving project completion dates, completing projects within budget and providing a high quality product. The frequency of the failure of projects to meet the expectations of clients in terms of one or all of these factors is a long-term and continuing cause for concern within the industry. The application of risk analysis and management provides a means of improving this situation. Risk management provides the opportunity to control the occurrence and impact of risk factors and provides clients with better information upon which to make decisions.

An indication of the range of risk management opportunity within the domain of the quantity surveyor is outlined below.

To assist in the cost management process

One area of perceived weakness in the service provided by the quantity surveyor is in the calculation of the contingency amount. In practice, the methods of determination of an appropriate contingency provision generally appear to be very crude, including the use of a standardised percentage addition to the estimated contract sum. It is apparent that in many situations, the contingency fund, which is intended to be an all-inclusive risk provision, is ill considered. For example, at tender stage for a project with an estimated cost of £15 million, the consultant quantity surveyor may include a contingency sum of £150000, perhaps with the general agreement of the client and architect, but generally on the basis of 'usual' practice, possibly accommodating some vague notion of the uniqueness of project circumstances.

The frequency and level to which the contingency provision is seen to be an inadequate device with which to protect clients against risk is such that many clients are likely to be dissatisfied at project outturn costs and may be critical of the level of cost advice provided by the quantity surveyor. This is irrespective of the likelihood that in some situations, clients themselves may be responsible for varying amounts of the additional costs. It is, of course, not only under-provision that is a concern. It is also possible that in some situations, due to an unwarranted contingency allowance, tendered project costs may be considered to be in excess of the client's budget resulting in unnecessary and costly post-tender remedial action.

Since the contingency allowance is seemingly the only consideration given to risk in many projects, the application of aspects of risk analysis could be used to greatly improve its accuracy. This view is supported by research by Mak *et al.* (1998) which found that '. . . the use of the ERA [estimating using risk analysis] approach has improved the overall estimating accuracy in determining contingency amounts . . .'.

Risk analysis skills appear to exist within quantity surveying organisations, although in practice they are rarely used. There is good opportunity for the practising quantity surveyor to utilise these skills for the benefit of clients and for professional credibility. If clients, in challenging the reasons for cost overruns, are advised that they were due to unforeseen events for which the contingency provision has proved inadequate, are they entitled to ask how the contingency premium was calculated? This seems a reasonable request of professional quantity surveyors; unfortunately in many situations the anticipated response is likely to be less reasonable.

The benefits that risk analysis may bring to the cost management process are not restricted to the client. Contractors, particularly in light of current procurement trends, are well used to dealing with more than construction risk, including that relating to design. Contingency provision should therefore be a key concern to contracting organisations, and methods of calculation should reflect professionalism rather than pure instinct, although commercial pressures demand a different perspective.

To assist in the decision-making process

Generally when we give cost advice to clients, the one thing that we are usually sure of is that our advice is not given with certainty. At feasibility stage, when we advise a client that a project will cost £5 000 000, this is given as an approximation. As consultants, our estimates are given on the basis of a forecast of a yet unknown party's forecast of costs (the contractors) usually in an uncertain timescale and with uncertain design information. It is therefore common practice to provide range estimates, e.g. £4 800 000 to £5 200 000, and to bring our uncertainty to the client's attention. Nevertheless, the decision to progress with a project will be made upon the basis of this information and will require interpretation by the client. Simulation, discussed in detail later in this chapter, is a technique which provides clients with a much more comprehensive view of project risk. A probabilistic cost forecast provides an open view on outturn cost likelihood and reduces the relative vagueness associated with traditional single point estimating. Although this may be regarded as a tool that would normally be restricted to dedicated project management, the quantity surveyor should understand its application and benefits.

To assist in bidding for construction work

The use of intuition in accommodating risk when bidding for construction work is prevalent amongst building contractors. This approach may be seen as the only practical way of winning work since, in the face of competition, the need to obtain work will ultimately be the major factor in determining price. This intuitive approach may be one cause of the high failure rate of contractors to achieve an acceptable profit that in extreme cases leads to business failure. It seems reasonable to assume that the intuition of contractors will be influenced by market conditions. Irrespective of this, a more rigorous approach to project risk would be beneficial in providing a clearer picture of project demands. This in turn would facilitate a more effective risk response.

The discussions above relate to cost risks; however, contractors are frequently faced with schedule risks. If liquidated and ascertained damages for late completion are set at say £50 000 per day, the risk to a contractor of late completion is very large. The risk management process can be used to assist in establishing a fuller understanding of the risks and can provide a means of improving their management.

Recent trends in construction procurement frequently interfere with the traditional risk balance. The progress of the Private Finance Initiative (PFI) is one such example whereby the contracting organisation may find itself as both promoter, either solely or jointly, and contractor in the same scheme. In such situations, the contractor is faced with a different risk outlook than traditionally encountered.

The application of risk management

In principle, risk management is a straightforward process in that it requires the evaluation of risk and the execution of a risk management strategy. The assessment of risk first entails risk identification, followed by the analysis of risks identified. This imparts a level of understanding that is needed to facilitate the adoption of a suitable risk management response.

Risk identification

The customary method of carrying out a risk analysis is by utilising a workshop at which participants 'brainstorm' recommended risks that they consider could have an impact upon a project. The workshop forum, which brings together specialists from a variety of relevant disciplines, promotes a wide project viewpoint, which, if managed well, will lead to meaningful debate and communication. This should be considered as an exercise that is beneficial in itself. Brainstorming activity is not the only approach to risk identification. Historical data may be used, possibly using the experience of the participants' records, formal or otherwise. Also, as with value management (see Chapter 8), the use of checklists may assist in providing structure to the thought processes used. An indicative example of such a checklist is provided in Fig. 9.1.

A checklist such as the one shown in Fig. 9.1 will be useful in directing attention to predetermined and recognised categories of risk and thus assisting with the identification of those which are project specific. The examples of categories given will incorporate a large range of risks; some categories in particular are wide in their potential scope. There is some danger with checklists that their use may limit deliberation to those categories contained in the list and it should be borne in mind that this could result in ruling out some major and possibly significant items.

The success of the risk identification process will depend on several factors including the level of experience and ability of the personnel concerned with the workshop, the amount of data readily available, the skill and experience of the analyst/facilitator, the time available and the timing of the workshop. It is important to realise that the process of risk identification is not likely to result in the discovery of all possible risks. This is not a practical objective.

When identifying risks, it is important to appreciate exactly what we are trying to establish. To facilitate the process of risk analysis and risk management, it is necessary to think about the possible sources of the risk, not merely the risk event. For example, if we consider the scenario of a basement excavation, one risk event could be that of the collapse of an adjacent road during the course of excavations. To allow proper risk analysis and risk management, the sources of this risk event should be understood. In this case, these may include inadequate direction of the workforce, accidental damage to an existing retaining wall, inadequate shoring design, inadequate shoring construction, or vandalism – all of which may be independently assessed and managed in some detail. Fig. 9.2 provides a summary of the existing risks and outlines possible considerations and actions with the assumption that this risk belongs entirely to the contractor.

Risk category	Indicative examples
Physical	Collapse of sides of trench excavations, surrounding infrastructure or striking existing services resulting in delays, additional cost and possible injury
Disputes	Disruption to a third party's business due to noise, dust, restricted access or construction traffic resulting in reduced sales, financial loss and possible litigation
Price	Increased inflationary pressures causing a severe increase in building costs and excessive financial loss; at present not a problem but consider 1970s and overseas
Payment	Delay in the payment by the main contractor to subcontractors causing a reduction in works progress and resultant programme delays
Supervision	Delays in the issue of drawings or instructions by the architect resulting in abortive work, delays, additional costs and contractual claims from the contractor
Materials	Non-availability of matching materials required in a refurbishment project resulting in possible redesign, programme delay and additional expense
Labour	Non-availability of labour due to the construction of another nearby major project which causes a regional shortage of specialist subcontractors
Design	Errors in the design due to lack of communications between structural engineers and architect resulting in abortive work, or possibly building failure

Fig. 9.1   Checklist of risk categories.

Risk analysis

To begin with, it is important to be aware that problems in construction do not restrict themselves to cost, although in due course all problems may have a cost effect. In numerous situations, time or schedule risk is of more significance than pure cost and, in some cases, quality may be the most important priority. Therefore, it is essential that risk analysis addresses the needs of a given situation and centres upon applicable areas of concern.

There is a range of risk analysis tools that may be used to evaluate the identified risks. The choice of the most appropriate approach will depend on project size, type and opportunity. Examples of some of the approaches, which may be categorised as qualitative, semi-quantitative and quantitative, are as follows.

Collapse of adjacent road during excavations		
Identified sources	Considerations	Possible actions
Inadequate direction of the workforce/ Accidental damage to existing retaining wall/ Inadequate shoring construction	All of these items more or less relate to one source, that of direction of the workforce, although each will require separate consideration. Potential high impact (in terms of time, cost and personal injury). Implications of Statutory Health and Safety transgression. Insurance?	Enforce quality control procedures. Allocate a reliable supervisor to the task. If subcontracted out, ensure the reliability and good standing of the subcontractor and that risk adequately transferred. Verify insurance provisions for such occurrences. Verify arrangements with municipal engineers as to condition of existing road, location of drains etc.
Vandalism	Potential high impact as above, also inner city location suggests above average likelihood.	Strictly enforce health and safety procedures including proper protection of the site and the works. Verify insurance provisions for such occurrences.
Inadequate shoring design	As above. The appointment of experienced engineers should reduce the likelihood.	Transfer the risk relating to the design of the temporary works, by appointing external consultants. Verify insurance provision.

Fig. 9.2   Scenario of a construction project – basement excavations to residential development.

The risk management workshop

The benefits of the risk management workshop have been highlighted earlier in this chapter. Without reference to any particular analysis technique, possibly the most simple and most effective aspect of risk analysis is the appraisal of risk that is possible during the course of structured workshop discussions. Workshops offer the means whereby risks may be identified, assessed and attended to; great advantage may be obtained without any element of more complex and demanding quantitative assessment.

Probability/impact tables (a 'semi-quantitative' approach)

One uncomplicated method of assessing risk is by the use of probability impact tables (P/I Tables). This simple process involves the weighting of a qualitative assessment and hence is termed 'semi-quantitative' analysis. An example of a P/I

Probability	Impact						
	Very low	Low	Medium	High	Very high		1. Existing road collapse
							2. Cut through gas mains
							3. Cut through power supply
							4. Labour dispute
							5. Excavation equipment
							breakdown
Very high		5	9				6. Delay in piling rig delivery
High			8				7. Collapse of large sewer
Medium		2, 3	6				8. Delay due to vandals
Low				7, 10	1		9. Delays due to hard rock
Very low				4			10. Planning delays

Fig. 9.3 Probability impact (P/I) table showing indicative consideration of schedule risk relating to the basement excavations outlined above.

table – indicating the consideration of risks relating to a possible schedule delay – is shown in Fig. 9.3.

To make the use of P/I tables more worthwhile a definition of the descriptors used in the table is required, for example, 'very high probability' equates to more than, say, a 75% chance of occurrence. It is clear that this technique should be designed and applied to meet the circumstances of specific project requirements. The example above, which examines aspects affecting time, may also be adapted and utilised to assess risks affecting cost and quality. Use of this simple technique permits risks to be positioned in terms of severity and therefore allows the management team to be more purposeful and focused upon the most significant issues. There is a need to concentrate upon those risks that have a high impact if they arise and have a high chance of occurrence. It is both inefficient and impossible to spend time on all risks, and matters that are inconsequential or of an extremely low incidence should be put to one side.

The case study example in Fig. 9.4 for a proposed lecture theatre shows the output from a combined value management/risk management workshop held in late 2000, relating to a short session focusing on project risk. Several key stakeholders, and the designated contractor and his designers, attended the workshop. The summary shows the identified risks, their evaluation in terms of simple P/I criteria and an outline response. Although the time allocated to the exercise was brief, it served a useful function in bringing to the fore several key issues in terms of both design and programming matters. It is important to appreciate that this is a summary; the workshop discussions allowed a wide range of views and concerns to be aired and resolved.

Risk registers

As part of the approach to risk analysis, a list of identified risks may be produced and expanded to contain important information relating to each item. This database may be used as management tool and will normally include key details of each identified risk with possible reference to:

Nr	Risk	Evaluation		Response
		Prob.	Imp.	
1	Noise from roof	M	VHi	Client/contractor research. Client to request satisfactory design and back-up information from contractor
2	Climbing on roof	Hi	VHi	Eliminate by design. Client to request contractor to design this risk out
3	Lecture theatres too small	Lo/M	Hi	Client research; confirmation of additional seats required/provided following incorporation of VM proposals. Client to request details of cost implications from contractor
4	Exceed delivery time (in terms of academic programme 2002)	Lo	Lo	Client review. If deadline 2002 October, risk low; however, uncertainty exists re funding
5	Is this the wrong building?	Lo/M	Hi	Because of faculty requirements/split, no alternative
6	An uncomfortably cold building if emergency exit used as circulation	Hi	VHi	Research by contractor. Client to request satisfactory design and back-up information from contractor
7	Budget compromises client satisfaction	Hi	Hi	Client/contractor review. PR/communication to be used to avoid dissatisfaction
8	Technology outstrips need for teaching	VLo	VLo	No action
9	Lack of user focus	Lo	VHi	No action
10	Future use/uncertainty	VLo	VLo	No action
11	Savings/income generation not met	Hi	Hi	Client considers targets to be realistic, therefore no action. Risk may be reduced by internal review
12	Planning	VHi	Hi	Client reduce/contractor review. No contract or physical activity until planning permission obtained. Building location may change. Client to monitor this carefully and to review when further details known. A change in position may jeopardise client satisfaction
13	Advice re incoming services	VLo	Hi	Client to eliminate by order of substation. Survey has reduced risks of problems associated with existing services
14	Cost uncertainty (if back to square 1)	M	M	Client/contractor review. Budget must be maintained

(*continued*)

Nr	Risk	Evaluation		Response
15	Duration of tender	Hi	M	Client/contractor reduce. Several doubts exist re legal situation with contractor. Client to promptly clarify
16	Uncertainty re client's detail design	M	Hi	Client to resolve subsequent to outcome of VM study
17	Lack of funds to complete			Not discussed. See 15
18	Risks from noise (e.g. exams)	Lo	M	Client/contractor review. Client accepts. Previously not a problem. No further action.
19	Subjective image	VHi	VLo	Client reduce by PR
20	Doubtful conference suitability	Hi	M	Client review. Considered not to be in doubt, provided adequate rain shelter provided
21	Under utilisation	Lo	Lo	Client accepts. No action
22	Programme			Previously discussed
23	Funding availability	VLo	VHi	Client review and monitor requirements of funding council and eliminate by compliance
24	Iterative design cost			Previously discussed. Reduce risks by obtaining fee quote for new design. NB. Not all VM outcomes are new designs, e.g. info/development re acoustics
25	Contractual disputes	Lo	VHi	Client/contractor review. Client to reduce by attention to contractual aspects, e.g. duration of tender
26	Lack of staff 'buy-in'	VHi	Lo	Client PR
27	Lack of faculty involvement and user 'buy-in'	VHi	Lo	See 26
28	Access to site/ designers' limitations	Lo	Lo	Considered low risk. No action
29	Lack of power back-up	Lo	Hi	Client accepts. Risks reduced by performance monitoring. Subdivision of boards to be considered by client to reduce risk
30	Unacceptable interior quality	Lo	Hi	Client accepts. PR as previous. Validity of design for conferences to be confirmed by client

Fig. 9.4 Example summary of risk evaluation and outline response – proposed lecture theatre.

- Description of the risk
- The predicted probability and impact
- 'P/I ranking' of the risk
- Identification of the owner of the risk
- Details of the strategy to be adopted to control both impact and probability
- Contingency provision
- An action window which identifies the period in the project when the risk may prevail.

The risk register acts as a control document and assists as a means of monitoring the management of risks throughout a project. For example, an identified risk in a refurbishment project may be the need to eradicate dry rot, which at the time of project design is not evident. Since, based on previous experience, the probability of its occurrence in the building is deemed high, and the potential impact large, as a precaution a contingency allowance is made for timber treatment throughout the building. In addition, the client's consultants include a contingency provision to allow for the replacement of timbers and additional work to the structure, where found necessary. The time that this possible expenditure would occur, if required, would be between month 3 and 4 of the contract. These and other details are retained in a risk register. At the closure of the 'action window', i.e. at the end of month 4, no dry rot had been found. At this stage, all areas of the building have been fully exposed and inspections carried out. No additional works are found necessary. This situation allowed the release of the contingency allowance, and the removal of this item from the monitoring process.

In addition, risk registers may be used as a reference tool for future project evaluation, with many of the included risks being relatively common in occurrence and very similar in content in most project situations.

Expected monetary value (EMV)

One method of bringing probability and impact together is by the use of the expected monetary value (EMV) technique. With this approach, possible outcomes are weighted by the expected probability of each occurring and combined to produce an aggregate result.

This approach can be used in several areas, for example, to produce an overall project outcome by application to independent elemental costs, or to calculate a sensitive rate in production of a tender. The EMV approach is demonstrated in Fig. 9.5, along with the limitations and dangers of its use.

Simulation (quantitative risk analysis)

On most construction projects, it is likely that quantitative risk analysis is neither sensible nor needed due to the relative payback from the extra time and know-how required to complete such an appraisal. Simulation does however present a

Example: Earthworks Sub-Contract

An earthworks subcontractor is preparing a bid for a large infrastructure project that incorporates 20 000 m^3 of excavation. The subcontractor's estimator is aware that there is a range of possible ground/weather conditions which will influence productivity. From previous experience and records of past projects, the estimator is able to broadly categorise four different types of working condition that may exist during the summer months, each type carrying a different cost of excavation and removal:

Very dry/hard	£3.00/m^3
Dry/firm	£2.50/m^3
Wet	£3.50/m^3
Very wet	£4.50/m^3

The question to be answered is – Which rate is applicable?

Previous records show that in the past 10 years the ground was very dry for 15% of the time, dry for 25% of the time, wet for 40% of the time and very wet for 20% of the time.

Using an expected monetary value (EMV) approach, the following calculation was used to produce a composite rate:

£3.00 * 0.15	=	£0.45
£2.50 * 0.25	=	£0.63
£3.50 * 0.40	=	£1.40
£4.50 * 0.20	=	£0.90
EMV	=	**£3.38**

With this more analytical approach, the contractor may choose to use the EMV, i.e. £3.38, resulting in the following tender (excluding all other components such as profit etc.)

Tender: 20 000 m^3 @ £3.38 = £67 600

Unfortunately, the real world is not quite so simple. For example, what happens in the event that the ground is permanently wet?

Cost: 20 000 m^3 @ £4.50 = £90 000

A heavy loss of £22 400!

Is the contractor prepared to take this risk? This is likely to depend on several factors (see section 'Willingness of a party to accept risk' later in the chapter). If the contractor is a major national organisation able to withstand this level of loss and, at the same time, spread the risk across many projects, then perhaps so. However, if the contractor is a small regional organisation, unable to suffer the consequences of a very wet site for the duration of the project, then perhaps not.

This example illustrates the dangers of prescribing the outcome of a purely mathematical approach to risk.

Fig. 9.5 Expected monetary value.

	Least cost (a)	Most likely (b)	Highest cost (c)
Substructures	150 000	170 000	245 000
External walls	325 000	335 000	345 000
Roof	185 000	195 000	240 000
External works	155 000	215 000	235 000
Totals	815 000	915 000	1 065 000

Fig. 9.6   Cost model showing a minimum, maximum and most likely cost (*Source*: Ashworth and Hogg 2000).

powerful and important method that may be appropriate to large and complex projects.

Construction is a complicated process involving a broad range of activities, each of which may go wrong during the course of a building project. As far as cost risk it concerned, Fig. 9.6 shows notional costs for a hypothetical project, which, for reasons of simplicity, has been reduced to four elements.

The costs shown in Fig. 9.6 have been produced following a workshop carried out in the early stages of a project's development. Discussions relating to the substructure element may have been recorded something like this:

> ...3.0   Substructure: The QS has produced a budget estimate amounting to £17 000. This includes the provision of £20 000 for excavation in bad ground (say hard rock removal). The architect advised that in a project constructed earlier in the year, on an adjacent site, no rock was encountered. The engineers agree with the architect that rock is unlikely, but are concerned that some piling may be required. No data from site investigations is available at present. The budget prepared by the QS is accepted as a reasonable provision for the substructures; however, a minimum cost and maximum cost are also identified.

Similar considerations have been made concerning each of the elements shown. A question to consider now is, 'Does the client have adequate finances to build the project?' Since this may depend on the amounts selected, which costs do we use? A negative client or client advisor may select the worst case scenario in each element resulting in a predicted total cost of £1 065 000. Whilst this approach to risk may be understandable in exceptional circumstances (where a 'fail safe' position is required), it is not in most construction projects. In the example used, there may be a 10% chance of the worst case in each element. The chances of each occurring simultaneously are thus $0.1 \times 0.1 \times 0.1 \times 0.1$ with a consequential probability of 0.0001 (i.e. 10 000 : 1).

Further consideration of the above cost model will reveal other weaknesses: there is an unmanageable range of 'what-if' scenarios; the values given are discrete (i.e. 'in between' values are not accommodated – the model does not allow a substructure cost of £160 000); no allowance is made for the fact that the minimum and maximum values are distinctly less probable than the 'most likely' value.

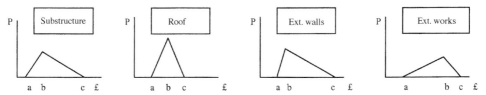

Fig. 9.7 Indicative probability distributions of the elements contained in the cost model (*Source*: Ashworth and Hogg 2000).

Simulation allows us to model each element in terms of cost likelihood (costs being used in the example but this can be applied to schedule risk also). It allows for continuous values (as opposed to discrete) and accounts for the possibility of each value by using probability density.

Figure 9.7 shows triangular probability density distributions representing each of the four elements considered, where probability (P) is on the vertical axis and minimum cost (a), most likely cost (b) maximum cost (c) are on the horizontal axis (Ashworth and Hogg 2000).

Following the construction of the elemental cost models in the form of probability density distributions, simulation software is used to randomly select elemental values that are collected to generate an estimate of total project costs. This exercise or iteration is repeated many times to produce, say, 500 estimates. Since, in each iteration, the selection of values is dependent upon each elemental probability density distribution, most of the elemental values will be selected about point 'b'. The frequency at which total project estimates will comprise elemental estimates tending toward values 'a' or 'c' is very low and the likelihood of an estimate being produced from the sum of four minimum elemental costs (point 'a') or four maximum elemental costs (point 'b') is mathematically unlikely. The output of the simulation exercise can be presented as a relative or cumulative probability distribution (Fig. 9.8). The information provides decision takers with a much clearer picture of the risks involved than by the provision of a single point estimate (Ashworth and Hogg 2000).

Information in this form allows clients and client advisors to see a full picture of possible project outcomes and therefore assists in decision-making. To illustrate the significance of this, consider the relative probability distributions of two projects that have been generated from a simulation exercise (Fig. 9.9). Both have similar 'most likely costs' (i.e. £5 million, which in normal conditions one would assume would be the basis of project estimates traditionally reported to clients) but the risk profile of project 'B' is considerably less attractive than project 'A' (Ashworth and Hogg 2000).

Risk management

Following the identification and evaluation of the risk, the way in which the risks should be managed needs to be determined. The successful management of risk requires:

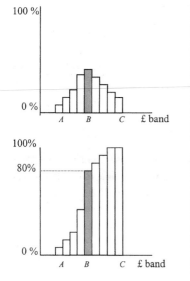

In this format, the **relative probability distribution** allows decision takers to see:

- that estimate 'B', although most likely, is unlikely to be exactly achieved
- the likelihood of occurrence of each cost band
- the minimum cost band 'A' and maximum cost band 'C'
- a 'risk profile' of project cost.

In this format, the **cumulative probability distribution** allows decision takers to see:

- that approximately 80% of the iterations produced a total cost estimate of value 'B' or less, which may be considered an acceptable degree of risk by a client (of a total of 500 randomly generated iterations, 400 resulted in a cost of 'B' or less).

Fig. 9.8   An illustration of output from a simulation exercise (*Source*: Ashworth and Hogg 2000).

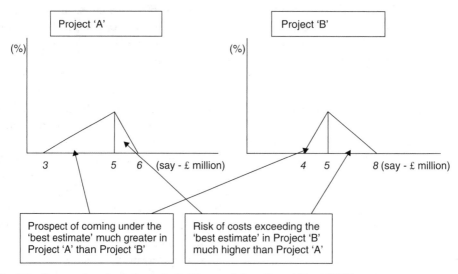

Fig. 9.9   Comparative simulation outputs (*Source*: Ashworth and Hogg 2000).

- Focus upon the most significant risks
- Consideration of the various *risk management options*
- An understanding of effective *risk allocation*
- An appreciation of the factors which may have an impact on a party's *willingness to accept risk*
- An appreciation of the *response* of a party if and when a risk happens.

Focus upon the most significant risks

Methods of deciding on the most important risks in a project have been outlined above. Whilst there are doubts about the need to artificially restrict the number of risks to be actively managed, it will be clearly advantageous to give attention to those risks which are considered to be high impact/high probability.

Risk management options

There are only a small number of risk management options available for consideration, which is helpful in simplifying the process. These may be categorised as follows:

- A risk can be *shrunk* or reduced by, for example, establishing more and better information about an unknown situation
- A risk can be *accepted* by a party as unavoidable and any alternative strategy may be considered as being inefficient or impossible to adopt
- A risk may be *distributed* to another party, for example, contractors usually distribute construction risk by selecting reputable subcontractors to carry out the work
- A risk may be *eliminated* by the rejection of a project or by the rejection of a particular part of the proposed works.

This may be presented to highlight a convenient Mnemonic (SADE): Shrink Accept Distribute Eliminate.

In so doing, there is no order of action suggested in the consideration of the four active risk management options. However, it is reasonable to consider the potential reduction of a risk before giving thought to further, perhaps more drastic action, since the level of the new risk may influence ensuing considerations.

There are two further passive alternatives in addition to the active options outlined above. It is possible to *monitor* risks without action (i.e. 'keeping an eye on the situation') or unintentionally *accept* the risk, the natural default situation if all other risk considerations are overlooked. Both of these alternatives should be considered as a poor response and both may result in disaster, particularly the latter.

To demonstrate the effect of the above risk strategy options, consider the following project scenario:

Project: the construction of a block of residential flats in which a double basement car park is required. The flats are positioned in a busy inner city location, surrounded on all sides by main roads/buildings in occupation. The ground is known to contain a significant amount of landfill, and existing services (including large Victorian sewers, gas mains and water mains) have been vaguely identified as close to the intended works. Clearly there is an abundance of potentially major risk events in constructing the basement car park, including earthworks collapse; cutting through existing services; damage to existing buildings and roads; problems of construction due to confinement of site and bad ground.

Risk management option	Possible action	Possible secondary risks
Shrink the risk	Obtain more accurate information about the nature of the site and location of existing services. Relocate the car park. Direct a specific construction method, e.g. contiguous piling	Additional information may be inaccurate. Relocated car parking may produce additional problems. Selected construction method may cause problems to contractor resulting in additional problems
Accept the risk	Allow a contingency to cover the eventuality. Insure against the risk. (Insurance in this case is seen as the contribution to a wider contingency fund. Some may classify this as distribution)	Contingency provision is inadequate. Risk acceptance by client promotes more 'carefree' attitude by contractor resulting in more risk eventualities. Insurance provision inadequate
Distribute the risk	Pass all associated risks to the main contractor. (Frequently it would seem, the natural default)	Excessive risk premium. Contractor loss resulting in aggressive attitude (e.g. claims) or insolvency. The latter will result in significant problems to the client
Eliminate the risk	Abandon the basement car parking. (This may not be a true option without the abandonment of the total project since planners may require the facility)	Location of new car parking causes new problems. Piling (bringing new risk) is a requirement of the new substructure design since the poor ground overburden is no longer removed during basement construction

Fig. 9.10   Identified risk – earthwork collapse say (high probability, very high impact) (*Source*: Ashworth and Hogg 2000).

Consider the indicative appraisal of risk management options (considered from the perspective of the client) in Fig. 9.10.

Appraisal of risk management options

It is important to recognise that when risk management action is taken, in each case (including that of the 'elimination' option) secondary risks should also be considered. These risks, which are identified in Fig. 9.10, arise as a result of the selected risk management strategy.

There seems to be an opinion held by many occupied in the construction industry that the best way to manage risk is to pass it to another party. This is prevalent throughout the construction hierarchy: clients to contractor/consultants; contractor to subcontractor. This may be in the belief that such action results in the removal of the risk, which as shown in Fig. 9.10, is not the case. It is possible to enlarge risk by distribution and this should be appreciated. This can be shown by consideration of the design and build procurement option. The key aim of this method of procurement is single point responsibility; this includes the transfer of design risk from the client (and his consultants) to the contractor. Since the design process in design and build is likely to be carried out speculatively and in a relatively short duration, it is realistic to presuppose that at the point of contractual agreement, the design will be less complete than at a similar stage with the traditional procurement option. If it is acknowledged that the design is less firm, it should also be evident that the risks relating to the design will be amplified. Thus, with design and build procurement, design risk is enlarged, not eliminated. This does not refute the existence of client advantages of this risk transfer but demonstrates the requirement to fully recognise the effects of risk allocation (Ashworth and Hogg 2000).

Considerations in risk allocation

As previously indicated in the consideration of the active risk management options shown in the above example (Shrink, Accept, Distribute, Eliminate), the most desirable method of management appears to be by distribution to a third party with acceptance by the existing owner being the least preferred. The effective distribution of risk is a key element of risk management and is a principal objective of contracts which should be arranged – from a client's perspective – to optimise his/her exposure to risk.

When taking into consideration the allocation of risk to another party, thought should be given to the following factors:

- The ability of the party to manage the risk
- The ability of the party to bear the risk if it eventuates
- The effect that the risk allocation will have upon the motivation of the recipient
- The cost of the risk transfer.

There are many examples of inappropriate risk allocation within the construction industry that occur due to the strong desire to minimise risk exposure at all costs.

Willingness of a party to accept risk

The readiness with which a party may be prepared to accept a risk will depend on several key factors including:

- Attitude to risk; a party who is risk averse is, in essence, someone less willing to accept risk than someone risk seeking. This fundamental will also be translated into the assessment of a risk premium.
- Perception of risk; a party who has recently experienced a serious injury on a construction site is quite likely to perceive the probability of a similar occurrence on a new project more highly than someone without the experience. This viewpoint may be translated into additional risk premium.
- Ability to manage risk; theoretically, a party unable to manage a risk due to lack of resources or experience should be less willing to accept a risk than someone with the necessary expertise. In practice, however, this may not occur on the basis that, occasionally, particularly where good management is lacking, 'fools rush in where angels fear to tread'.
- Ability to bear risk; theoretically, a party unable to bear a risk due to the lack of the necessary financial back-up should be unwilling to accept.
- The need to obtain work; this factor is likely to be the most significant of all. When the construction industry is in recession, a party is more willing to accept risk as a necessary means of business survival. Alternatively, when work abounds, a full consideration of project risk can be allowed for. Risk acceptance is therefore market sensitive.

Response when a risk eventuates

If and when things ultimately go wrong, and in construction they appear to do so a great deal of the time, a contracting party who has accepted the corresponding risk will be accountable for the related losses. Whilst this is understood, it is also quite possible that the party suffering the loss will be moved toward recovery in some way. In some situations, this may be manifested in contractual claims or a decrease in the quality of construction. This reality should be accepted and the mitigation of risk should be seen as desirable irrespective of who is in possession.

Merging risk management and value management opportunity?

In several important respects, the approach to risk analysis is comparable to that of value management. It typically involves a workshop which is managed by a risk analyst/facilitator; brainstorming and other techniques which help with the decision-making process; an ordered approach; and multi-discipline involvement. This relationship is recognised by some practitioners and organisations who now combine the activities of risk and value management into one comprehensive workshop. The use of project workshops committed to the consideration of value improvement and risk management appears to offer a good opportunity for adding value to the service provided to clients. This action should not be confused with the regular activities of design team meetings.

Discussion topic

Scenario: On behalf of the client, at project inception, you are required to carry out a risk management exercise for the construction of a rail tunnel from Portsmouth to Ryde (Isle of Wight), which will be approximately 6 miles in length.

(a) Using the checklist of risk categories shown in Fig. 9.1, identify some of the key risks indicating their possible sources.

(b) Analyse the risks using a probability/impact table (see Fig. 9.3) and with regard to Risk 1 (explosion at the face of excavations), discuss your rationale. Rank the risks in order of significance.

Please note that a major civil engineering project such as this would, in practice, warrant significant risk analysis and the involvement of a full range of relevant experts. Notwithstanding this, most practitioners involved in the construction sector would be able to envisage many of the problems likely to be encountered.

(a) Identifying risks

Checklist	Identified risk	Risk sources
Physical	Explosion at the face of the excavations	i) Negligent working practices in terms of Health and Safety provision – say, due to inappropriate fuel storage ii) Naturally occurring gas iii) Unsettling an unexploded WW2 mine iv) Terrorism
Disputes	Delay due to anti-tunnel protesters	Protests from environmentalist lobby groups about the loss of unique sea bird habitat resulting in significant disruptive activity and high profile media coverage
Price	Exceptional tunnel boring problems due to rock	Inadequate survey information
Payment	Exchange rate volatility relating to equipment and materials imported to the UK	Relatively high levels of inflation in the UK economy resulting in a fall in the value of the pound sterling

(continued)

Checklist	Identified risk	Risk sources
Supervision	Internationally, limited availability of specialist engineering expertise	Competition arising from major projects around the world, for example in China and India
Materials	Delay in the delivery of tunnel boring equipment	Late completion of the manufacture of specialist equipment
Labour	Shortage of experienced steel erectors	The work clashes with preparation for the 2012 Olympic Games and the continued escalation of a European high-tech construction boom
Design	Failure to design adequate escape facilities	Incorrect design criteria provided by consulting engineers

Fig. 9.11   Risks and risk resources.

	Impact					
Probability	Very low	Low	Medium	High	Very high	
Very high						
High						
Medium		7	3	2		
Low		4	5	6	1	
Very low					8	

1. Explosion at boring face
2. Delay due to anti-tunnel protestors
3. Delays due to hard rock
4. Exchange rate volatility
5. Limited availability of expertise
6. Delay in boring equipment delivery
7. Shortage of steel erectors
8. Inadequate escape facilities

Fig. 9.12   Probability impact (P/I) table showing indicative consideration of schedule risk in a tunnel project.

(b) Analysing risk

As we have seen, a useful tool to use when evaluating risk is the use of probability impact tables (P/I tables) which allow us to record the relative importance of different risks. The P/I table below shows consideration of the risks identified above. When using this approach it is important to have a common understanding of the values implied by the descriptors used. For example, what is meant by 'high probability'? This should be established at the outset of a risk management workshop and is considered in more detail below.

Figure 9.12 is a probability/impact (P/I) table showing indicative consideration of schedule risk in a tunnel project.

Rationale

In appraising the risks as indicated in Fig. 9.12, detailed analysis of each item will be required, utilising any opinion and evidence available. As an example, take Risk

1; Explosion at the face of the excavations, the considerations for which could reasonably include:

i) The severity of the event; looking at the risk event overall, it is clear that any form of explosion, particularly underground, will have very serious consequences. Workers are likely to be killed or injured, work is likely to be severely disrupted and costs arising will be very significant. In addition, the adequacy of the health and safety provision will be questioned and if any negligence is established, prosecution will follow with severe consequences for those held responsible. A very high impact indeed!

ii) The likelihood of occurrence. The potential damage from this event is such to make all participants vigilant, nevertheless, the scale of the works and previous experience of the risk workshop participants makes the probability of this low, rather than near to impossible.

The different sources of risk have not been considered in this evaluation. Without an awareness and understanding of these, it is impossible to determine relative importance and a proper management response. Each of the four causes shown has a different likelihood of occurrence and even a different potential impact. For example, the scale of explosion and its degree of predictability will affect impact; an act of terrorism could cause more damage than an exploding fuel tank.

To be more precise with the evaluation it is also possible to examine different aspects of probability and impact, for example, by looking at time, cost and quality as separate components (Fig. 9.13). In doing this, it will be necessary to establish definitions for each of the descriptors for each of the performance outcomes as shown in Fig. 9.14.

Figure 9.13 is a P/I table showing consideration of time, cost and quality.

Impact
Risk 1: Explosion at face of excavations
(Risk 5: 'Engineering expertise' in parentheses)

Probability	Very low	Low	Medium	High	Very high
Very high					
High					
Medium					
Low		(T,C)		£ (Q)	T
Very low	Q				

In this example, the impact on time and cost would be very similar; however, the probability/impact of this event in terms of quality would be very low – the design would hopefully not be compromised by such an accident. Comparatively, if we consider Risk 5, the availability of engineering expertise, the risks relating to time and cost may be less than that to quality (shown in parentheses).

Fig. 9.13 Probability impact (P/I) table showing consideration of time, cost and quality.

	Probability		Impact					
			Time (delay in weeks)		Cost (% increase)		Quality	
	%	Scale	Criteria	Scale	Criteria	Scale	Criteria	Scale
Qualitative descriptor								
Very high	>80	0.9	>12	0.8	>20	0.8	Major remedial works preventing occupation/use	0.8
High	60–80	0.7	8–12	0.4	15–20	0.4	Major remedial works interfering with occupation/use	0.4
Medium	40–59	0.5	4–8	0.2	5–10	0.2	Extensive remedial works	0.2
Low	20–39	0.3	2–4	0.1	1–5	0.1	Minor remedial works	0.1
Very low	<20	0.1	<2	0.05	<1	0.05	Minor snags	0.05
Nil	0	0	0	0	0	0	0	0

Fig. 9.14 Defining qualitative descriptors: scaling factors and values.

However, in practice, dependent on the situation, it is common to rationalise and to rely on a single assessment for all risk components, which allows comparison in one P/I table. Although this may result in a loss of detail, it nevertheless accommodates meaningful discussion, is pragmatic, and will achieve the same end by identifying those risks which are seen to be significant.

Ranking the risks

From the above analysis shown in Fig. 9.12, we can rank the risks in order of significance. This can be done relatively simply by ascribing a value to each grade of impact and probability and thereafter calculating the product of the relevant scores. Figure 9.14 tabulates some possible performance criteria and scaling factors; these would need to be reviewed for each project situation.

Therefore, using the table below (Fig. 9.15) we have a scoring/ranking of the risks identified. Although the values used are notional, in terms of risk management, a judgement of the relative importance of the risks has been made. In terms of managing a project this is important since either the score or the ranking may be

Risk	Impact score		Probability score		Notional aggregate	Importance ranking
	Descriptor	Value	Descriptor	Value		
1	Very high	0.8	Low	0.3	0.24	1
2	High	0.4	Medium	0.5	0.20	2
6	High	0.4	Low	0.3	0.12	3
3	Medium	0.2	Medium	0.5	0.10	4
8	Very high	0.8	Very low	0.1	0.08	5
5	Medium	0.2	Low	0.3	0.06	6
7	Low	0.1	Medium	0.5	0.05	7
4	Low	0.1	Low	0.3	0.03	8

Fig. 9.15   Ranking of risks.

used to determine an appropriate management action. In this example, a value of over 0.20 could be designated as requiring an immediate risk reduction and so forth.

References and bibliography

Abrahamson M.W. *Risk Management.* International Construction Law Review. 1998.

Ashworth A. and Hogg K.I. *Added Value in Design and Construction.* Pearson. 2000.

Byrne P. *Risk, Uncertainty and Decision Making in Property Development.* E & F N Spon. 1996.

Chapman C.B., Ward S.C. and McDonald M. *Roles, Responsibilities and Risk in Management Contracts.* SERC Research Grant Report; University of Southampton. 1989.

Dallas M. *Value and Risk Management. A Guide to Best Practice.* (In association with the CIOB.) Blackwell Publishing. 2005.

Hogg K.I. The role of the quantity surveying profession in accommodating client risk. *Journal of Financial Management of Property and Construction* 8 (1): 49–56. March 2003.

Hogg K.I. and Greenwood D.G. *Evaluating Construction Project Risk: The differences between groups and individuals.* 4th International Conference on Construction Project Management. Singapore. 2004.

Hogg K.I. and Morledge R. Risks and design and build: keeping a meaningful perspective. *Chartered Surveyor Monthly.* RICS. May 1995.

Institution of Civil Engineers and the Faculty and Institute of Actuaries. *Risk Analysis and Management for Projects (RAMP).* Thomas Telford. 1998.

Kelly J.R. and Male S.P. *Value Management in Design and Construction: The economic management of projects.* E & F N Spon. 1993.

Latham Sir M. *Constructing the Team.* The Stationery Office. 1994.

Mak S., Wong J. and Picken D. The effect on contingency allowances of using risk analysis in capital cost estimating: a Hong Kong case study. *Construction Management and Economics*, 16. 1998.

Raftery J. *Risk Analysis in Project Management.* E & F N Spon. 1994.

RICS. *Improving Value for Money in Construction; Guidance for Chartered Surveyors and Clients.* University of Reading for RICS. 1995.

Smith N.J., Merna T. and Jobling P. (eds) *Managing Risk in Construction Projects.* Blackwell Publishing. 2005.

Vose D. *Quantitative Risk Analysis: A Guide to Monte Carlo Simulation Modelling.* John Wiley & Sons. 1996.

10 Procurement

Introduction

Procurement is the process that is used to deliver construction projects. The dictionary definition states that procurement is 'acquiring or obtaining by care or effort'. Clients who have made the major decision to build are faced with the task of procuring the construction works that they require. This may be a daunting prospect, given the level of financial commitment and other risks associated with the venture, the complex nature of construction and the possible perception of the construction industry as one that frequently under-performs.

A little over 35 years ago the clients of the construction industry had only a limited choice of procurement methods available to them for commissioning a new construction project. Since then there have been several catalysts for change in procurement, such as:

- Government intervention
- Pressure groups being formed to create change for the benefit of their own members, for example, the British Property Federation
- International comparisons, particularly with the USA and Japan, and the influence of developments relating to the Single European Market
- The apparent failure of the construction industry and its associated professions to satisfy the perceived needs of its customers in the way that the work is organised
- The increase in PFI projects and planned expansion of PFI policy
- The influence of developments in education and training
- The impact from research studies into contracting methods
- The response from industry, especially in times of recession, towards greater efficiency and profitability
- Changes in technology, particularly information technology
- The attitudes towards change and the improved procedures from the professions
- The clients' desire for single point responsibility
- The publication of headline reports: in 1994, the Latham Report, and in 1998, the Egan Report.

This has resulted in a significant shift in methods of procurement used by clients.

Although traditional procurement systems are frequently used, in recent years there has been a significant shift towards alternative strategies. Forms of Design and Build procurement now appear to be as commonly encountered as traditional procurement. Similarly there is continued use of management contracting which, by value, represents a significant proportion of procurement overall.

General matters

The wide range of procurement systems now available, and the understanding that procurement choice may have a significant bearing on the outcome of a project, signify both the opportunity and importance of meeting the procurement challenge with a well-considered strategy. The selection of appropriate contractual arrangements for any but the simplest type of project is difficult because of the diverse range of views and opinions that are available. Much of the advice is conflicting and lacks a sound base for evaluation. Individual experiences, prejudices, vested interests and familiarity, together with the need for change and the real desire for improved systems, have all helped to reshape procurement options available to us at the commencement of the twenty-first century. The proliferation of differing procurement arrangements has resulted in an increasing demand for systematic methods of selecting the most appropriate arrangement to suit the particular needs of clients and their projects. In recent years, the amount of research and publications within the field of construction procurement has also grown, as the bibliography at the end of this chapter indicates.

Whilst the main issue is that of satisfying the client's objectives, a matter examined in detail later, at an implementation level the following are the broad issues involved.

Consultants or contractors

These issues relate to whether to appoint independent consultants for design and management or to appoint a contractor direct. The following should be considered:

- Single point responsibility
- Integration of design and construction
- Need for independent advice
- Overall costs of design and construction
- Quality, standards and time implications.

Competition or negotiation

There are a variety of different ways in which designers or constructors can secure work or commissions, such as invitation, recommendation, speculation or reputation. However, irrespective of the final contractual arrangements that are selected, the firms involved need to be appointed. Evidence generally favours some form of competition in order to secure the most advantageous arrangement for the client.

There are, however, many different circumstances that might favour negotiation with a single firm or organisation. These include:

- Business relationship
- Early start on site
- Continuation contract
- State of the construction market
- Contractor specialisation
- Financial arrangements
- Geographical area.

Also, the advent, development and promotion of partnering has changed the view of some clients toward the need for competition (see later in this chapter). In determining the need for competition, it must not be assumed that the choice between that and the option of negotiation is clearly defined, as each case must be decided on its own merits.

Measurement or reimbursement

There are in essence only two ways of calculating the costs of construction work. The contractor is either paid for the work executed on some form of agreed quantities and rates or reimbursed the actual costs of construction. The following are the points to be considered between the alternatives:

- Necessity for a contract sum
- Forecast of final cost
- Incentive for efficiency
- Distribution of price risk
- Administration time and costs.

Traditional or alternative methods

Traditionally, most projects built in the twentieth century in the UK have used single-stage selective tendering as their basis for contracting. With a wider knowledge of the different practices and procedures around the world, and some dissatisfaction with this uniform approach, other methods have evolved to meet changing circumstances and aspirations of clients. The following factors should be considered (these are examined in more detail later in this chapter):

- Appropriateness of service
- Length of time from inception to completion
- Overall costs inclusive of design
- Accountability
- Importance of design, function and aesthetics
- Quality assurance
- Organisation and responsibility

- Project complexity
- Risk apportionment.

Standard forms of contract

There is a wide variety of different forms of contract in use in the construction industry. The choice of a particular form depends on a number of different circumstances, such as:

- Client objectives
- Private client or public authority
- Type of work to be undertaken
- Status of the design
- Size of proposed project
- Method used for price determination.

Local authorities use different forms of contract from central government departments, while some of the larger manufacturing companies have developed their own forms and conditions. These often place a greater risk on the contractor and this is in turn reflected in the contractor's tender prices. The different industry interests continue to develop a plethora of different forms for their own particular sectors. A trend toward greater standardisation would be welcomed. As long ago as 1964, the Banwell Report recommended the use of a single form for the whole of the construction industry as being both desirable and practicable. The message has largely gone unheeded owing to the variety of interested parties involved.

Although the general layout and contents of the various forms are similar, their details and interpretation may vary immensely. Different forms exist for main and subcontracts, for building or civil engineering works, and according to the relationship between the client, consultants and contractors. The different versions of the JCT (Joint Contracts Tribunal) forms of contract are used on the majority of building contracts. These were amended and published as a comprehensive set of alternative forms in 2005.

An alternative to the contracts published by the JCT is the New Engineering Contract (now known as NEC3) which provides a family of standard contract forms. The New Engineering Contract is increasingly used for major projects, for example, it has been selected by the Olympic delivery authority for the production of new buildings and infrastructure in connection with the 2012 London Olympics. It is a substantially different contractual approach to those provided by the JCT suite of contracts and further meaningful commentary is considered to be beyond the scope of this text.

Practice notes

The Joint Contracts Tribunal from time to time issue 'practice notes', which express their view on some particular point in practice. While due account should be taken of such opinions, they do not affect the legal interpretation of the terms of the contract and are thus not finally authoritative. They are similar to a discussion in Parliament of the interpretation of an Act.

Methods of price determination

Building and civil engineering contractors are paid for the work that they carry out on the basis of one of two methods:

- *Measurement*   The work is measured in place, i.e. in its finished quantities, and paid for on the basis of quantity multiplied by rate. Measurement may be undertaken by the client's surveyor, in which case an accurate and detailed contract document can be prepared. With this method, the risk relating to measurement is carried by the employer and that for the rate by the contractor. Alternatively, measurement may be undertaken by the contractor's surveyor or estimator, in which case it will be detailed enough only to satisfy the contractor concerned. With this situation, the risk relating to both measurement and rate is carried by the contractor.
- *Cost reimbursement*   The contractor is paid the actual costs based on the quantities of materials purchased and the time spent on the work by operatives, plus an agreed amount to cover profit. Elements of measurement contracts may be valued on the same basis by the adoption of dayworks.

Measurement contracts

The alternative forms of measurement contract which may be used in the construction industry are as follows.

Drawing and specification

This is the simplest type of measurement contract and is really only suitable for small or simple project work. There has, however, in the past, been some use of this method on inappropriately large projects, often based on misconceived ideas. Each contractor measures the quantities from the drawings and specification and prices them in order to determine the tender sum. The method is thus wasteful of the contractor's estimating resources, and does not really allow for a fair comparison of tender sums. The contractor also has to accept a greater risk, since in addition to being responsible for the pricing they are also responsible for the measurements. In order to compensate for possible errors, contractors will tend to overprice the work.

Performance specification

This method results in an even more vague approach to tendering. The contractor is required to provide a price based upon the client's brief and user requirements alone. The contractor must therefore choose a method of construction and type of materials suitable for carrying out the works. The contractor is likely to select the least expensive materials and methods of construction that comply with the laid down performance standards. Some design and build contracts (see section on procurement options later in this chapter) may be based on performance specification.

Schedule of rates

In some projects it is not possible to predetermine the nature and full extent of the works. In these circumstances a schedule is provided that is similar to a bill of quantities, but without the quantities. Contractors then insert rates against these items and these will be used to calculate the price based on remeasurement. This procedure has the disadvantage of being unable to provide a contract sum, or any indication of the likely final cost of the project. On other occasions, a comprehensive schedule already priced with typical rates is used as a basis for agreement. The contractor in these circumstances adds or deducts a percentage adjustment to all the rates. This standard adjustment can be unsatisfactory for the contractor, as some of the listed rates may be high prices and others low prices.

Bill of quantities

This provides the best basis for estimating, tender comparisons and contract administration. The contractors' tenders are therefore judged on price alone as they are all using the same measurement data. This is an efficient approach to the measurement of the works, since only one party is responsible rather than each individual contractor. As a negative, this type of documentation relies on the production of working drawings before tender stage and is time consuming to prepare. It is therefore not a practical approach where time is in short supply. Also, the employer's acceptance of the risk relating to measurement and design contributes to uncertainty of final cost.

Bill of approximate quantities

In some instances it may not be possible to measure the work accurately. In this case a bill of approximate quantities would be prepared and the entire project measured upon completion. This is a useful approach where an early start on site is required; approximate bills can be measured from incomplete drawings.

Cost-reimbursement contracts

These types of contract are not favoured by many of the industry's clients as there is an absence of a tender sum and a predicted final cost. This type of contract also often provides little incentive for the contractor to control costs. It is therefore only used in special circumstances, for example:

- Emergency work projects, where time cannot be allowed for the traditional process. For example, following a fire in the departure hall of a major airport, the key priority of the client would be to ensure passengers could use the airport

without disruption as quickly as possible. Any possible additional cost arising due to the nature of the contract would probably be insignificant relative to the possible loss in revenue.
- When the character and scope of the works cannot really be determined. For example, this is frequently the case with elements of measurement contracts for which the works are valued on a dayworks basis.

Cost reimbursement contracts can take several forms. The following are three of the types that may be used in the above circumstances. Each of the methods pays contractors' costs and makes an addition to cover profit. Prior to embarking on this type of contract, it is important that all the parties concerned are fully aware of the definition of contractors' costs as used in this context.

Cost plus percentage

The contractor is paid the costs of labour, materials, plant, subcontractors and overheads, and to this sum is added a percentage to cover profits. The percentage is agreed at the outset of the project. A disadvantage of this method is that the contractor's profit is related directly to expenditure. Therefore, the more time spent on the works, the greater will be the profitability. In other words, lower efficiency leads to greater profit.

Cost plus fixed fee

In this method, the contractor's profit is predetermined by agreeing a fee for the work before the commencement of the project. There is therefore a possible incentive for the contractor to attempt to control the costs, because it will increase the rate of return. However, to counter this, it is also possible that since the fee is fixed, the contractor can improve his profitability only by reducing management costs, precisely what the client is seeking. In practice, because it is difficult to predict cost accurately beforehand, it can cause disagreement between the contractor and the client's professional advisers when trying to settle the final account, if the actual cost is much higher than that which was estimated at the start of the project.

Cost plus variable fee

The use of this method requires a target fee to be set for the project prior to the signing of the contract. The contractor's fee is made up of two parts: a fixed amount and a variable amount depending on the actual cost. This method provides an even greater incentive to the contractor to control costs, but has the disadvantage of requiring the target cost to be fixed on the basis of a very rough estimate.

Contractor selection and appointment

There are essentially two ways of selecting a contractor: through competition or by negotiation. This will apply to any working arrangement, including strategic partnering, which in the first instance requires the appointment of a contractor partner. Competition may be restricted to a few selected firms or open to almost any firm that wishes to submit a tender. The contract options described later are used in conjunction with one of these methods of contractor selection.

European legislation imposes restriction on tendering arrangements for the procurement of public goods and works. Where government related expenditure is involved above a prescribed contract value, it is necessary to invite tenders from member states by advertising the project in the *Official Journal of the European Communities*. The subdivision of contracts into smaller 'projects' with the intent of falling within the threshold value is not allowed. At present, it seems that this European initiative is not having a major effect upon competition for construction work in the UK. It does, however, impact significantly on securing publicly funded projects.

The Construction Industry Board (CIB) has developed a Code of Practice for the Selection of Main Contractors. This replaces the previous Code of Tendering Procedure issued by the National Joint Consultative Committee (NJCC). This publication, although not mandatory, does provide guidance and suggested good practice on the selection of contractors and the awarding of construction contracts. With reference to the CIB document, the key principles of good practice to be adopted when appointing contractors in competition (either by single or two-stage tendering) are:

- 'clear procedures should be followed that ensure fair and transparent competition in a single round of tendering consisting of one or more stages'
- 'the tender process should ensure receipt of compliant, competitive tenders'; where contractors feel it necessary to attach conditions to tender submissions due to their inability to fully comply with the tender documents, tender evaluation becomes more complex
- 'tender lists should be compiled systematically from a number of qualified contractors'

  As stated, it may be necessary to consider European procurement law when compiling lists of tenderers. Considerations to be made when selecting the preliminary list of firms include: the firm's financial standing and record; its recent experience of building over similar contract periods; the general experience and reputation of the firm for similar building types; the adequacy of its management; and its capacity to undertake the project. Although in some respects, the inclusion of some tenderers may appear to be automatic due to their size and previous record, consideration should be given to regional standing, resources and reputation which may differ from the national or another regional position. Generally, with a prequalified list of tenderers, there should be no doubt as to the ability of any of the tenderers to satisfactorily complete the contract.
- 'tender lists should be as short as possible'

  The code is prescriptive on this matter and the recommendations are shown in Fig. 10.1 below. Whilst the rationale for this guidance is not in doubt, in practice, clients must consider the possibility of collusion and breaches in confiden-

Contract type	Preliminary list	Tender list
	Number invited to respond	Number invited to tender
Design and construct	Maximum 6, ideally 3 or 4	Maximum 3
Construct only	Maximum 10, ideally 4–6	Maximum 6, ideally 3 or 4

Fig. 10.1 Recommended number of tenderers (*Source:* Code of Practice for the Selection of Main Contractors (CIB 1997)).

Contract type	Preliminary enquiry	Tender
	Time to return preliminary enquiry	Time to return tender
Design and construct	Minimum 3 weeks	Minimum 12 weeks
Construct only	Minimum 3 weeks	Minimum 8 weeks

Fig. 10.2 Recommended tender periods (*Source:* Code of Practice for the Selection of Main Contractors (CIB 1997)).

tiality. The risks of this are increased where the list of tenders is very small, three for example. (See section on 'Selective competition' below.)

- 'conditions should be the same for all tenderers'
- 'confidentiality should be respected by all parties'
- 'sufficient time should be given for the preparation and evaluation of tenders'

The code is also prescriptive on this matter and the recommendations are shown in Fig. 10.2. These times clearly depend on situation and procurement type.

(It should be noted that both the preliminary list and preliminary enquiry in Figs 10.1 and 10.2, relate to a 'first' tender list and enquiry which will be subsequently refined on receipt of details relating to willingness to tender, available capacity, relevance of skills and experience, proposed project team, and understanding and attitude towards the project. This is not to be confused with two-stage tendering discussed later in this chapter.)

- 'sufficient information should be provided to enable the preparation of tenders'
- 'tenders should be assessed and accepted on quality as well as price'
- 'practices that avoid or discourage collusion should be followed'
- 'tender prices should not change on an unaltered scope of works'
- 'suites of contracts and standard unamended forms of contract from recognised bodies should be used where they are available'
- 'there should be a commitment to teamwork from all parties'

This is very much the essence of the way ahead for improving the construction industry, central to the partnering approach and desired in all forms of contracting.

Selective competition

This is the traditional and most popular method of awarding construction contracts. In essence a number of firms of known reputation are selected by the project team. The guidelines outlined above should be adhered to where relevant.

Open competition

Open competition, whereby details of the project are first of all advertised in local or trade publications inviting requests for tender documents, is not an efficient practice, and in consideration of the guidelines above, is not a recommended practice. Whilst the approach allows new contractors, or those who are unknown to the project team and the client, the possibility of submitting a price, the costs of tendering are high and the process of pricing the items lengthy. Where selective tendering is adopted with a list of six tendering contractors, the law of averages indicates that tendering costs for six projects will be absorbed in one successful submission. In open competition, this waste is much greater, depending on the total number of interested contractors, and is of course borne by clients. Some very reputable contractors may not be interested in tendering in such conditions.

The use of open tendering may relieve the client of the obligation of accepting the lowest price. This is because firms are generally not vetted before the tenders are submitted. Factors other than price must therefore be taken into account when assessing tender bids. It is generally accepted that a lowest price tender will be obtained by the use of this method.

Negotiated contract

The negotiated contract method of contractor selection involves the agreement of a tender sum with a single contractor. The contractor will offer a price using the tender documentation, and the client's surveyor then reviews this in detail. The two parties then discuss the rates that are in contention, and through a negotiation process a tender acceptable to both parties can be agreed. Owing to the absence of any competition or other restriction other than the acceptability of price, this type of contract procurement is not generally considered to be cost advantageous. It can be expected to result in a tender sum that is higher than might have been obtained by using one of the previous methods, although substantiation of the order of this is difficult to assert. Because of the higher sums incurred, public accountability and the possible suggestion of favouritism, local government does not generally favour this method.

A negotiated contract should result in fewer errors in pricing. It also accommodates contractor participation during the design stage, and this may result in savings in both time and money. It should also lead to greater cooperation during the construction period between the designer and the contractor. Where an early start on site is required, it clearly offers distinct advantages to a drawn out competitive tendering approach.

Two-stage tendering

The main aim of two-stage tendering is to involve the chosen contractor on the project as early as possible. It therefore tends to succeed in getting the person who knows what to build (the architect or engineer) in touch with the firm that knows how to build it (the contractor) before the design is finalised. The contractor's expertise in construction methods can thus be used in the architect's design. A further advantage is that the selected contractor will be able to start on site sooner than would be the case with the other methods of contract procurement.

In the first instance, an appropriate contractor must be selected. This can be achieved by inviting suitable firms to price the major items of work from the project. A simplified bill of quantities is therefore required that will include the preliminary items, major items and specialist items, allowing the main contractor the opportunity of pricing for profit and attendance sums. The guidelines published by the CIB are applicable to this stage in the process.

The contractor will also be required to state their overhead and profit percentages. The prices of these items will then form the basis for subsequent price agreement that will be achieved through negotiation.

Serial tendering

Serial tendering is a development of the system of negotiating further contracts, where a firm has successfully completed a contract for work of a similar type. Initially contractors would tender against each other, possibly on a selective basis, for a single project. There is, however, a legal understanding that several other similar projects would automatically be awarded using the same bill of rates. The contractors would therefore know at the initial tender stage that they could expect to receive a number of contracts, which could provide them with continuity in their workload. As an alternative, it may be that a 'series' of projects are awarded to a contractor who successfully tenders in competition on the basis of a notional 'master' bill of quantities that will include a comprehensive range of items and will be used to price each future project in a defined series. In either situation, conditions would be written into the documents to allow further contracts to be withheld where the contractor's performance was unsatisfactory.

Serial contracts should result in lower costs to contractors since they are able to gear themselves up to such work by, for example, purchasing suitable types of plant and would generally benefit from economies through the increased total contract size. Serial contracts are appropriate to buildings such as housing and schools in the public sector. This method may also be usefully employed in the private sector in the construction of industrial units. It has been successfully used with industrialised system buildings.

This arrangement for letting contracts, although having several advantages including that of promoting a good working relationship, is not to be confused with the concept of partnering which is discussed in detail later in this chapter.

Procurement options

There are several procurement options available to the client and within each broad type there are several variants, each of which may be possibly refined to accommodate particular client needs and project specifics. For example, within a traditional arrangement, it is normal to have some of the works carried out under a cost plus or remeasurement arrangement and possible also to let a portion of the works on a design and build basis. An appreciation of the operation and application of each of the procurement options is essential to developing a sound procurement strategy.

Traditional

In this approach, the client commissions an architect to take a brief, produce designs and construction information, invite tenders and administer the project during the construction period and settle the final account. If the building owner is other than small, the architect, traditionally the first point of client contact, will advise the client to appoint consultants such as quantity surveyors, structural engineers and building services engineers. Other consultants, particularly the quantity surveyor, may also be the client's first port of call. The contractor, who has no design responsibility (unless particular portions of the work are so identified), will normally be selected by competitive tender unless there are good reasons for negotiation. The design team are independent advisers to the client and the contractor is only responsible for executing the works in accordance with the contract documents. Fig. 10.3 shows the relationship of the parties.

The key feature of this form of procurement is the separation of design and construction. The client appoints a team of consultants, frequently led by the architect, to design the building and prepare tender documentation. The main advantages and disadvantages of this procurement option are as follows.

Advantages

- A high level of price certainty for the client. Since cost is known before construction commences, and providing the design process has been completed fully in the pre-contract stage, a high degree of price certainty exists.
- A low tender price.
- Accommodates design changes and aids the cost management process.
- Relatively low tender preparation costs. In addition, subject to the status of the tender documents, high tender quality.

Disadvantages

- A relatively lengthy time from inception to start on site.
- Problems relating to design error. The risk relating to the design lies with the client. Post-contract design changes are frequently abundant and resultant delays and disputes are common.
- Lack of involvement of the constructor in the design process.

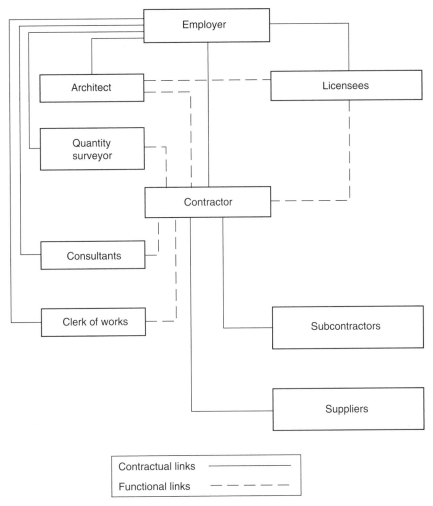

Fig. 10.3 Traditional contract.

Design and build

It has been suggested that the separation of the design and construction processes, which is traditional in the UK construction industry, has been responsible for a number of problems. Design and build (Fig. 10.4) can often overcome these by providing for these two separate functions within a single organisation. This single firm is generally the contractor. The client, therefore, instead of approaching an architect for a design service, chooses to go directly to the contractor. With this method of procurement, the contractor therefore accepts the risk for the design element of a project.

It is common for the client to initially appoint design consultants to develop a brief, examine feasibility and prepare tender documents that will include a set of

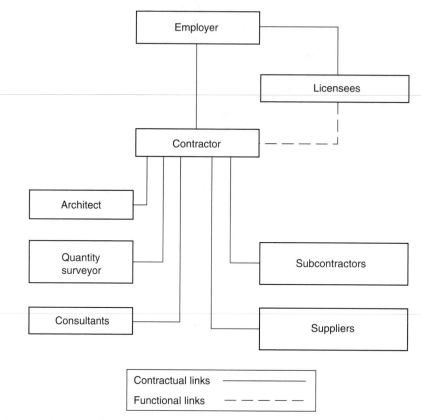

Fig. 10.4 Design and build.

employer's requirements. Contractors are invited to tender on the basis that they will be responsible for designing and constructing the project and will submit a bid, which will incorporate design and price information. The contractor's proposals will be examined by the client and the project subsequently let. An issue to be considered in the early stages of the project is the nature of the employer's requirements, which may vary significantly in terms of detail. Clients may need to balance their conflicting desires to both direct the design and transfer full design risk to the contractor.

Develop and construct is an approach whereby the client's consultants prepare a concept design and ask contractors to develop that design and construct the works. Frequently, the client attempts to transfer the risks associated with the early design by novation. With this approach, the client's architect and pre-contract designs are 'transferred' to the contractor who accepts the related liability as part of the contractual arrangement. This procedure is unpopular with contractors who are inclined to believe that in such situations the allegiance of the architect remains with the client.

There are several advantages and disadvantages to the use of design and build that should be considered, including the following.

Advantages

- *Single point responsibility*   If problems arise during the works, the contractor is unable to place blame with the client's consultants and will be motivated toward the reduction of design problems and their mitigation when they arise.
- *Price certainty prior to construction*   Provided there are no client changes, a high level of price certainty exists.
- *Reduced project duration*   This is made possible due to the overlap of the design and construction phases.
- *An improved degree of buildability*   This may be achieved since the contractor has a greater opportunity to influence the design, although in practice, this will depend on the level of design direction contained in the employer's requirements.
- *Fitness for purpose*   The potential exists to extend design liability beyond reasonable skill and care to include fitness for purpose. Insurance difficulties regarding the increased design liability and the possible questionable ability of contractors to withstand the impact of a large claim in the event of failing to achieve the fitness for purpose requirement are such that it will normally be preferable to forego this opportunity (Morledge and Sharif 1996).

Disadvantages

- *Client's reduced ability to control design*   The scale of this concern depends on the design and build approach adopted.
- *Commitment prior to full design*   In essence the client contracts to buy a building that is yet to be fully designed.
- *Difficulty in comparison of tenders*   The evaluation of the differing design alternatives contained within the contractors' proposals may add a significant complexity to the normal tender review process.
- *Cost management difficulties*   A much reduced level of price information is available to the client's surveyor, creating significant cost management problems.

Management based contracts

There are several possible variants to a management based procurement strategy. Each shares the main characteristic of the appointment of a party – usually a contractor – to manage the construction works (and in the case of design and manage, the design of the works also) in return for a lump sum or percentage fee. The scope of the management provision and methods of establishing contractual links in the construction supply chain provide differentiation between specific types. This approach to procurement is particularly suited to large, complex projects whereby it may be beneficial to reduce risk for the contractor (Murdoch and Hughes 2000).

Management contracting

Management contracting is a term used to describe a method of organising the project team and operating the construction process. The management contractor

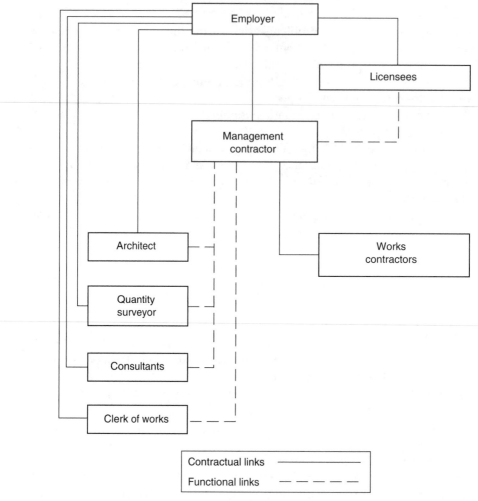

Fig. 10.5   Management contracting.

acts in a professional capacity, providing the management expertise and buildabil-ity advice required in return for a fee to cover overheads and profit. The contrac-tor does not therefore participate in the profitability of the construction work itself and does not employ any of the labour or plant, except for the possibility of the work involved in the setting up of the site and the costs normally associated with the preliminary works. The works are let on a package basis, usually by competi-tive tender (although other methods of procuring individual works packages may be adopted), and contracts are between the management contractor and works con-tractors. Fig. 10.5 shows the relationship of the parties.

Advantages

- *Early involvement of the management contractor* Because the contractor is employed on a fee basis, the appointment can take place early during the design stage. The contractor is therefore able to provide a substantial input into the practical aspects of the building technology process.

- *Reduced project duration*   Construction work may start as soon as sufficient work has been designed.
- *Accommodates later design decisions*   Some design decisions relating to work which may be sequentially at the end of the construction phase may be deferred due to the letting of work in packages.

Disadvantages

- *Commitment prior to full design*   Design is incomplete at the time of commencement and therefore aspects of price, quality and programme are uncertain when the client decides to proceed.
- *Increase in client risk*   The risks relating to additional costs arising as a result of the faults of the works contractors (e.g. delays, defective work, claims from other works contractors) lie with the client.

Construction management

Construction management offers an alternative to management contracting and has been adopted on a number of large projects over recent years. Construction management shares a close similarity to management contracting; however the main difference is that the individual trade contractors are in a direct contract with the client.

The client appoints a construction manager (either consultant or contractor) with the relevant experience and management expertise. The construction manager, if appointed first, would take the responsibility for appointing the design team, who would usually also be in direct contract with the client. The construction manager is responsible for the overall control of the design team and trade contractors throughout both the design and construction stages of the project.

Because of the direct contract arrangements, additional client involvement is required and this approach is therefore not recommended for those without adequate experience and resources.

Design and manage

Design and manage is similar to management contracting in that a management contractor is paid a fee for managing the construction of the works; however, as the term may suggest, it extends the role to incorporate the management of the design of the project. The benefits and pitfalls of the system are also similar to those experienced in management contracting. The major difference relates to the additional design responsibility of the management contractor. To this extent the characteristics of this form of procurement bear similarity with design and build – the gain of single point responsibility and the loss of design control.

Contract strategy

In deciding on a contract strategy, recognition should be given to client priorities (and where possible, desires) in terms of project duration, cost and quality objectives. In order to determine an appropriate method, it is necessary to match the needs of the client with the most suitable approach available. To enable this, an understanding of both the client's objectives and the operation and relative attributes of the available procurement/contract types and methods of appointment, is required. The following points should be borne in mind.

Project size

Small schemes are not suitable for more elaborate forms of contractual arrangement. Such procedures are also unlikely to be cost effective. Small schemes will more likely use either selective or open tendering or a type of design and build. Medium to large projects are able to use the whole range of methods, with the very large schemes more advanced, and complex forms of procurement may be necessary.

Client type

Clients who regularly carry out construction work are much better informed, develop their own preferences and will not require the same level of advice as those who only build occasionally. They may, however, need to be encouraged to adopt more suitable and appropriate methods, and the quantity surveyor will need to convince such clients that the adoption of such suggestions will lead to improved procedures. Experienced clients may have a working knowledge of the construction industry, may retain the services of in-house construction advisors and will be able to contribute to the process throughout. Inexperienced clients may have unrealistic expectations and the tendency to inappropriately interfere rather than contribute.

There is a range of client types which, although each client will be unique, may be categorised by key characteristics. The categories of client that are likely to be encountered in practice include: public bodies, including local authorities, who are experienced and have a large and wide portfolio of construction needs; large commercial developers who build for profit; and large and small companies who build to improve and extend their business and are thus owner occupiers. These different client types will have different procurement needs.

The priorities of these groupings will generally differ in terms of the balance of time, cost and quality objectives, accountability and certainty of output. For example, public sector clients are likely to be driven more by the need for low cost, cost certainty and accountability than are private sector clients.

Client procurement needs

Time

Project duration or completion dates may be critical to the success of a project, and in some situations if not met could lead to total failure in meeting a client's objectives. Whilst most clients are likely to have a desire for an early building completion, it is important to distinguish between this and true need since attempting to meet the objective or early completion is likely to have consequences on other project requirements. The choice of procurement strategy can have a large impact on the duration of a construction project.

Cost

In the event that a limited capital budget is the prime consideration of the client, quality, in the form of a reduced specification, is likely to be restricted and project duration will be the optimum in terms of construction cost rather than client choice.

It is generally believed that open tendering will gain the lowest price from a contractor, although, as previously stated, this brings several possible problems and is therefore not recommended. Negotiated tendering supposedly adds about 5% to 10% to the contract sum although it is difficult to ascertain an exact premium. Projects with unusually short contract periods tend to incur some form of cost penalty. The introduction of conditions that favour the client, or the imposition of higher standards of workmanship than normal, will also push up costs.

The degree to which cost certainty is required prior to commitment to construction or at project completion, restricts procurement choice considerably. The risks associated with abortive design fees are also a factor of which clients should be aware.

Quality

The quality of a building is influenced by several factors including: the briefing process; the suitability of materials, components and systems and their interrelationships within the total design; and the quality control procedures that are in place during both design and construction. The choice of procurement strategy can affect the design process and means of control by which the client and his advisors can monitor both specification and construction activity. It should be noted that quality is a subjective issue, sometimes difficult both to define and identify; it may not necessarily mean a more complex building or a higher specification.

Accountability

Whilst organisations receiving public funding will naturally be concerned with accountability since they are subject to public scrutiny, it is often assumed that accountability is less of a concern for companies in the private sector. This assumption may be misplaced. Research carried out by Masterman in 1988 showed that 'private, experienced secondary clients, i.e. major and active manufacturers, retailers, service organisations etc . . .' indicated accountability as the most important criterion in ensuring project success (Masterman 1992).

Certainty of project objectives

Some forms of procurement (e.g. management contracting) incorporate an inherent facility to accommodate design development throughout a project and others are particularly unsuited to design changes (e.g. design and build).

Market conditions

The selection of the process to be used will vary with the general state of the economy and should also take into account the predicted changes to it within the project's life. When there is ample work available, contractors may be reluctant to enter what they perceive to be unsatisfactory relationships. When the market is at a low point, contractors will be more willing to accept risk, a situation which should be understood by the client for its negative effects as well as positive, since liquidations and claims are more likely to ensue in such conditions.

The relative strengths and weaknesses of the available options may be evaluated in terms of the above factors. Since each strategy will contain a differing balance of the various attributes required by clients, a prioritisation of key objectives is necessary to enable the most suitable choices to be made.

Partnering

Many of the problems that exist in construction are attributed to the barriers that exist between clients and contractors. In essence, partnering is about breaking down these barriers by establishing a working environment that is based on mutual objectives, teamwork, trust and sharing in risks and rewards.

Within the UK construction industry, partnering activity is relatively recent and has been given significant impetus by the Latham Report (1994) and the report by the Construction Task Force (1998): *Rethinking Construction (The Egan Report)*. Further attention was given to the need to increase the use of the partnering approach in 'Accelerating Change' (a strategic forum for construction chaired by Sir John Egan, 2002). One of the forum's six headline targets stated that: '20% of construction projects (by value) should be undertaken by integrated teams and supply chains by the end of 2004, rising to 50% by the end of 2007.' The prominent recognition given to partnering is a clear indication of the strong belief that the partnering approach can make a major contribution to improving value within the construction industry. However, it should also be recognised that significant negative opinion on the practice of partnering exists. A review of articles relating to partnering indicates the level of conflicting views.

Partnering relies on the principle that co-operation is a more efficient method of working than the approach resulting from traditional contracting in which each party is driven toward looking after their own independent objectives.

There is no universally accepted version of partnering and the range of definitions clearly demonstrates this. The version here (which it is important to add, has been subsequently revised by the authors) provides a relatively simple view, which should be easily understandable by all:

'Partnering is a management approach used by two or more organisations to achieve specific business objectives by maximising the effectiveness of each participant's resources. The approach is based on mutual objectives, an agreed method of problem resolution and an active search for continuous measurable improvements.'

(Bennett and Jayes 1995)

It can be seen from this definition that there are three main components to a partnering arrangement: mutual objectives; an agreed method of problem resolution; an active search for continuous and measurable improvements.

The nature of partnering is such that it may take several different forms depending on the situation and the objectives of the parties involved. However, it is possible to broadly classify a partnering arrangement as either 'project partnering' or 'strategic partnering'. Although the differences between these two, which relate to scale and level of relationship, are significant, the essence of the partnering concept is the same in both.

Project partnering

As the name would suggest, project partnering relates to a specific project for which mutual objectives are established and the principles are restricted to the specified project only. The great majority of partnering opportunity is of this type since:

- It can be relatively easily applied in situations where legislation relating to free trade is strictly imposed
- Clients seeking to build on an occasional basis may use it.

Strategic partnering

Strategic partnering takes the concept of partnering beyond that outlined for project partnering to incorporate the consideration of long-term issues. The additional benefits of strategic partnering are a consequence of the opportunity that a long-term relationship may bring and could include:

- Establishing common facilities and systems
- Learning through repeated projects
- The development of an understanding and empathy for the partners' longer term business objectives.

The partnering process

Once the decision to partner has been made, the procedure of partnering principally involves a selection procedure, an initial partnering workshop and a project review.

The selection of a good client or contractor partner who is trustworthy and committed to the arrangement is fundamental to the process. Partnering can be used with traditional procurement methods in the initial selection stages. This is followed by a workshop that will be attended by key stakeholders and usually results in the production and agreement of a partnering charter that will be signed by all participants.

During the project implementation stage, performance will be regularly reviewed. This will incorporate all relevant project matters including quality, finance, programme, problem resolution and safety.

Advantages and disadvantages of partnering

The adoption of partnering – at a strategic level or for a specific project – is considered to bring major improvements to the construction process resulting in significant benefits to each partner. These may include:

- Reduction in disputes
- Reduction in time and expense in the settlement of disputes
- Reduction in costs
- Improved quality and safety
- Improvement in design and construction times and certainty of completion
- More stable workloads and income
- A better working environment.

However, in considering the use of a partnering approach, it is necessary to acknowledge the existence of some important disadvantages and concerns regarding its use. These may include:

- Initial costs
- Complacency
- Single source employment: strategic partnering could result in either party becoming very dependent on one client and thus becoming extremely vulnerable should this source of work be threatened.
- Lack of confidentiality of client/contractor processes and systems: disputes could occur in the event that information, which may be regarded as commercially confidential, were to be withheld from one or more of the partners.
- The lack of competition: the accepted route to securing a 'good value' price is through the competitive tendering process.
- Absence of partnering through the supply chain: the benefits of partnering seem to be absent from the contractor–subcontractor relationship.
- Concerns about legal issues: there is some doubt as to the legal status of the Partnering Charter.

The future of partnering is unclear. The prevailing culture that underlies the construction industry contains strong elements of mistrust, cynicism and resistance to change. These cannot be easily overturned. Market realities should not be ignored, nor can the importance of the need for the delivery of a well balanced education to future construction professionals (Ashworth and Hogg 2000).

The Private Finance Initiative (PFI)

It is important to note the impact in recent years of the Private Finance Initiative (PFI) and its extensive application in the provision of capital projects, for example, those relating to the Department of Health. PFI, as a method of procurement, is a means of funding major capital investments, without the need for government funding. It is expedited through private consortia who are given the opportunity to design, build, and increasingly, manage, new projects. This one-stop approach, frequently contractor led, is complex, incurs high costs in tendering and carries with it the risk of poor quality of service in facility operation. However, assuming the client has made clear the required outputs, it is seen by many to offer an attractive alternative on large government projects, with an inbuilt incentive to design in a manner that makes operation more efficient. There is extensive debate about the merits and difficulties surrounding PFI and it is, to say the least, a subject of great controversy.

The role of the quantity surveyor

It is of fundamental importance to clients who wish to undertake construction work that the appropriate advice is provided on the method of procurement to be used. The active selection of an inappropriate form of procurement, or the passive acceptance of a regularly used and hence 'comfortable' practice, can have a major impact on the success of a project.

Advice must be relevant and reliable and should be given independently, without the intrusion of individual bias. The advice provided should exclude considerations of self-interest. A client should not be faced with unwittingly selecting an inappropriate form of procurement as a consequence of selecting a particular consultant or profession as the first port of call.

Quantity surveyors are in an excellent position as procurement managers with their specialist knowledge of construction costs and contractual procedures. They are able to appraise the characteristics of the competing methods that might be appropriate and to match these with the particular needs and aspirations of the employer. Procurement management may be broadly defined to include the following:

- Determining the employer's requirements in terms of time, cost and quality
- Assessing the viability of the project and providing advice in respect of funding and taxation advantages

- Recommending an organisational structure for the development of a project as a whole
- Advising on the appointment of the various consultants and contractors in the knowledge of the information provided by the employer
- Managing the information and coordinating the work of the different parties
- Selecting the methods for the appointment of consultants and contractors.

Procurement procedures are dynamic activities that are evolving to meet the changing needs of society, the industry and its clients. There are no longer standard solutions; each individual project needs to be separately evaluated on its own individual set of characteristics. A wide variety of different factors need to be taken into account before any sound advice or implementation can be provided. The various influences, at the time of development, need to be weighed carefully and always with the best long-term interest of the client in mind.

Discussion topic

Discuss the procurement advice you would give to clients in the following situations:

(a) Six weeks before Easter a fire occurs in the booking hall at Newtown Airport. The airport authority is heading towards one of its busiest times of the year and could suffer a loss of profit, somewhere in the order of £50 000 per day if facilities are not able to cater for the holiday demand. (Anticipated contract value: £1.5 million.)

(b) In the midst of a property boom, a developer purchases an old Victorian cotton mill and wishes to refurbish the building, converting it into high quality residential accommodation. (Anticipated contract value: £8 million.)

(c) The mid-shire local authority is to develop a new community facility to provide locally available social care. (Anticipated contract value: £2 million.)

Before we look at the particular requirements of the scenarios listed above, it is necessary to reflect upon varying client types and the range of needs they may have. It is essential to provide procurement advice objectively and this will require an analysis and appreciation of the client's position. To determine this, the following factors will need to be considered: time constraints; cost constraints; quality expectations; the need for price certainty; the need for accountability; market conditions. (These factors are detailed elsewhere in this chapter.)

Scenario (a) (fire damaged airport)

In this situation, time is clearly of the essence. The client will almost certainly have the view that whatever it needs, whatever it costs, it must be operational within, say, three weeks. The consultants will be required to appraise the feasibility of this requirement and, assuming that it is feasible in terms of scale and the short time available, would be wise to appoint a contractor on the basis of cost reimbursement with responsibility for both design and build. Given the urgency of the situation, a cost plus percentage agreement will be the most likely variant to interest and motivate the building contractors. In this situation, it will also be necessary to forego any of the normal preliminary steps in contractor selection and immediately discuss with the contractor or subcontractors known to the client/consultants and deemed the best choice for speedy completion, irrespective of cost. Ideally, companies with a vested interest in supporting the client, say from the opportunity of further work, should be given first consideration. In this respect, it is very likely that the airport will have ongoing or recent relationships with suitable contractors which will be most helpful in this position.

The negative aspects of this procurement approach, i.e. high cost and uncertainty, are outweighed by the key requirement of speed to construction and completion. Whatever the amount of excess profit the constructors may receive, the client will find it to be insignificant relative to the potential short-term business loss (£50 000 per day) and the potential longer term effect due to lost allegiance from airlines and passengers.

Scenario (b) (£8 million refurbishment of a cotton mill in the midst of a property boom)

The approach to this project is more complex than the last. The best way to begin is to think of the client's objectives and in this respect the situation becomes clearer. Profitability!

If this is the main driver we need to address certain key factors:

- **Time.** Although, unlike the last scenario, time isn't the main factor, perhaps timing is. First, property booms don't last indefinitely. In the surging market of 1989–1992, some developers started a little too late and suffered the consequences of releasing residential property into a collapsing market. Second, competitors may benefit from delayed completion, for example, by enabling them to bring a development to the market first. Also, time will have an impact on building costs and financing.
- **Cost.** This is fundamental to profitability and costs will need to be controlled and, as far as possible, known in advance in confidently determining project feasibility. The nature of the construction doesn't help in this regard since the refurbishment of any Victorian building has a substantially higher element of risk than new build.
- **Quality.** The location of the development demands high quality, particularly in the type and standard of finishes. Notwithstanding the obvious emphasis on the quality of appearance, a project of this type, in the centre of a major city and with a national corporate profile to protect, needs to be high quality throughout.
- **Certainty of price.** This is an important aspect at feasibility stage and should be maximised to minimise risk.

- **Accountability.** The developer will no doubt be accountable to shareholders, however, this is unlikely to be an issue at an individual project level, provided the company is profitable and provides good dividends. In comparison to the next scenario, accountability is thus not an issue as such.

In this scenario there are several competing factors. In summary, the client is likely to appreciate single-point responsibility (most clients are likely to prefer this, given the choice), reduced project duration, an improved degree of buildability and a reasonable amount of price certainty. This points toward a design and build approach, however, there are difficulties with this in that the construction of the building is unlikely to be straightforward and due to the nature of the project there will be significant design and development variations. Also, it is necessary to achieve a good quality level and the additional risks relating to the existing building fabric may have an impact on this. In the event that the design and build contractor was to encounter such difficulties and was liable for associated costs, the desire to retrieve this loss by a reduction in quality would be great.

Alternatively, traditional procurement would solve some of these key problems, allowing more control over design and quality. The downside to this approach is the potential additional time penalty and the loss of single-point responsibility. Also, in any event, irrespective of design input, there is bound to be substantial uncertainty with regard to work to the existing building, preventing meaningful pre-contract measurement.

As we can see, there are inevitable compromises to be made when deciding upon a best-fit procurement option; the objective is to select the most appropriate rather than the perfect. In this case, this can be achieved by adopting more than one strategy; perhaps at the core, a traditional approach, supported by the extensive use of re-measurement and the use of approximate quantities. This will provide good central design control, provide reasonable certainty of price and accommodate the risks of refurbishment fairly, thus reducing the likelihood of confrontation. Projects of this nature are inevitably more risky than new build schemes and will benefit from an experienced and cooperative design team.

Scenario (c) (mid-shire local authority community project)

With a government funded body there is inevitably more emphasis on accountability and price certainty, and, subject to particular circumstances, likely to be less concern about time. In this project, for example, unlike a school with possible demands for completion prior to the start of a term, completion deadline is unlikely to be a primary requirement, although there may be unique local circumstances. Accountability and the community nature of the project mean that both cost and quality will be dominant factors.

As we have seen, traditional procurement, if the building is well designed, provides a high level of price certainty, low tender price, a high level of accountability and good opportunity to manage quality.

However, there are other factors that may be important to consider. For example, this may be one relatively small project out of several that the local authority is planning over a period of two to three years. If this is the case, a partnering approach may be considered, which would also possibly accommodate facilities management requirements relating to the entire client estate. Provided the initial appointment was made in competition, this would not go against the requirement for accountability.

Summary

The discussion above is based on an understanding of the various procurement options and experience of project development. Since this is a variable, there are possible alternative views to those outlined. However, what is clear is the need to approach each project with a clear and open mind since each situation will have some unique characteristics. There is a danger amongst the QS profession that advice may be restricted as a consequence of out of date knowledge and misperception. An analytical approach will negate against this. It is certainly unsound to advise clients to use any form of procurement if this is based on bias and personal preference as opposed to professionalism and situation demands.

References and bibliography

Ashworth A. *Contractual Procedures in the Construction Industry.* Pearson Education. 2001.

Ashworth A. and Hogg K.I. *Added Value in Design and Construction.* Pearson Education. 2000.

Bennett J. *Partnering in Action.* SBIM Conference Proceedings. 1999.

Bennett J. and Jayes S. *Trusting the Team.* Centre for Strategic Studies in Construction. University of Reading. 1995.

Bennett J. and Jayes S. *The Seven Pillars of Partnering; a guide to second generation partnering.* Reading Construction Forum Ltd; Thomas Telford Publishing. 1998.

CIB (Construction Industry Board). *Code of Practice for the Selection of Main Contractors.* Thomas Telford. 1997.

Clients Construction Forum. *Survey of Construction Clients' Satisfaction 1998/99 – Headline Results.* Clients Construction Forum. 1998.

Construction Industry Board. *Constructing Success: Code of practice for clients of the construction industry.* Thomas Telford. 1997.

Construction Task Force. *Rethinking Construction (The Egan Report).* Department of the Environment, Transport and the Regions. 1998.

Coulter S. Most clients yet to try partnering. *Building,* June, 1998.

Davis, Langdon and Everest. *Contracts in Use: A Survey of Building Contracts in Use during 1998.* RICS. 2000.

Eggleston B. *The NEC 3 Engineering and Construction Contract – A Commentary,* Second Edition. Blackwell Publishing. 2006.

Franks J. *Building Procurement Systems – A client's guide.* The Chartered Institute of Building and Longmans. 1995.

Latham M. *Constructing the Team.* The Stationery Office. 1994.

Long P. Partnering agreements: the legal knot. *Construction Law Bulletin.* Cameron McKenna. 1998.

McGeorge D. and Palmer A. *Construction Management – New Directions.* Blackwell Publishing. 2002.

Masterman J.W.E. *An Introduction to Building Procurement Systems.* E & FN Spon. 1992.

Morledge R. and Sharif A. *The Procurement Guide: A guide to the development of an appropriate building procurement strategy.* RICS. 1996.

Morledge R., Smith A. and Kashiwagi D.T. *Building Procurement.* Blackwell Publishing. 2006.

Murdoch J. and Hughes W. *Construction Contracts: Law and Management,* 3rd Edition. E & FN Spon. 2000.

Strategic Forum for Construction. *Accelerating Change.* A report by the Strategic Forum for Construction, chaired by Sir John Egan. 2002.

Thomas G. and Thomas M. *Construction Partnering and Integrated Teamworking.* Blackwell Publishing. 2006.

11 Contract Documentation

Contract documents

The contract documents for a building project (JCT 2005, clause 2.9) comprise:

- Contract drawings
- Contract bills (or specification)
- Articles of agreement
- Conditions of contract
- Appendix.

In addition, immediately after the signing of the contract the architect will provide the contractor with other descriptive schedules and documents that are necessary for carrying out the works. The contractor will also provide a copy of the master programme.

On civil engineering projects, using the ICE (Institution of Civil Engineers) Conditions of Contract, the following typically represent the contract documents:

- Conditions of contract
- Specification
- Drawings
- Bills of quantities
- Tender
- Written acceptance
- Contract agreement.

The range of contract forms have been referred to in Chapter 10. When the choice of form of contract has been decided, the next step is the preparation of the documents that will accompany the signed form of contract.

The contract documents for any construction project will include, as a minimum, the following information (Ashworth 2006):

- *The work to be performed* This usually requires some form of drawn information, including plans, elevations and cross-sections. Additional details will also be prepared depending on the complexity and intricacy of the project. This will provide information for the client and even a non-technical employer is usually able to grasp a basic idea of the architect's or engineer's design intentions. The drawings will also be required for planning permission and building regulations approval, where appropriate. On some projects drawings may be used as a basis for three-dimensional models and computer graphics.

- *The quality of work required*   The quality and performance of the materials to be used and the standards of workmanship must be clearly conveyed to the constructor, the usual way being through a specification or preamble clauses.
- *The contractual conditions*   In all but the simplest projects some form of written agreement between the employer and the constructor is essential. This will help to avoid possible misunderstandings. It is recommended that one of the standard forms of contract should be used as discussed in Chapter 10. It is preferable to adopt the use of a standard form of contract rather than to devise separate conditions of contract for each project.
- *The cost of the finished work*   Wherever possible this should be predetermined by a firm estimate (tender) of cost from the contractor. However, it is recognised that on some projects it is only feasible to assess the cost once the work has been carried out. In these circumstances the method of calculating this cost should be clearly agreed. Whilst value is now an important criterion, cost nevertheless forms a part of this calculation.
- *The construction programme*   The length of time available for the construction work on site will be important to both the client and the contractor. The client will need to have some idea of how long the project will take to complete in order to plan arrangements for the handover of the project. The contractor's costs will be affected by the time available for construction. The programme should include progress schedules to assess whether the project is on time, ahead or behind the programme.

Coordinated project information

Research undertaken by the Building Research Establishment in the mid-1980s identified that the biggest single cause of quality problems on building sites is unclear or missing project information (Allott 1984). Another significant cause of disruption of building operations on site has been highlighted as shortcomings in drawn information. Much of the site management time is devoted to searching for missing information or reconciling inconsistencies in that which is supplied. This, together with a lack of compatibility in project information generally between the drawings, specifications and bills of quantities, is a major concern. The difficulty is partially due to the fact that the information is sourced from several different professional disciplines involved in the project.

In order to improve this situation, the Coordinating Committee for Project Information (CCPI) was established by the major bodies in the construction industry. After consultation they produced the following documents:

- Common Arrangement of Work Sections for Building Works (CAWS 2000)
- Project specification

- Production drawings
- Code of Procedure for Measurement of Building Works
- SMM7.

CCPI as a working committee has been disbanded but its work is now monitored by the Building Project Information Committee (BPIC). The purpose of CAWS is to define an efficient and generally acceptable identical arrangement for all documents. The main advantages are:

- *Easier distribution of information, particularly in the dissemination of information to subcontractors* One of the prime objectives in structuring the sections was to ensure that the requirements of the subcontractors should not only be recognised but should be kept together in relatively small tight packages.
- *More effective reading together of documents* Use of CAWS coding allows the specification to be directly linked to other documents. This reduces the content in the latter while still giving all the information contained within the former.
- *Greater consistency achieved by implementation of the above advantages* The site agent and clerk of works should be confident that when they compare the drawings with the contract bills they will no longer ask the question 'Which is right?'.

CAWS is a system based on the concept of work sections. To avoid boundary problems between similar or related work sections, CAWS gives, for each section, a list of what is included and what is excluded, stating the appropriate sections where the excluded item can be found. CAWS has a hierarchical arrangement in three levels, for instance:

- Level 1 R Disposal systems
- Level 2 R1 Drainage
- Level 3 R1O Rainwater pipes/gutters.

CAWS includes some 300 work sections commonly encountered in the construction industry. Although very dependent on size and complexity, no single project will need more than a fraction of this number, perhaps as a general average 25% to 30%. Only level 1 and level 3 are normally used in specifications and contract bills of quantities. Level 2 indicates the structure, and helps with the management of the notation. New work sections can be inserted simply, without the need for extensive renumbering.

Form of contract

This is the principal contract document and will generally comprise one of the preprinted forms of contract available to the industry. Such forms have the general agreement of the different parties involved. The form of contract, under JCT contracts, takes precedence over the other contract documents. The conditions of contract establish the legal framework under which the work is to be undertaken.

Whilst the clauses aim to cover for any eventuality, disagreement in their interpretation frequently occurs. When disputes arise these should be resolved quickly and amicably by the parties concerned. Where this is not possible, it may become necessary to refer the disagreement to a form of alternative dispute resolution, adjudication or arbitration. The parties to a contract usually agree to take any dispute initially to adjudication rather than to litigation. This can save time, costs and adverse publicity which may be damaging to both parties. Where the dispute cannot be resolved, it is taken to court to establish a legal opinion. Such opinions, if held, eventually become case law and can be cited should similar disputes arise in the future. The majority of the standard forms of contract comprise, in one way or another, the following three sections.

Articles of Agreement

This is the part of the contract which the parties sign. The contract is between the employer (building owner) and the contractor (building contractor). The blank spaces in the articles are filled in with the:

- Names of the employer, contractor, architect and quantity surveyor
- Date of the signing of the contract
- Location and nature of the work
- List of the contract drawings
- Amount of the contract sum.

If the parties make any amendments to the Articles of Agreement or to any other part of the contract, then the alterations should be initialled by both parties.

In some circumstances it may be necessary or desirable to execute the contract as a deed (formerly under seal). This is often the case with local authorities and other public bodies.

Conditions of Contract

The Conditions of Contract include, for example, the contractor's obligations to carry out and complete the work shown on the drawings and described in the bills to the satisfaction of the architect (or contract administrator). They cover matters dealing with the quality of the work, cost, time, nominated suppliers' and subcontractors' insurances, fluctuations and VAT. Their purpose is to attempt to clarify the rights and responsibilities of the various parties in the event of a dispute arising.

Appendix

The Contract Particulars for the Conditions of Contract include that part of the contract which is peculiar to the particular project in question. It includes key information, for example, on the start and completion dates, the periods of interim payment and the length of the rectification period for which the contractor is responsible. The Contract Particulars include recommendations on some of this information.

Contract drawings

The contract drawings should ideally be complete and finalised at tender stage. Unfortunately this is seldom the case, and both clients and architects rely too heavily on the clauses in the conditions allowing for variations. Occasionally the reason is due to insufficient time being made available for the pre-contract design work or, frequently, because of indecision on the part of the client and the design team. One of the intentions of the SMM7 was to only allow contract bills to be prepared on the basis of complete drawings. To invite contractors to price work that has yet to be designed is not a sensible or fair course of action. Tenderers should be given sufficient information to enable them to understand what is required in order that they may submit as accurate and realistic a price as possible. The contract drawings will include the general arrangement drawings showing the site location, the position of the building on the site, means of access to the site, floor plans, elevations and sections. Where these drawings are not supplied to the contractors with the other tendering information, they should be informed where and when they can be inspected. The inspection of these and other drawings is highly recommended, since it may provide the opportunity for an informal discussion on the project with the designer.

The contractor, upon signing the contract, is provided with two further copies of the contract drawings. These may include copies of the drawings sent to the contractor with the invitation to tender, together with those drawings that have been used in the preparation of either the contract bills or specification. The contract drawings are defined in the Articles of Agreement as those which have been signed by both parties to the contract. It will be necessary during the construction phase for the architect to supply the contractor with additional drawings and details. These may either explain and amplify the contract drawings, or because of variations identify and explain the changes from the original design. An information release schedule is to be provided and further drawings and details are the norm and are expected to be provided.

Schedules

The preparation and use of schedules is particularly appropriate for items of work such as:

- Windows
- Doors
- Manholes
- Internal finishings.

Schedules provide an improved means of communicating information between the architect (or engineer) and the contractor. They are also invaluable to the quantity surveyor during the preparation of other documents. They have several advantages over attempting to provide the same information by way of either correspondence or further drawings. The checking for possible errors is simplified, and the schedules can also be used for the placing of orders for materials or components.

During the preparation of schedules the following questions must be borne in mind:

- Who will use the schedule?
- What information is required to be conveyed?
- What additional information is required?
- How can the information be best portrayed?
- Does it revise information provided elsewhere?

The designer should supply the contractor with two copies of these schedules that have been prepared for use in carrying out the works. This should be done soon after the signing of the contract, or as soon as possible thereafter should they not be available at this time. The architect should supply the information release schedule.

Contract bills

Some form of contract bills or measured schedules should be prepared for all types of building projects, other than those of only a minor or simplified nature. The bill, or schedules, comprises a list of items of work to be carried out, providing a brief description and the quantities of the finished work in the building. The bill may include firm or approximate quantities, depending upon the completeness of the drawings and other information from which it was prepared.

Purpose of contract bills

The main purpose of contract bills is for tendering. Each contractor tendering for the project is able to price the work on precisely the same information with the minimum amount of effort. This avoids duplication in quantifying the construction work, and allows for the fairest type of competition. Despite the predicted demise of contract bills a large proportion of all contracts awarded in the UK still use some form of quantification. However, contract bills are not appropriate for all types of construction work, and other suitable methods of contract procurement should be used. For example, for minor works drawings and a specification may be adequate, or where the extent of the work is unknown, payment may be made by using one of the methods of cost reimbursement.

In addition to tendering, contract bills have the following uses. These should be borne in mind during their preparation:

- Valuations for interim certificates
- Valuation of variations
- Ordering of materials if used with caution and awareness of possible errors and future variations (the risks relating to this lie with the contractor)

- Cost planning
- Planning and progressing by the contractor's site planner
- Final accounting
- Quality analysis by reference to the trade preamble clauses
- Subcontractor quotations
- Cost information.

Preparation of contract bills

Contract bills are prepared by the quantity surveyor, adopting best practice procedures. The items of work are measured in accordance with a recognised method of measurement. Building projects are generally measured in accordance with SMM7. Other methods of measurement are also available to the construction industry (Fig. 11.1). On mass housing projects it may be preferable to use a simplified version such as the Code of Measurement of Building Works in Small Dwellings. Separate methods also exist for work of a civil engineering nature and petrochemical plants. An international version is available from the RICS. Descriptions and quantities are derived from the contract drawings and SMM7 and are on the basis of the finished quantities of work in the completed building. The contents of contract bills are typically as follows.

Preliminaries

This covers the employer's requirements and the contractor's obligations in carrying out the work. SMM7 provides a framework for this section of the bill. It includes, for example:

- Names of parties
- Description of the works
- Form and type of contract
- General facilities to be provided by the contractor.

In practice, although the preliminaries may comprise over 20 pages of the contract bills, only a small number of items (10–15) are priced by contractors. The remainder of the items are included for information and contractual purposes only. The value of the preliminary items may account for between 8% and 15% of the contract sum.

Preambles

The preamble clauses contain descriptions relating to:

- The quality and performance of materials
- The standard of workmanship
- The testing of materials and workmanship
- Samples of materials and workmanship.

1915	Scottish SMM for Building Works.
1922	Standard Method of Measurement of Building Works, 1st Edition.
1927	SMM Building, 2nd Edition.
1933	Standard Method of Measurement of Civil Engineering Quantities.
1935	SMM Building, 3rd Edition.
1945	Code for the Measurement of Building Works in Small Dwellings, 1st Edition
1948	SMM Building, 4th Edition.
1953	SMM Civil Engineering Quantities (Revision).
1956	Code for Small Dwellings (Revision).
1958	Scottish Mode of Measurement.
1963	SMM Civil Engineering Quantities (Revision).
1963	Code for Small Dwellings, 2nd Edition.
1963	SMM Building, 5th Edition.
1964	SMM Building, 5th Edition (Amended).
1968	SMM Building, 5th Edition (Metric).
1969	Method of Measurement for Roads and Bridgeworks.
1970	SMM Building, 5th Edition (with fitted carpet amendment).
1972	Standard Method of Measurement of Construction Engineering Works.
1976	Civil Engineering Standard Method of Measurement.
1979	SMM Building, 6th Edition (with Practice Manual).
1979	Code for Small Dwellings, 3rd Edition.
1979	Principles of Measurement (International) for Works of Construction.
1984	Standard Method of Measurement for Industrial Engineering Construction.
1984	SMM6, with amendments.
1985	CESMM, 2nd Edition.
1987	SMM7 Measurement Code.
1988	SMM7.
1991	CESMM 3rd Edition.
1998	SMM7 with amendments.
1999	SMM7 Measurement Code, second edition.

Fig. 11.1   Methods of measurement.

This section of the contract bills has in some cases now been replaced by a set of standard and comprehensive preamble clauses. The contents of the preambles section are usually extracted from a library of standard clauses. Contractors rarely price any of this section.

Measured works

This section of the bill includes the items of work to be undertaken by the main contractor or to be sublet to subcontractors. There are several different forms of presentation available for this work, the most common of which are:

- *Trade format*   The items in the bill are grouped under their respective trades. The advantages of this format are that similar items are grouped together, there is a minimum of repetition, and it is useful to the contractor when subletting.

- *Elemental format*   This groups the items according to their position in the building, on the basis of a recognised elemental subdivision of the project, e.g. external walls, roofs, wall finishes, sanitary appliances. In practice contractors tend to dislike it since it involves a considerable amount of repetition.

Named subcontractors

Some aspects of a building project are not measured in detail but are included in the bills as lump sum items. These sums of money are intended to cover specialist work not normally undertaken by the general contractor or work which cannot be entirely foreseen, defined or detailed at the time that the tendering documents are issued. They are separately described as for defined or undefined work.

Appendices

The final section of the bill includes the tender summary and a list of the main contractors and subcontractors.

The form of tender is the contractor's written offer 'to undertake and execute the works in accordance with the contract documents for a contract sum of money', and will also state the contract period and any price adjustment. The tenders are submitted to the architect/contract administrator who will then make a recommendation regarding the acceptance of a tender to the client. If the client decides to go ahead with the project, the successful tenderer is invited to submit a copy of the priced bills for checking. The form of tender usually states that the employer:

- May not accept any tender
- May not accept the lowest tender
- Has no responsibility for the costs incurred in their preparation.

Methods of measurement

The quantifying of construction works is best done using an agreed set of rules or method of measurement. It is then clear to all users how the work has been measured and what has not been measured (deemed to be included in SMM terminology).

In 1912, a committee was formed to prepare rules for the measurement of construction works, resulting in the first edition of The Standard Method of Measurement of Building Works (SMM) in 1922 (Fig. 11.1). However, it should be noted that this was preceded by a SMM in Scotland published in 1915. A unified SMM for the whole of the UK was not agreed until 1965.

Historically, prior to this time, quantity surveyors or measurers as they were then described, prepared quantities and remeasured work using their own ideas and preferences. This often caused ambiguity, confusion and doubt, not to mention disagreement, on the part of the contractor's staff, particularly the estimators and surveyors. Some surveyors would measure the work in detail, whereas others would adopt a practice of measuring only the cost-significant items. When SMM7 was being developed, everyone wanted simplicity. This meant to those in private practice, fewer items to measure. Contractors wanted more precise, and probably more detailed, information.

The main aim of the different SMMs is to provide a clear set of rules that can be used for measuring construction work. The rules apply equally to work that is proposed or executed. Some words and phrases in contract bills have developed implied meanings, trade customs and practices. Standard phraseologies have also been developed to standardise and clarify meanings of bill descriptions. However, the introduction of SMM7 largely reduced the need for these, since it was written with the dual purpose of measuring and billing in mind. SMM7 also has the advantage of being readily suited to the use of information technology.

SMMs have not been without their critics. Some have contested the technical adequacy of their rules, the ease of application and the overall value of measured analysis. Others have argued that it is inappropriate to value varied work in the way intended and recommended by the forms of contract. Some have suggested that the limitation of contract bills is as a direct result of the rigidities of SMMs. There has in the past also been criticisms of the time involved in measuring cost-insignificant items. The introduction of SMM7 has largely remedied this.

In countries where quantity surveyors have been long established they have developed their own methods of measurement for building works. These have often been based on a UK SMM, and have been adapted to suit local conditions, such as different methods of construction and its associated technologies.

At the beginning of the twentieth century, no standard method of measurement existed by which construction work could be adequately measured. By the end of that century a multiplicity of different methods had emerged. The methods that are acceptable include the following facets:

- Technical adequacy of the rules
- Lacking in ambiguity
- Ease by which the rules can be applied
- Measurement only of items of cost significance
- Consideration of a wide variety of applications.

Contract specification

In certain circumstances it may be more appropriate to provide documentation by way of a specification rather than contract bills. The types of project where this may be appropriate include:

- Minor building projects
- Small-scale alteration projects
- Simple industrial shed type projects.

The specification provides detailed descriptions of the work to be performed, to assist the contractor in preparing the tender. A specification is used during:

- Tendering, to help the estimator to price the work that is required to be carried out
- Construction by the designer in order to determine the requirements of the contract, legally, technically and financially, and by the building contractor to determine the work to be carried out on site.

With the introduction of Coordinating Project Information (CPI), as described above, the specification has become more important than ever. BPIC in their publications make it clear that the specification is the key document from which all other information, either for drawings or contract bills, will flow. The writing and use of specifications is a subject in its own right and as such warrants separate study (Willis and Willis 1997).

National Building Specification

The National Building Specification (NBS) is not a standard specification. It is a large library of specification clauses from which to select relevant clauses. They often require the insertion of additional information. NBS thus facilitates the production of specification text specific to each project, including all relevant matters and excluding text that does not apply.

NBS is available only as a subscription service, and in this way it is kept up to date by issue of new material several times a year. It is available on-line, via disk and hard copy for insertion into loose-leaf ring binders. NBS is prepared in CAWS matching SMM7, and complies fully with the recommendations of the CPI Code of Procedure for Project Specifications. There are three versions of NBS, the Standard Version, an abridged Intermediate Version and a Minor Works Version.

Schedules of work

A schedule of work is a compromise between a specification and contract bills. It is more like contract bills and does not include any quantities for the work to be carried out. Its main purpose is therefore in valuing the items of work once they have been completed and measured. A schedule of rates may be used on:

- Jobbing work
- Maintenance or repair contracts
- Projects that cannot be adequately defined at the time of tender
- Urgent works
- Painting and decorating.

Master programme

It is the contractor's responsibility to provide the master programme. This shows when the works will be carried out. Unless otherwise directed by the architect, the type of programme and the details to be included are to be at the entire discretion of the contractor. If the architect agrees to a change in the completion date because of an extension of time, then the contractor must provide amendments and revisions to the programme within 14 days.

Information release schedule

This schedule informs the contractor when information will be made available by the architect. The schedule is not annexed to the contract. However, the architect must ensure that the information is released to the contractor in accordance with that agreed in the information release schedule. In practice the architect needs to coordinate the information release schedule with the contractor's master programme for the works.

Discrepancies in documents

JCT 05 clause 2.15 requires the contractor to write to the architect if any discrepancies between the documents are found. The form of contract always takes precedence over the other documents. The contractor cannot therefore assume that the drawings are more important than the contract bills. However, drawings drawn to a larger scale will generally take precedence over drawings that have been prepared to a smaller scale. JCT 05 clause 2.15 expects the contractor to adopt a similar course of action should a divergence be found between statutory requirements and the contract documents. Where discrepancies or differences result in instructions to the contractor requiring variations, these will be dealt with under section 5. The contractor should not knowingly execute work where differences occur within the various documents supplied by the design team. However, it is not the contractor's responsibility to discover differences should they arise.

Discussion topic

Since the role of the quantity surveyor is changing, is there any real need for an understanding of the measurement of construction works?

The historical developments of quantity surveying have been described by many, including Seeley and Winfield (1999) and Ashworth and Hogg (2000). These both describe how the quantity surveying profession evolved and the important part that measurement contributed towards this growth and then subsequent diversification.

The role of the measurement of construction works in the work of the quantity surveyor has undeniably reduced over the past two decades. Different practice and procedures for the assessment of cost and value have been introduced and the use of electronic programs have contributed to its supposed demise. However, the ability to measure construction works in an organised and useful way remains a core skill of the quantity surveying profession. The ability to measure and to analyse engineers' and architects' drawings in this way is one of the distinctive skills of the quantity surveyor and has influenced the way that quantity surveyors carry out their varying roles today.

The practice of measurement is supposedly something that is British or done only in those countries that are part of the Commonwealth. This is clearly false and a study of methods of measurement around the world will demonstrate this and also suggest that there is some agreement on how construction work should be measured.

In the USA, for example, one of the main services provided by construction cost consultants is the production of pretender estimates to assist clients in their planning and later evaluation of bids (Picken and Jagger 2005). Such estimates are produced by measuring and valuing work from drawings prior to construction on site taking place. In the People's Republic of China, there are very formal procedures for measuring and valuing construction work (Lu 1994). In more recent years there has been much anecdotal evidence that China is very interested in the ways in which construction work can be quantified for costing purposes.

The debate continues to take place as to whether the subject of measurement has a place anymore in the quantity surveying curriculum. The subject remains important in most undergraduate programmes throughout the world. Where it is missed out is very much the exception. Recent correspondence in *RICS Business Matters* has also suggested that this is an important skill for those hoping to join the profession. A discussion in 2005, by the author, of students on a much sought after non-cognate quantity surveying programme indicated that three-quarters, at least, were measuring as a normal part of their work. A majority of these worked in some of the very largest quantity surveying practices. These quantity surveying programmes must also be assumed to reflect and meet the needs of practice, where university departments use an advisory board of practitioners to advise and guide them about the needs of practice, which happens in the majority of cases.

However, it is recognised that measurement is not as important in the curriculum as it was forty years ago. In the then RICS external examinations it was described as a 'typical' subject, meaning that students needed to gain higher marks in order to pass it. Fortune and Skitmore (1995) take a more holistic view of measurement by identifying the qualities which underpin measurement skills and identifying the complex and sometimes intangible process that is taking place in its practice.

The argument that measurement has been replaced by automated electronic systems is nothing new. Such systems have been around for a significant number of years and many practices now make excellent use of the packages that are available. Also, whilst in the past measurement was often undertaken by the most senior personnel, today it is often viewed as more of a technician role, with the

	1985 (%)	1987 (%)	1989 (%)	1991 (%)	1993 (%)	1995 (%)	1998 (%)	2001 (%)
By number of projects	45.5	38.3	42.6	30.5	36.8	41.3	32.7	21.3
By value of projects	64.7	55.5	55.9	50.8	45.7	46.1	30.1	23.1

Fig. 11.2   Percentage of projects where bills or approximate bills are used as tender documents in the UK (*Source*: RICS 2002).

professional quantity surveyor taking up a more strategic position. But practitioners still require their staff to be able to measure and to have a detailed understanding of the techniques that are involved in order to carry out the higher-level functions.

Studies of the use of bills of quantities by the RICS have indicated a considerable decline in the use of this document since 1985, but their decline seems to have further accelerated since the turn of this century (Fig. 11.2). This information has been derived from surveys on the use of alternative documentation that is being used. The survey may of course be biased depending upon the size and types of project examined (Ashworth 2006). This has varied because of the nature of work and respondents to the survey. It should also not be forgotten that a considerable amount of pre-measurement is now done on behalf of contractors by quantity surveyors as a result of changes in procurement practices.

However, at some point in the procurement process, someone has to quantify the construction works. The best person to do this is undoubtedly the quantity surveyor (RICS 2003).

Picken and Jagger (2005) further identify that measurement leads to acquisition of the following range of skills:

- Ability to understand complex three-dimensional relationships expressed in a two dimensional form
- Understanding of the technology required in construction projects and the relationship of such technologies to other technologies
- Understanding of context and significance in terms of the resource and financial implications involved in construction
- Understanding of rudimentary arithmetic, geometry and trigonometry
- Ability to set out information in a concise and orderly manner
- Understanding of the need for quality assurance in order to ensure accuracy and consistency in the presentation of information
- Awareness of how information technology can play a key role in the production of cost information
- Awareness of the need to be able to construct accurate and timely models describing construction projects in financial terms
- Awareness of the role of measurement in the estimating process and its relevance in design cost management.

Measurement is sometimes a misunderstood activity by those who have never carried it out in practice. It is also a much under-valued skill.

References and bibliography

Allott A. *et al. Common Arrangement for Specifications and Quantities.* Coordinating Committee for Project Information. 1984.

Ashworth A. *Contractual Procedures in the Construction Industry.* Pearson Education. 2006.

Ashworth A. and Hogg K. *Added Value in Design and Construction.* Longman. 2000.

Building Project Information Committee. *Common Arrangements for Works Sections for Building Works* (CAWS). 1987.

CAWS (Common Arrangement of Work Sections) Committee for Project Information. 2000.

CIOB *Code of Estimating Practice.* Chartered Institute of Building. 1997.

Fortune C. and Skitmore R.M. Quantification skills in the construction industry. *Construction Management and Economics* 12. 1995.

Lee S. and Trench W. *Willis's Elements of Quantity Surveying.* Blackwell Publishing, Oxford. 2005.

Lu Y. Building economics research in the PRC – a review. In: *Proceedings of CIB W.55 (Building Economics Working Commission)* (edited by D.H. Picken and J. Raftery). Hong Kong. September 1994.

Picken D. and Jagger D. Whither measurement? *Construction Bulletin*, RICS. 2005.

RICS *Contracts in Use Survey: A Survey of Building Contracts in Use During 2001.* RICS Construction Faculty. 2002.

RICS *Measurement Based Procurement of Buildings.* RICS QS and Construction Faculty. 2003.

Seeley I.H. and Winfield R. *Building Quantities Explained.* Palgrave. 1999.

Willis C.J. and Willis J.A. *Specification Writing for Architects and Surveyors.* Blackwell Science. 1997.

12 Preparation of Contract Bills

Appointment of the quantity surveyor

The appointment of the quantity surveyor is likely to have been made at an early stage when early price estimates were under consideration. This may be before any drawings are available, in order to provide some cost advice to the client. Only on very small projects will a quantity surveyor not be required at all. Whilst contractors should not be asked to submit tenders in competition without quantities being provided this is now becoming commonplace. Quantity surveying firms are then sometimes involved in preparing quantities directly for such firms. These may take several different formats today. There may, of course, be special considerations that would justify such invitations for a larger contract and, equally well, similar considerations that could call for quantities being provided for a contract of a lesser value.

It should not easily be forgotten that the abortive work involved is considerable if six or eight contractors are each asked to prepare their own quantities for estimating. The costs involved are high and must be borne by the industry and ultimately passed on to the client. The number of firms that might be involved in open tendering makes this point even more forcefully.

The EU Directives on Public Procurement aim to encourage competitive tendering for public contracts throughout the European Union. Irrespective of the source of funding, public bodies must comply with these Directives, which lay down strict procedures for placing written contracts exceeding the threshold values. Failure to comply with the Directives may result in legal action on the part of unsuccessful bidders. They may claim damages to cover not only the cost of preparing the bid, but also opportunity costs such as loss of profits. They need to show not that they would have won the contract, only that a breach of the Directives has occurred. The sterling equivalent of the threshold value is recalculated every two years.

Under the Directives, bids must be sought by one of the following procedures:

- Open Procedure: a notice is placed in the *Official Journal of the European Community (OJEC)* giving all interested suppliers the opportunity to tender

- Restricted Procedure: a notice is placed in the *OJEC*, allowing interested suppliers to respond. Invitations to Tender are sent only to selected tenderers
- Negotiated Procedure: one or more firms of choice are approached, terms of contract negotiated with them, and a Contract Award notice placed in the *OJEC*.

This latter procedure may be used where open or restricted procedures have been used and produced no acceptable tenders, where only one supplier is available for technical, artistic or copyright reasons or in cases where extreme urgency exists for unforeseeable reasons.

Receipt of drawings

The quantity surveyor will usually collect the drawings and any specification notes from the architect's office and at the same time discuss the job. General notes will be made of any verbal instructions given, but the more detailed questions will arise once some detailed examination of the documents has been made. A timetable for the completion of the contract bills will be agreed, along with dates when additional detailed drawings and information can be expected. It is likely that drawings of some sections of the work are incomplete, and a good deal of time may be wasted if the taking-off of that section is begun on the drawings available. A 1:20 detail may quite likely alter the 1:100 drawing, and an alteration, however slight, may affect a lot of items in the dimensions. If further drawings are to follow, it is helpful if the order in which they are being prepared can be agreed, having regard both to the architect's office procedure and the surveyor's requirements.

Project title: Southtown School			Project Nr: 00
Consultant: Architect			Sheet Nr: 1
Drawing identification		Revisions	
Nr	Title	Scale	
1	Block plan	1:1000	
2	General plan	1:100	
3	Sections, elevations	1:100	
4	Classroom plans	1:20	
5	Cloaks, toilets plans	1:20	
6	Assembly hall plan	1:20	
7	Heating chamber	1:20	
8	Door details	full size	
10	Metal window schedule	1:50	
15	Drainage	1:200	
17	Classroom store fittings	1:20	

Fig. 12.1 Sample drawing register.

Study of documents

The drawings received should be stamped with the surveyor's name and date of receipt except, of course, originals that have to be returned. On a job of any size, a register of drawings should be completed, giving their reference number, scale and brief particulars (Fig. 12.1). The advantage of separate sheets is that each taker-off can have a copy, which will assist quick reference until thoroughly acquainted with the drawings. Particulars of any further drawings received should be added to the list when they come in.

   The documents will then be examined by a principal or senior assistant in charge of the project and the takers-off will be allocated. The first things to be done are to:

- See that all the necessary figured dimensions are given, both on plans and sections
- See that the figured dimensions are checked with overall dimensions given
- Insert any dimensions that can be calculated and may be useful in the measurement
- Confirm any errors in figured dimensions with the architect, so that the originals can be amended accordingly.

Except on the smallest jobs the drawings should be supplied in duplicate, and on larger jobs there may be three or more copies. It is advisable to number the sets so that it can be seen at a glance to which set a drawing belongs.

   It will make the rooms stand out clearly if the walls are coloured in on plan and section in the surveyor's office; moreover, the act of colouring them will give an early indication of the general construction. This can be done quickly with coloured pencils and will be found well worthwhile. There may also be manuscript notes, or even alterations in plan, made by the architect at the last moment on one copy, and these should be transferred to the other copies.

   It will also be found useful on a job of any size to mark on the general plan (usually 1:100) the positions of the parts that are detailed. A cross-reference in coloured pencil in both cases will stand out. There may, for instance, be a number of 1:20 details spread over several sheets: for example, entrance doors, bay windows, or particular points of construction. Sections can be referenced by normal section lines superimposed on the plan in a distinctive colour with the drawing number and section reference given. Elevations can be referenced in a similar way. It may be found frequently that sections on detail sheets are not given a letter reference by the architect, but have a title such as 'Section through kitchen'. The surveyor can give them a letter reference for the special purpose. The marking of the general plans in this way makes them serve as a key, so that a taker-off working on some particular part of the building can see at a glance what details are available.

   A careful perusal should then be made of the specification or specification notes. By following through systematically the sections of the taking-off, gaps may be found in the specification which require information. These may be quite numerous when only notes are supplied, as 'notes' vary considerably in quality, thoroughness and extent. When a standard specification, such as NBS (National

Building Specification) in one of its versions, is used then the opportunity can be taken to check that the correct alternatives have been chosen, that superfluous matter has been deleted, and that all the gaps have been filled in.

Schedules

Schedules are useful, both for quick reference by the taker-off and for eventual incorporation in the specification for the information of the clerk of works and site agent. They may be supplied by the architect with the drawings, or it may be necessary for the surveyor to draft them. Internal finishings should certainly be scheduled in a tabulated form, so that the finishes of each room for ceiling, wall and floors can be seen at a glance, with particulars of any skirtings, dadoes or other special features. Schedules for windows and doors would include frames, architraves and ironmongery. Those for manholes would give a clear size, invert, thickness of walls, type of cover and any other suitable particulars.

Some of the material on schedules may be otherwise shown on drawings, but the schedule brings the parts together and gives a clear view of the whole. The schedule of finishings, if not supplied in the form of a drawing, should be copied and given to each taker-off, as each will at some time want to know what finish comes in a particular place, when it is necessary to make deductions or allow for making good.

Taking-off

Query sheets

After drawings and specification have been examined a first list of queries for the architect or engineer will be prepared. These should be written on the left-hand half of A4 sheets, numbered serially and dated. The use of printed headed sheets can be prepared for the purpose. When a sufficient batch of queries has been collected, they can be communicated to the architect or engineer, asking for the return of replies, as soon as possible. It is often more convenient and expedient for the surveyor to meet the architect or engineer to discuss the queries, in which case the responses can be noted. On return to the office, details should be sent to the architect or engineer for confirmation and clarification.

The queries should be given serial numbers, carried from one set to another, and the sets filed together as they are received so that the series is complete. Each taker-off should, if possible, have a copy and record the queries as they are dealt with. A final check should be made when the taking-off is completed to see that there are no gaps or overlaps.

It often happens that some proprietary material, of which the surveyor has never heard, is mentioned in the specification. If the name and address of the maker are not provided, the surveyor should find them out and obtain a manufacturer's catalogue. At the same time material prices should be obtained and if possible the expected fixing or installation costs. In suitable cases it may be prudent to ask for a small sample, as the sight and handling of a piece of the material is often helpful in disposing of difficulties that may arise in describing the fixing. Lists of trade names with the names and addresses of manufacturers will be found in many reference and price books. The internet can also be used effectively to obtain such information.

The surveyor should *never* accept an unknown name without investigation. A material specified has, before now, been found to be obsolete, and to describe it in the contract bills not only looks foolish and reveals ignorance, but involves queries being raised by tenderers. Even when the surveyor knows a material, if there has not been an occasion to refer to it for some time, up-to-date particulars should be sought. Specifications may vary, new developments will occur and prices will fairly certainly have changed.

Project Nr	**Southtown School** **Queries**	Sheet Nr . . . .
Date		
Ref *Query*	*Reply*	
1. Finish to floor of entrance hall specified wood block, coloured as tile?		*Date*
2. Should not dimension between piers on north wall be 5.08 not 5.03? (to fit the overall 56.85)		
3. Dpc not mentioned in the spec. notes? ?lead cored felt		
4. Should brick facing to concrete beams be tied back?		
Etc. etc.		

Fig. 12.2 Sample query sheet.

References to merchants' catalogue numbers or numbers of British Standards and Codes of Practice should be verified. An incorrect figure may appear in the specification either through error or through the writer not realising that the reference has become obsolete. Queries to an architect could take the form shown in Fig. 12.2.

Division of taking-off

For a small project it is most satisfactory if one person does the whole of the taking-off. For larger buildings the amount of subdivision will depend on the time allotted for the job and the availability of takers-off. Where two are made available a subdivision might be:

- Carcase of the building
- Internal finishes, windows, doors and fittings.

Such sections as sanitary plumbing, drainage and roads are more or less independent of other sections, and could be allotted as one or other of the takers-off becomes available. When three takers-off are available, the third could start on these sections, and on buildings involving a lot of joinery fittings they might take responsibility for these. There are few projects on which it would be practicable to use more than four takers-off. The measurement of the carcase of a building would not normally be subdivided. The superstructure might be subdivided for a steel or reinforced concrete frame building. The frame, with floor, roof slabs and beam casings, would be the charge of one person, while the brickwork and roof coverings would be dealt with by another. Such a section as roofs could, if necessary, be separated. If possible, however, one person should see the whole structure through.

In the same way, the measurement of the windows and external doors can hardly be subdivided, as they are sometimes structurally combined and have similar finishings. It would be preferable for internal doors to be done by the same person, as many items such as lintels and plaster reveals will occur in both sections. Internal finishings can be done by someone else if a careful schedule has been prepared and there is close cooperation between the takers-off measuring finishings and openings. The Scottish method alleviates this problem by measuring the work on a trade basis.

The more the taking-off is subdivided the greater the risk of duplicating items or assuming that someone else will have measured specific items. There are certain items in which practice differs with different offices. For example, some surveyors measure skirtings with floor finishings and others with wall finishings. Some measure them net, others adjust for openings in the doors section. It is important, therefore, that clear rules are adopted and care is necessary to see that new or temporary takers-off, not used to the office custom, are informed.

A set of 1:100 and 1:20 or other general layout drawings should be available for each taker-off, and it is worth the surveyor paying for the extra prints if they are not supplied. Single copies of special drawings are usually sufficient, such as details of joinery fittings or layout of plumbing services and drainage. These are usually the concern of only one taker-off. If a second copy of the specification can be supplied it will be found to save a lot of time when two or more takers-off are involved. On a large contract it may be worth having copies of the specification or substantial extracts photocopied, particularly when, as may be the case, it is supplied in draft. The amount of supervision of the taking-off necessarily varies with the expertise of the individual. A junior taker-off just starting to do this work may require a large amount of supervision and answering of queries. This is part of the training required. The beginner must learn but it must not be at the expense of serious mistakes. Quite a casual query may reveal unexpected ignorance, and with the inexperienced one must always be alert.

Computerised systems

Computerised methods of measuring vary from the traditional process of writing to the computer screen to the use of digitised or electronic measuring from computerised pad, to the wider use of computer-aided design (CAD) systems. These systems have now been in existence for over 25 years and there is the capability to generate automatic contract bills. Electronic data interchange (EDI) also allows the process to be taken even further by transferring the bills and other documentation directly into the contractor's own computer system. This allows for a form of automatic estimating, using a contractor's own pricing information. This can then be analysed and adjusted to suit the particular project conditions.

Buildings erected in stages

When buildings are to be erected in stages, separate prices may be required for each stage. If so, each stage will require an entirely separate bill or set of bills. It may be, however, that the division into stages is only for organisation of the work, for example, when part of a building must be completed before another part can be commenced. As far as the client is concerned only one price is required, but it would be much more convenient, both for interim certificates and final accounting, if each section were separately billed. Provisional sums, for specialist work, would be allocated accordingly.

Similar circumstances arise for housing estates where houses are completed and handed over one at a time or in batches. Though it is not necessary to have a separate bill for each type, this is valuable. In any case, some idea of the value of each type must be calculated for interim certificates and for the release of retention monies that may become due. Clients will also be interested in the different costs.

Elemental bills

Where the billing is to be by elements instead of by trades, the taking-off must be subdivided accordingly. The Building Cost Information Service (BCIS) publishes an agreed standard list of elements which can be used. The peculiar character of the job may mean that all in the list are not used, and it may be that a new element must be introduced. In the interest of more accurate cost analysis the standard list should be adhered to so far as possible. If it is found practicable to make the sections of the taking-off correspond with the elements, it will probably be possible to save time in writing the bills.

Until someone is used to taking-off by elements there is also the risk of something being missed or measured twice. There is nothing to prevent the normal order and classification of taking-off being followed, as long as it is clear to which element each item is assigned. It is useful to give code letters to each element, such as FDN for foundations, EXTLW for external walls or P for partitions. Dimensions can be marked up with these code letters prior to billing.

Numbering dimension sheets

It is advisable to mark every page at the top with the section of the dimensions to which it belongs and the serial number of the page, such as Roofs 24 or Windows 38. On reference to any sheet one can then see quickly to what section it belongs. Sheets are also usually numbered serially right through, either on each page or on each column. It is important to ensure that sheets do not go missing. It is advisable to keep a running index to the dimension sheets as a check, giving at the same time, whenever referred to, an overall view of the status of the taking-off.

Alterations in taking-off

Alterations are sometimes made by the taker-off after the dimensions have been squared and perhaps even later when they have been carried through to the bill stage. It is important that such alterations are at the same time taken through all the stages involved. If the taker-off marks the alterations with a pencil cross and hands the sheet personally to whoever is responsible for the next stage this should ensure that the corrections are made. Alterations should not be made to the taking-off by others. There may be a reason for an apparent error. Often, after making what appears to be a correction, it is found that the dimensions were right the first time.

Contract bills

Standard descriptions

To the casual onlooker all bills of quantities may look the same, but a more detailed examination will reveal personal differences and preferences. Most quantity surveying practices have developed their own house style for their bills of quantities. In an attempt to overcome misinterpretation by contractors and to ease the process of bill preparation, the use of standardised bill descriptions was developed in the 1960s. Also the effective use of computers for this purpose requires a standard library of descriptions. The bill items, instead of being written as a single description, are frequently composed of a group of phrases. The descriptions are built up by using levels of phrases or words; by combining these, an ordered description is compiled.

The introduction of SMM7 signalled the end of standard phraseologies since SMM7, as well as being a set of rules for measuring, also provided the key phrases for writing the bills. The use of this method has resulted in a much greater uniformity than otherwise existed for contract bills. This is help-ful to contractors with their own processes of tendering, estimating and final accounting.

Preliminaries and preamble clauses

If possible, before billing of the measured work commences, the preambles to each section covering materials and workmanship should be drafted from the specifica-tion, amplified if necessary by reference to previous bills. This should be done by one of the takers-off, who will by then have a comprehensive knowledge of the job. By writing the preamble clauses first, the biller can see how far the descriptions are already covered and make notes of any clauses which it is thought should be added.

If the project is for a public authority, the surveyor may be supplied with their typical preliminaries items clauses. Otherwise the surveyor will use a previous bill as a guide. Care should be taken to ensure that everything is applicable to the par-ticular project. There may be clauses, parts of clauses, or even single words, which were inserted specially for the previous job and which, being in type like the rest, are not now evident as insertions. In the same way, owing to some particular circumstances, omissions may have been made previously that now need to be reinstated.

Everything that concerns price must be included in the contract bills. The un-necessary duplication of descriptions in specification and bill can be avoided by reference in the bill to clause numbers and headings of the specification, where there is a full specification prepared and issued with the bills. Where, however, the specification is not part of the contract, it is more satisfactory for the tenderer to have everything in the bills to save the extra effort spent in cross-reference.

Numbering items

Items in the contract bill can be serially numbered from beginning to end. Alter-natively, each page can have items referenced from A onwards in order that an item might be referenced, as for example, item 50 C or 94 B. The latter is the preferred method, since if new items have to be inserted at a late stage the whole sequence of numbers does not need to be revised.

Schedule of basic rates (Fluctuation Option B)

If the contract alternative is used, requiring recovery of fluctuations to be based on the rise or fall in prices of materials (an unlikely event in current practice), an appendix will be required to the contract bills, in which the contractor can set out the rates on which the tender is based. The appendix can contain a list of the principal materials (which the contractor can supplement if necessary) or it can be left to the contractor to prepare the list. Alternatively, a fixed priced list could be given of the principal materials, but it is preferable to let contractors include their own rates, as one contractor may be better placed than another through being a large buyer, or for some other reason.

If the priced contract bills are not being returned with the tender, it is advisable to reprint the schedule of basic rates of materials as an appendix to the form of tender as well, so that it can be considered when tenders are compared.

Schedule of allocation

When the price adjustment formula for calculating variation of price claims is to be used, it is necessary when the contract bills are complete, and before they are sent out to tender, to prepare an allocation of all the items in the bills. This is done by preparing a schedule of all work categories and then allocating all the bill items to the work categories. Preliminaries and certain provisional sums are included in a balance of adjustable items section.

When this allocation is complete it is included in the bills, and in submitting a tender a prospective contractor is deemed to have agreed to the allocation chosen by the quantity surveyor. It is unlikely that a tendering contractor will challenge the allocation, but if they do they must do so when submitting the tender. When the tender has been accepted it is a straightforward exercise to complete the schedule by filling in the appropriate amounts. The whole process is self-checking in that the end result must be the contract sum.

Completing the contract bills

The draft bills should receive a final careful read through, an overall checking of the quantities and editing prior to their issue. An examination should be made to see that all gaps in the text have now been completed, such as cross-references to item numbers or summary pages. The bills should have their quantities compared by cross-checking against similar quantity related items. This aspect of work is usually carried out by the principal or a senior assistant whose experienced eye should be able to detect any major discrepancies, if these are present. Someone not directly involved in the project will often be able to spot possible errors.

A final careful examination should also be made of the drawings and specification. All notes on the drawings should be examined in case any have been missed. 'Of course, you've taken this' may produce the answer, 'No, I thought *you* had!' The specification can be run through in pencil, clause by clause, in confirmation that each has been included, either by the takers-off measuring or by those who drafted the preliminaries or preambles sections. At some stage between completion of the draft bill and passing of the proof the whole of the dimensions and the billing process should be examined to ensure that arithmetical checks have been carried out and the process audited.

The number of copies required of the finished document will include the number sent out to tender and copies for the client, architect, quantity surveyor and contractor. Two copies are also usually required for the contract.

Proof reading

The reading through of a proof copy of the contract bills is an important task. One person comparing a draft and a proof will very easily miss differences and mistakes will occur. The best way is for two to check, one reading from the draft and the other following in the proof. Periodically the duties and documents should be changed to avoid the soporific effect of listening to the reading for too long. It is usual to read from the draft copy because this is slower than reading from the proof copy. A good way to simplify the reading through is to go through the quantities columns first to ensure:

- Correctness of figures, both of item numbers and quantities
- That they are in the right column
- That the m or m^2 or m^3 are correct.

Copyright in the bills of quantities

Copyright is established by the Copyright Act 1988 in every 'original literary, dramatic, or musical work'. It is not clearly established whether there is copyright in contract bills. The RICS has taken the opinion of counsel, and were advised that in his opinion copyright existed, on the ground that the bill was an original literary work within the meaning of the Act of 1911, which used the same wording. This is an expert opinion and it should be remembered that experts sometimes differ. This is a reason for actions in law. There are inevitably many clauses in a bill of quantities that are more or less standard and used in very similar form by many surveyors, but the quantities are undoubtedly original. There will be in all bills a number of items that are original and peculiar to that particular bill.

There might be those who would take advantage of an opportunity to reuse a surveyor's bill. The best protection is to have a specific reservation by surveyors in their agreements with their clients of the rights of reprinting and reusing bills. If, of course, the surveyor is advised, when instructed, that it is proposed to use the bill again as and when required, and the condition is accepted, there can be no redress. The difficulty is avoided if the RICS Form of Agreement, Terms and Conditions for the Appointment of a Quantity Surveyor is adopted without amendments. In clause 8 of that agreement, copyright in the contract bills is retained by the quantity surveyor.

The dimensions and other memoranda from which the contract bills are prepared are in a different category. These are the surveyor's own means to an end. It has been held that the surveyor is entitled to retain these documents, unless, of course, as is sometimes the case, the contract with the client provides otherwise.

Invitation to tender

The project team will usually prepare, often in consultation with the client, a list of firms to be invited to tender, in some instances by way of some form of pre-qualification such as interview. Guidance is provided in the Code of Practice for the Selection of the Main Contractor. The firms selected will be sent a letter of invitation in advance of the documentation to confirm that they are on the list of proposed tenderers. This letter will include the following:

- Name of client and architect
- Title and location of the job
- Approximate date when the contract bills will be issued
- The time to be allowed for tendering
- Where the drawings may be inspected or some description of the works
- The form of contract to be used.

A typical letter for such a case is given in Fig. 12.3, based on the CIB Code of Practice for the Selection of Main Contractors. Occasionally some of the requirements of the Code of Procedure are not necessary. The requirement that tenderers should have two copies of the bill is important if they must send a priced copy with their tender or even forward it on advice that their tender is under consideration. If, however, a blank copy is sent after receipt of tenders, the extra expense of two copies to all tenderers does not seem justified. The offer of additional copies of the bill, or of sections of it, which the above-mentioned code stipulates, is only likely to be necessary in large contracts.

Form of tender and envelopes

Except for clients who have their own standard tender forms, the surveyor will probably be required to draft one to be issued with the contract bills. Tenders are sent either to the client or the architect according to the designated arrangements made for the delivery, which must be adhered to by all tenderers. They are normally to be sent in an envelope marked with the name of the job so that they can be put aside on receipt and all opened together.

Some public and private authorities require the priced bill to be returned with the tender. Where this is required, a separate addressed envelope of a suitable size should be provided for its return. The priced bills will be delivered sealed, and only opened if they are to be considered for acceptance. The envelope must, therefore, be marked with the name of the project in the same way as that enclosing the tender

Dear Sirs
Southtown Church of England Primary School

We are authorised to prepare a preliminary list of tenderers for the construction of the works described below.

Your attention is drawn to the fact that apart from the alternative clauses to the Standard Form of Building Contract as detailed below, further amendments to the Standard Form of Contract are annexed and will be incorporated in the tender documents.

Will you please indicate whether you wish to be invited to submit a tender for these works on this basis. Your acceptance will imply your agreement to submit a wholly bona fide tender in accordance with the principles laid down in the *Code of Practice for the Selection of Main Contractors*, and not to divulge your tender price to any person or body before the time for submission of tenders. Once the contract has been let we undertake to supply all tenderers with a list of the tender prices.

Please state whether you would require any additional unbound copies of the contract bills in addition to the two copies you will receive. A charge may be made for extra copies. You are requested to apply by [insert date]. Your inability to accept will in no way prejudice your opportunities for tendering for further work under our direction; neither will your inclusion in the preliminary list at this stage guarantee that you will subsequently receive a formal invitation to tender for these works.

Yours faithfully

- Job: Southtown Church of England Primary School
- Employer: Blankchester Diocesan Board of Finance
- Architect: LMN Chartered Architects
- Quantity Surveyor: RS&T Chartered Quantity Surveyors
- Consultants: None
- Location of site: Site plan enclosed
- General description of work: New Primary School
- Approximate cost range: £2–3 m
- Form of contract: JCT 2005
- Examination and correction of bills: overall price to be dominant (confirm or withdraw)
- The contract is to be: under hand
- Anticipated date for possession: [date]
- Period of completion of the works: 65 weeks
- Approximate date for dispatch of all tender documents: [date]
- Tender period: 5 weeks
- Tender to remain open for: 4 weeks
- Liquidated damages value: £5000 per week
- Details of bond requirement: None

Fig. 12.3   Preliminary enquiry for invitation to tender.

TENDER FOR: **Southtown Church of England Primary School**

TO: **LMN Chartered Architects**

Dear Sirs

Southtown School

We having read the conditions of contract and the bills of quantities delivered to us and having examined the drawings referred to therein do hereby offer to execute and complete in accordance with the conditions of contract the whole of the works described for the sum of £2 726 517.00 and within 65 weeks from the date of possession.

We agree that should obvious errors in pricing or errors in arithmetic be discovered before acceptance of this offer in the priced bills of quantities submitted by us, these errors will be dealt with in accordance with the CIB Code of Practice for the Selection of Main Contractors – overall price to be dominant.

This tender remains open for 28 days from the date fixed for the receipt of tenders.

Dated this . . . . day of . . . . 20. . . . . .

Name

Address . . . .

. . . .

. . . .

Signature . . . .

Fig. 12.4 Form of tender. *Note:* If used in Scotland a paragraph needs to be added confirming the tenderer's willingness to enter into a formal contract, and the signature requires two witnesses.

form, and the tenderers should be instructed that each must put their name clearly on the outside. (See Fig. 12.4 for an example form of tender.)

Issue of drawings

The contractors who are invited to tender should each be issued with copies of the location drawings, the site plan and the main plans and elevations of the project. Where appropriate, relevant component details and drawings should also be provided. During the preparation of the bill of quantities, if errors are found in the architect's and engineer's drawings these should be corrected. The architect or engineer should be advised of these in time to allow the necessary corrections to be made prior to issuing them to the tenderers.

Dispatch of finished bills

A covering letter (Fig. 12.5), will be issued with the drawings and contract bills to the tenderers. It should state:

- What documents are enclosed
- The date, time and place for delivery of tenders, and that tenders are to be delivered in the envelope supplied
- What drawings are enclosed and where and when further drawings can be inspected

---

[Date]

Dear Sirs

Southtown Church of England Primary School

Following your acceptance of the invitation to tender for the above, we now have pleasure in enclosing the following:
- Two copies of the contract bills
- Two copies of the general arrangement drawings indicating the general character and shape and disposition of the works, and two copies of all detailed drawings referred to in the contract bills
- Two copies of the form of tender
- Addressed envelopes for the return of the tender and priced bills of quantities together with instructions relating thereto.

Will you please also note:
- Drawings and details may be inspected at the offices of the architect
- The site may be inspected by arrangement with the architect
- Tendering procedure will be in accordance with the principles of the Code of Practice for the Selection of Main Contractors
- Overall price to be dominant (confirm or withdraw).

The completed form of tender is to be sealed in the endorsed envelope provided and delivered or sent by post to reach the architect's office not later than 12.00 noon on [date]. Please acknowledge receipt of this letter and enclosures and confirm that you are able to submit a tender in accordance with these instructions.

Yours faithfully

Quantity surveyor

Fig. 12.5 Formal invitation to tender. *Note:* In Scotland it is mandatory for the priced bills to be returned with the tender, elsewhere it is a desirable alternative to adopt the option of delivery within four days.

- What arrangements the tenderer must make for visiting the site, with whom an appointment should be made, or where the key can be obtained. If the site is open for inspection this should be stated
- A request for an acknowledgement of the tender documents.

It is also advisable to state that the client (employer) is not bound to accept the lowest or any tender or to pay any expenses incurred by the tenderer in preparing the tender.

Care must be taken in arranging the documents for each contract, so that all have them complete and correct. They are probably best laid out in piles with their envelopes and checked as they are put into them. For a comprehensive checklist of the information to be provided with the tender information, refer to Appendix 1 of the Code of Practice.

Correction of errors

Once the bills are dispatched to the contractors for tendering purposes, a copy should be examined. Mistakes may still be found, even after careful checking has taken place, perhaps made in the rush to send out the bill in time. Queries may arise, too, from contractors tendering. Unless errors are of a minor nature, they should be circulated to all contractors in time for them to correct their copies before tenders are completed. An acknowledgement should be requested to ensure that all tenderers have incorporated the corrections. During the examination of the priced bills of the successful contractor it should be verified that these corrections have been made. A typical letter is shown in Fig. 12.6.

[Date]

Dear Sirs,

Southtown School: New Extensions

Will you please incorporate the following corrections in the contract bill:

Item 24D For 16 m^3 read '66 m^3'.
Item 35C For 'm^2' read 'm'.

Please acknowledge receipt of this letter.

Yours faithfully,

RS&T

Fig. 12.6   Letter to contractors: corrections to the contract bills.

Receipt of tenders

Delivery and opening

In public authorities, tenders will probably be addressed to the secretary or principal chief officer. With private clients they are usually forwarded to the architect or the quantity surveyor. On the due date for receipt of tenders, the envelopes received will be counted to check that they have all been received, prior to being opened. After opening, the official concerned or the architect or quantity surveyor will prepare a list of the tendered amounts.

The practice sometimes adopted of not giving contractors the list of tenders, or giving the figures only without the names, is to be discouraged. Publication of the result is the least that can be done in return for the time, effort and cost that is involved in preparing tenders. It is recommended that tenders delivered late, after the due time, should be returned, unopened, to the contractors. Contractors do share information in a variety of ways and an unscrupulous contractor might take advantage of this. If the postmark showed the tender to have been dispatched before the time for delivery, this might be an exception to the rule.

Reporting of tenders

In considering the tenders received, factors other than price may be of importance and these should form part of the assessment criteria provided to tenderers. The time required to carry out the work, if stated as a requirement on the form of tender, may be compared. Time is frequently an important matter to the client. Although there may be reasonable excuses for failing to keep to the time agreed, and even justification for avoiding the liquidated damages provided for by the contract, the time stated by a reputable contractor may be taken as a reasonable estimate, having regard to the prevailing circumstances. The *Code of Procedure for Selective Tendering*, however, recommends that contractors tender only against price, since it is not known what value a client may place on time.

If the contract is subject to adjustment of the price of materials via the fluctuation option B method, the schedule of basic rates of materials must also be considered. The question should be asked, 'Has the tenderer assumed reasonable basic prices for materials?' If they are too low there can be an excessive increased cost on a rising market or too little in reduced costs on a falling market. Where tenders are very close, the schedules of basic rates may be compared, since the lower tenderer may have less favourable prices. Only a preliminary examination will be made at this stage to ascertain which tender or tenders should be considered for acceptance. A fuller report will be made later by the quantity surveyor.

The architect will rely extensively on the quantity surveyor for advice on these matters. A report will be prepared for the client or committee concerned, setting out clearly the arguments in favour of acceptance of one tender or another.

When tenders are invited from a limited number of contractors the lowest, or potentially lowest, should generally be accepted, although consideration must be given to other assessment criteria stated in the invitation to tender. However, tenderers are usually informed that the client has no legal obligation to accept the lowest or any tender or be involved in the costs of their preparation. But, in a limited invitation to tender (selective tendering) there is a moral obligation to accept the lowest, if any.

In open tendering, which is not recommended, there may be clear justification for rejecting the lowest tender on grounds other than cost. However, when the expenditure of public money is involved, there may be repercussions and the grounds for such rejection will need to be clearly demonstrated.

Examination of priced bill

Before acceptance of a tender, the tenderer whose offer is under consideration is required to submit a copy of the priced bills to the quantity surveyor for examination. If this has not been delivered with the tender, an additional copy should be sent to the contractor for this purpose. Sometimes, to save time, the original bills may be requested. However, the original bills are often marked with the estimator's pricing notes which would be injudicious to disclose. There is no justification for the statement that is sometimes required that the copy of the bill has been compared and checked with the original. The tender is a lump-sum tender and the sole purpose of obtaining the pricing is to provide a fair schedule for the adjustment of future variations.

The first check is an arithmetical check. Clerical errors are common. It is important that the contract bills are as correct as they possibly can be prior to the signing of the contract. If, for example, an item has been priced at £0.50 per m and extended at £0.05, it will not be fair that either additional quantity or omission of the item should be priced out at the incorrect rate in adjusting accounts. All clerical errors should be corrected in the contract copy of the bills. The amount of the tender will of course not normally be altered. Any difference will be shown as a rebate or addition as an addendum to the summary (see Fig. 12.7). This addendum will be used for interim valuations and adjustment of variations, to all rates except prime cost and provisional sums.

In addition to the arithmetical check, a technical check is also made of the pricing by examining the contractor's rates and prices. Deliberate or accidental errors may be found. Items may accidentally have been left unpriced. Items billed in square metres may have been priced at what is obviously a linear metre rate, or vice versa. An obvious misunderstanding of a description may be noted. Corrections should be made so that a reasonable schedule of rates for pricing variations results.

A secondary reason for examination of the priced bill is to ensure that the tenderer has not made such a serious mistake that they would prefer to withdraw the tender. Under English law contractors may do this at any time prior to the acceptance of a contract. When such a serious error is detected, it is always advisable to bring this to the attention of the tenderer. The error will sooner or later be discovered, resulting in a risk that constant attempts will then be made to recover the loss, to the detriment of the client's interest. The contractor should, however, be recommended to stand by the tender that has been submitted.

Summary	£
Preliminary items	18 662.00
Named subcontractors	55 408.00
Groundwork	17 479.06
	8 131.16
In situ concrete	~~8 031.16~~
Masonry	19 083.69
Structural carcassing	4 789.88
Cladding/covering	18 103.97
	4 392.36
	~~4 393.36~~
	1 884.38
Linings/partitioning	~~1 884.28~~
Window/doors/stairs	8 616.01
Surface finishes	1 036.77
Paving/fencing	4 302.74
Disposal systems	3 728.98
Engineering services	16 445.98
	£182 064.98
	~~£181 965.88~~
Water and insurance 3.5%	6 367.57
	£188 432.55
	~~£188 333.45~~

Tender submitted £188 300.00

Fig. 12.7   Corrected summary in contract bills.

Notwithstanding the guidance provided in the Code of Practice for the Selection of Main Contractors, if the correction of the error does not bring the tender under consideration above the next highest, the architect may feel that the client should be advised to allow amendment of the tender. Otherwise the next lowest tender should be considered. If a tender has been accepted and a contract formed, tenderers strictly cannot withdraw. However, this may not be a good policy to adopt for the above-mentioned reason. There are cases where a successful tenderer realises that the tender submitted is well below that of the other competitors. There is the temptation to adjust the bills, knowing that these will still be below other contractors' sums. To avoid this temptation occurring, the practice of submitting priced bills with the tender is recommended.

If priced bills are delivered with the tender by all contractors, only the bills of tenderers under consideration for acceptance should be opened. All others should be returned unopened. In the summary in Fig. 12.7, alterations have been made correcting clerical and other errors found in the priced bills. The increased total means that all rates except PC and provisional sums (which the contractor has no power to reduce) will be subject to a percentage rebate. To calculate this percentage, extract PC and provisional sum amounts, as shown in Fig. 12.8.

		£
Provisional sums		10000
Mechanical services		14850
Electrical services		8223
Wood flooring		2200
External staircase		4050
Metal windows		6500
Water mains		200
Ironmongery		550
Sanitary fittings		890
Daywork		2610
		£50073

The rebate to be expressed as a percentage is:

Errors	182064	
	181966	98
Rebate in tender	188433	
	188300	133
		£231

(This total equals the total difference between £188433 and £188300.)

The percentage is calculated as follows:

Corrected total without insurance, etc.	182064
Less PC and provisional sums	50073
	£131991

$$\text{Percentage} = \frac{231}{131991} \times \frac{100}{1} = 0.18\%$$

Fig. 12.8 Calculation of percentage rebate.

The effect of this calculation is as follows. In the variation account all rates will be subject to addition of 3.50% for water and insurances. The water and insurance percentage can be converted into a percentage on the contractor's own work instead of on the whole total as appears in this example. In that case the two percentages can be combined into a single percentage. However, in this case, since the contractor has expressed water and insurances as a percentage of the whole, they are so treated.

Correction of errors

Two alternatives are described for dealing with genuine errors in rates and prices. Where overall price is dominant the contractor either confirms or withdraws the tender. Where rates are dominant the contractor is allowed to confirm or correct. The difficulty is to decide what is a genuine error, particularly as, under the JCT

Standard Form, four days can elapse before the priced contract bills must be produced – a further reason to request that they are submitted at the same time as the tenders. The option to be adopted must be made prior to the tenders being invited.

Examination of schedule of basic rates

Where a contract is to be subject to price adjustment and exceptionally the formula method is not being used, a basic price list of materials must also be provided. This is usually requested at the same time as the bills for checking. The basic price list only covers those materials for which an adjustment is required. All rates not supported by a bona fide quotation must be carefully checked. Some materials, such as Portland cement, have standard prices. Other, such as steel tubing or stoneware drainage goods, are often quoted at a percentage on or off a standard list. Current rate comparisons can be obtained from the trade association concerned. The majority of the basic rates will probably be compared with those in price books or from local builder's merchants. It is usual for only the major items to be included, to reduce the unproductive paperwork of all concerned.

The prices will usually vary with the amount purchased. Quotations should be examined within a context of an expert knowledge of prices. It is not unknown for merchants to make mistakes against themselves in the price quoted. It will probably be some months before the quotation is accepted, and then only after further tenders have been obtained by the contractor in an attempt to reduce costs. The mistake being discovered, the contractor may claim the correction as increased cost. Under the normal forms of contract fluctuation in market price must be proved, so recovery of the amount of the error cannot be made.

The introduction of the price-adjustment formula, which is now generally preferred, negates this process and instead the quantity surveyor will be required to complete the schedule of allocation and the reduction of the work categories to the agreed number if required.

Addendum bills

The lowest tender is sometimes for a higher amount than the client is prepared to spend. It is usually possible to reduce the tender sum by changing the design in terms of either quantity or quality. However, the correct application of cost planning procedures should avoid this problem. An addendum bill is prepared in a similar way to the variation account referred to in Chapter 14. Changes to the design are measured and priced using the rates from the contractor's priced bills. The addendum bill is prepared prior to the contract being signed and the contract sum is hence based on the revised tender figure. The bill modifies the original quantities, and the quantities so modified become part of the contract. The adjustments are mostly omissions, but balancing additions are also required. Where there are no rates or prices for these in the bills, they are agreed with the contractor through a process of negotiation. Sometimes, if a variation is complex but its value can be estimated fairly accurately, the adjustment can be made on a lump sum agreed by the parties concerned.

Preparation of contract bill of quantities

A fair copy of the priced bills is required for signature with the contract. If a blank copy has been sent to the contractor to be completed, this can be used as the surveyor's office copy after making any alterations necessitated by the checking process. A corrected copy, for the contract, is made by the quantity surveyor. If necessary it should contain the schedule of basic rates of materials or the allocation of the items, so making them part of the contract.

The prices in the contractor's priced bill of quantities are confidential. They must be used solely for the purpose of the contract (JCT). Though they naturally contribute to the quantity surveyor's knowledge of current rates and prices, they can only be referred to for the surveyor's own information. They should not, for example, be discussed with another contractor. Where contractors have submitted priced bills with their tenders, those not considered for acceptance are returned unopened. Their prices remain unknown to all but the contractor. It is not unknown for contractors to submit tenders out of courtesy, because they do not like to risk offending a client or architect. Their rates and prices will be kept unknown if this procedure is followed correctly.

Preparing the contract

The duty of preparing the contract by completing the various blank spaces in the articles of agreement is the responsibility of the architect, although the quantity surveyor is often asked to complete it. It is sometimes necessary to add special clauses to the conditions of contract and to amend other clauses. Where this is required they must be written in, and the insertion or alteration must be initialled by both parties at the time of signing the contract.

Extreme care must be taken if clauses within standard conditions of contract are to be amended, as specific alterations can affect other clauses. It is not recommended. It is usual to seek legal advice if it is intended to make substantial amendments. Any portions to be deleted must be ruled through and similarly initialled. All other documents contained in the contract, i.e. each drawing and the contract bills, should be marked for identification and signed by the parties; for example:

- This is one of the drawings
- This is the bill of quantities
- referred to in the contract signed by us this day of 20 . . .

For the contract bills, this identification should be on the front cover or on the last page, and the number of pages can be stated. If the standard form with quantities is used the specification does not form a part of the contract. It will thus not be signed by the parties. If there are no quantities the specification is a contract document and must be signed accordingly. All the signed documents must be construed together as the contract for the project.

Contracts are either signed under hand, when the limitation period is six years, or when a twelve years period is required the contract is completed as a deed, formerly under seal. In the latter case it is important to ensure that this is duly recognised, as failure to do so could have serious implications.

Case law exists that illustrates the importance of ensuring that all the contract documents are in agreement with each other. In *Glesson* v. *London Borough of Hillingdon* (1970) EGD 495, there was a discrepancy between completion dates set out in the contract bills and completion dates in the Appendix to the JCT form of contract. Delays had occurred, and the question was raised with regard to the correct date to be used when calculating liquidated damages. The court held that under the relevant clause of the form of contract in use at the time (JCT 63), the date in the Appendix prevailed. Litigation would not have occurred had the contract documents been properly checked for any inconsistencies.

E-tendering

E-tendering provides a framework where both clients and tenderers are able to reduce their costs, remove unnecessary administration and streamline the overall tendering process. The tenth edition of this book (1994) made reference (p. 88) to the electronic data exchange in construction to reap benefits and efficiencies from the use of information technology especially in the context of exchanging information between different companies and during the process of tendering. The use of standard practices, both in terms of presentation and content, will help clients and tenderers benefit from a consistency in approach as well as the avoidance of ambiguities and technical incompatibility. A number of issues need to be examined in the tendering process from the initial preliminary enquiry through to tender acceptance or withdrawal. These include:

- Tendering methodologies
- Different electronic formats and their impact upon information exchange
- Benefits and constraints that different technologies provide
- Security issues, tendering procedures and workflow
- Assessment of tenders and notification of results.

The RICS carried out a survey in 2004 to find out the extent of e-tendering amongst members of the quantity surveying profession. At that time the survey found that whilst e-tendering was being developed most of the documentation for tendering purposes was still using paper copies, in some cases as back-up information to electronic sources. An overwhelming majority of those involved in the survey expected that the RICS should prepare a code of practice for e-rendering. The then Construction Faculty (now Quantity Surveying and Construction) has published a helpful guide for members of the profession in respect of e-tendering. This provides standards, guidance, promotes awareness and offers legal and contractual advice.

The RICS survey revealed the information shown in Fig. 12.9 from those involved in a telephone survey and amongst those who had attended a conference under the title, *E-tendering – What are we doing with it?*

Have you used e-tendering?		68%
What technology did you use?	Simple disk or email	53%
	Proprietary ExtraNet	47%
Size of projects		£0.50–60.00m
Positive experience?		79%
E-tendering as an opportunity?		85%
Who does e-tendering benefit?	All parties	100%
	Clients	76%
	Contractors	35%
	Subcontractors	18%

Fig. 12.9   RICS survey on e-tendering.

Those who responded that e-tendering was not a positive experience cited client expectations not being met, use of conflicting IT systems, no opportunity for innovation, incomplete enquiry information and that contractors still requested hard copies as being some of the main reasons. Some also had a fear that such systems might replace them and this was seen by them, a small minority, as a threat.

Discussion topic

Describe the development of information technology for quantity surveying practice making particular reference to the preparation of bills of quantities.

The use of computers in society at large is a relatively modern day phenomenon. There are now hardly any areas of life that they do not touch and we accept their positive use as if they have been around for all time. They are used in medicine and surgery, shopping and in the arts through the media by producers and users. They have removed the routine and drudgery from many activities and today they have high standards of reliability.

Less than fifty years ago they were unknown to the profession of quantity surveying. Perhaps the larger quantity surveying practices were beginning to consider their use, but the high costs of the hardware, the absence of software, their poor reliability and the need for air-conditioned space prohibited any practice from owning their own machines.

This is what some of the leading thinkers have had to say about the use of computers in the past (references in RICS 2000):

- Computers in the future may weigh no more than 1.5 tons (Popular Mechanics 1949)
- There is no reason why anyone would want a computer in their home (Digital Equipment Corporation 1977)
- Everything that can be invented has been invented (US Office of Patents 1899)
- 640k ought to be enough for anybody (Bill Gates, Microsoft 1981)
- I think that there is a world market for maybe five computers (IBM 1943).

There are two surprising observations from the above. The first is that these comments were made by individuals who were *in the know*, and second that some of the comments were made in recent times. The statements, as we now know, were all incorrect forecasts of the future.

It should also be remembered that less than fifty years ago, the photocopier was still in the early stages of development with the use of wet processes borrowed from photography. Calculators were at the stage of large and noisy machines and the hand-held versions did not commonly appear until about 1970. At the time these cost almost a week's earnings. As for the mobile telephone, it would be 1995 before they became commonplace.

Dent (1964), in his book *Quantity Surveying by Computer*, describes a process of automation in the office and the need for quantity surveyors to gain some understanding of computers. At about the same time some computing firms were realising that there might be a market in the use of computers for quantity surveying, since the billing process is largely a routine process and this is something that computers are good at carrying out. This would not be carried out in the quantity surveyor's own office but outsourced to the firms who had access to computers but more importantly access to computer programs for this purpose. Alvey (1976) much later produced *Computers in Quantity Surveying*. The RICS was also very much at the forefront in advising its members about current and future developments. In 1980 it issued *An Introduction to Information Systems and Technology* and this was followed closely by *Chartered Quantity Surveyors and the Microcomputer* which was a joint publication by the RICS and the Computers in Construction Association (CICA) (RICS 1981). Brandon (1984) at the RICS 13th triennial conference provided a clear incisive view in his paper *Computers – Friend or Foe*. In this he clearly showed the relationship between computing costs and computing capability. As the former was falling rapidly the latter was increasing at an even faster rate. Since 1984 these two factors, and especially the latter, are the reasons why computers themselves still do not have a real life much beyond three years. Evans (1979) in *The Mighty Micro* makes all sorts of fascinating predictions about the future, i.e. today, and it is surprising just how many of his predictions are wide of the mark. This is a trait of forecasters since the future is unknown and often cannot even be imagined. A chapter in Evan's book describes the decline of the professions but, since 1979, it can be argued that rather than becoming weaker the professions have actually become stronger and more important.

The use of computers in quantity surveying initially has focused on measurement and bills of quantities since these two have been the predominant core activities of work. However, as we are all aware today, the use of computers can be applied to many different situations. This is evident from the contents of Chapter 5 on Research and Innovation in this book.

In 1991, the RICS published, *QS2000* (Davis *et al.* 1991). This suggested that the biggest impact of IT on quantity surveying practice had been on improving the speed and efficiency of professional services. It also suggested that information flows in construction would increasingly be made electronically. Earlier editions of *Practice and Procedure for the Quantity Surveyor* (Willis and Ashworth 1987) also indicated that a wide range of measurement, estimating and costing procedures would in the future be done through the use of computer packages.

Specifically in terms of measurement and bill production, Seeley and Winfield (1999) make reference to two systems that have been extensively used by quantity surveying organisations, namely Masterbill and CATO (computerised applications for taking-off). The taking-off of quantities with the assistance of computers can be carried out in a variety of different ways. In some cases this can be done through a direct input of dimensions and descriptions through the keyboard, whereby dimensions and descriptions are done electronically. The descriptions generated comply with SMM7 and little inputting is required for the system to generate full descriptions in a similar way that Word is able to automatically correct spelling. Alternatively a separate digitiser package is used for calculating quantities directly from drawings in the way commonly used by CATO. Similar systems are employed in the USA to generate quantities by contracting firms for estimating purposes. Some of these use a light pen for measuring directly from drawings as an alternative to the digitiser. An alternative method is to use AutoCAD whereby measurements and descriptions can be imported directly as a part of the AutoCad function. The Masterbill system allows traditional approaches to the task of taking-off to be adopted. This ensures that, even in today's world, less computer literate surveyors will not be reluctant to use such systems. Electronic data exchange (EDI) has become more widely used where bills can be issued to contractors in an electronic form and this then allows them to price the work directly for tendering purposes. This practice helps to eliminate the wasted effort of re-keying information into a separate system.

The University of Salford is the centre for information technology in the construction industry. More information about the work of this centre can be found at www.construct-it.org.uk.

References and bibliography

Alvey R. *Computers in Quantity Surveying*. Macmillan. 1976.

Brandon P.S. *Computers – Friend or Foe*. Proceedings of the 13th Triennial Conference. The Royal Institution of Chartered Surveyors. 1984.

CIB. *Code of Practice for the Selection of Main Contractors*. Thomas Telford. 1997.

Construction Faculty. *E-Tendering*. RICS Books. 2005.

Davis, Langdon and Everest. *QS2000*. The Royal Institution of Chartered Surveyors. 1991.

Dent C. *Quantity Surveying by Computer*. Oxford University Press. 1964.

Evans C. *The Mighty Micro*. Hodder and Stoughton. 1979.

Hackett M., Robinson I. and Statham G. *The Aqua Group Guide to Procurement, Tendering, and Contract Administration*. Blackwell Publishing. 2006.

Lee S. and Trench W. *Willis's Elements of Quantity Surveying*. Blackwell Publishing. 2005.

Martin J. *E-tendering: What are we doing about it?* RICS. 2004.

Nisbet J. *Called to Account: Quantity Surveying, 1936–1986*. Stoke Publications. 1989.

RICS *An Introduction to Information Systems and Technology*. The Royal Institution of Chartered Surveyors. 1980.

RICS *Chartered Quantity Surveyors and the Microcomputer*. The Royal Institution of Chartered Surveyors. 1981.

RICS *Appointing a Quantity Surveyor*. Royal Institution of Chartered Surveyors. 1999.

RICS *Victorian to Virtual*. The Royal Institution of Chartered Surveyors. 2000.

Seeley I.H. and Winfield R. *Building Quantities Explained*. Palgrave. 1999.

Willis C.J. and Ashworth A. *Practice and Procedure for the Quantity Surveyor*, Ninth Edition. Collins. 1987.

13 Cost Management

Introduction

The scope of the surveyor's involvement during the post-contract stage of a project will generally require the preparation of interim valuations, the preparation and agreement of the final account and the management of project costs throughout. Final accounts are discussed in the next chapter. It is usual for both the client and the contractor to employ a surveyor or team of surveyors during the post-contract phase. The successful execution and completion of the post-contract procedures and the final account very much depend on cooperation between the client's appointed surveyor and that of the contractor. Whilst the responsibilities of these differ, there are areas of involvement common to both and it is important that each side has an understanding of the process and possible approaches. This is further underlined by the developments in procurement and contract choice, which continue to emerge.

Depending upon the size and nature of the project, the post-contract administration may be undertaken by site-based staff involved on a full time or intermittent basis. Nevertheless the duties to be performed will be somewhat similar. Likewise, the degree of involvement may vary according to the type of main contract. For example, if the contract is awarded on an approximate quantities basis requiring re-measurement on site, there will be a need for additional surveyors to be involved to carry out the site measurement. Alternatively, if the project is design and build, it is probable that the demands upon the time of the client's surveyor will be much reduced.

The issue of contract choice and the impact it may have upon practice is somewhat of a difficulty to overcome in a book of this type. There are now many contract options available and it is impossible to accommodate the specifics of each. Furthermore, there will undoubtedly be additional choice by the time this book reaches many of its readers. Also, at the time of writing, the industry has just seen the introduction of many new JCT programmes (JCT 2005) which will take over from the 1998 versions they are intended to replace in the forthcoming years. Therefore, in the interest of clarity and pragmatism, the emphasis of this chapter is on the practical rather than contractual. Where reference to the contract becomes a necessary part of the explanation, the main form which will be assumed to be in use is the Standard Building Contract 2005 (JCT 2005 SBC). Despite the decline in the use of this traditional approach over recent years, it is still considered to be the most commonly understood, and knowledge of it is a fundamental requirement for students and practitioners. Although only occasional reference to other forms of contract is made, much of the explanation and associated practice and procedure relating to JCT 2005 can be transferred to these with little or no amendment. However, although strong similarities may exist there are also subtle and distinct differences which the reader should be wary of and research further where practice demands.

Since the contractual basis of this chapter generally relates to JCT 2005 SBC, it accommodates the provisions of the Housing Grants, Construction and Regeneration Act (the Construction Act), which became effective in May 1998.

For those readers familiar with JCT 98, it may be useful to note that the new form, JCT 2005, has pulled together the several amendments that have occurred since 1998 and is presented in a new format. Although one of the objectives in JCT 05 is to make the contract easier to use, readers will need to acquaint themselves with the new format and detail.

Valuations

The construction industry survives on cash flow and the role of the surveyor is of key importance in this regard. JCT 2005 makes clear the duty of the client's quantity surveyor in this respect (clause 4.11):

> 'Interim valuations shall be made by the Quantity Surveyor whenever the architect/contract administrator considers them necessary for the purpose of ascertaining the amount to be stated as due in an Interim Certificate . . .'

In addition, the contract states that where the use of the price adjustment formula applies (fluctuations option C), in accordance with clause 4.11, the quantity surveyor must prepare the valuation.

Most construction projects encountered by the surveyor will have contractual provision for the payment of the contractor for work done, at regular intervals during the contract period. The amounts involved in the construction of major works and the duration of most projects warrant this approach. Therefore, the contractor has a regular cashflow, which is vital to profitability within the construction industry. This issue of regular payment is a major feature of the Housing Grants, Construction and Regeneration Act 1996 (the Construction Act), the contents of which are reflected in JCT 2005 SBC.

Between the date of the first interim certificate, subject to agreement between the parties, and the issue of the 'practical completion certificate', interim certificates must be issued at the regular intervals stated in the contract particulars. It should be noted that the contract does not include a definition of practical completion and that this should be clarified prior to start, thus avoiding potential disagreement at a later date. The most common period of payment, and default situation if none is stated, is at one monthly intervals. If requested by the architect, which is the norm, it is the responsibility of the surveyor to calculate the amount of such interim payments. Following the issue of the 'practical completion certificate', the architect/contract administrator may issue further interim certificates 'as and when further amounts are ascertained'.

Certificates and payments

In the course of a construction project for which interim valuations apply, an architect or contract administrator will be called upon to make decisions and issue a series of certificates that must be issued in accordance with the provisions of the particular contract. In so doing when acting in an arbitral situation, an architect is under a duty imposed by law to act fairly and impartially between the parties.

Where an interim certificate is required by the contract, and certainly under JCT 2005, its issue is a condition precedent to payment. If the certificate is improperly withheld, entitlement to payment may be enforced in the absence of the requisite certificate.

The employer must pay the contractor no later than 14 days from the issue of the interim certificate. Failure to do so may result in the application of one or more of the remedies available to the contractor:

- Claim for interest on the late payment. In addition to the amount due to the contractor, simple interest will be due on the amount of late payment, calculated at a rate of 5% above the Base Rate of the Bank of England, current at the due date, for the duration of the delay in payment.
- Suspension of the work (clause 4.14). If the employer fails to pay in accordance with the contract, the contractor may issue a notice to the employer, copied to the architect/contract administrator, to the effect that the works will be suspended after a further 7 days, allowing time for delivery. Once the employer has paid the amount due, the work should recommence within a reasonable time thereafter.
- Determination in accordance with clause 8.9 of the standard form (JCT 05).

The surveyor should be alert to the above conditions and advise the client and architect where necessary.

In addition to the certificate, the contractor is entitled to a 'Notice of Payment' which should be issued within 5 days of the date of issue of an interim certificate, showing the basis of the valuation. Where the employer intends to 'withhold and/or deduct an amount from the payment due', the contractor should be provided with a 'Notice to withhold' which should be issued no later than 5 days before the final date of payment. This should show the amount to be withheld/deducted, stating the grounds for such action. Whilst neither the surveyor nor the architect should exclude amounts from the certificate, when there are grounds for withholding, they should advise the employer of the entitlement to do so, including provision of the required detail.

Accuracy

The valuation for certificates should be made as accurately as is reasonably possible. In preparing or verifying the valuation, the surveyor has two opposing concerns:

- The contractor is entitled under the contract to the value of work done, less a specified retention sum. If the valuation is kept low, the retention sum is in effect increased. To a contractor having a number of contracts in hand, these excessive amounts of 'retention' will mount up and demand additional capital, which may have serious consequences.
- The client must be protected against the possible insolvency of the contractor. When an overpayment resulting from an excessive valuation cannot be recovered from the contractor, as in the case of insolvency, the additional payment will be effectively 'lost' and may become an additional expense to the client. In turn, this additional amount may become the liability of the surveyor.

It is reasonable to assume that to the client's surveyor, the latter consideration will be given more significance than the former since self-preservation is a powerful incentive. It appears that some surveyors translate their understandable caution into the under-valuation of contractors' work. This tendency, which may be greater amongst less experienced surveyors, should be resisted for the reasons outlined above. Unfortunately, in cases where the financial reputation of the contractor becomes in doubt, the reasonable inclination of the surveyor is to become more cautious, which in turn may result in greater financial difficulties.

In determining the need for accuracy, the surveyor should appreciate that an interim valuation is merely a snapshot of the progress of the works, which, depending on the work stage, contractors' programme, deliveries to site, etc., may change significantly within 24 hours of the assessment.

Timing

If, as is usually the case, the contractor is entitled to certificates at regular monthly intervals, (in terms of JCT 2005, this should strictly be adhered to), it will be found convenient to arrange the dates at the beginning of the contract: say, the last Thursday in every month. Since the timing of the first interim certificate determines the dates of the remainder, it is important that this is done with some forethought. Sometimes the dates must be fixed to suit the client's convenience, particularly when payment is passed at a board or committee meeting held at fixed intervals. Similarly, where there are monthly site meetings which the surveyor needs to attend, it may be possible to arrange the valuation for the same day. This is particularly beneficial where the site may be some distance from the surveyor's office. In any event all parties should be aware of the implications of failure to certify and pay correctly in accordance with the conditions of contract.

In the past, the client's surveyor found it very helpful if the contractor submitted a statement with supporting invoices, delivery notes, etc. as a basis for the valuation. This approach saved a great deal of time and was also possibly more accurate since the contractor was able to provide more detail than may otherwise have been possible to establish during a site visit. JCT 2005 now reflects this previously accepted practice by means of clause 4.12. This allows the contractor to prepare and submit an application for interim payment, 'not later than 7 days before the date for issue of an Interim Certificate'. It is important to emphasise that this action does not remove the responsibility for the valuation from the client's surveyor who, in normal circumstances, will be liable for its correctness and in the absence of the contractor's application is still obliged to prepare a valuation in accordance with clause 4.11. (Although the correctness of the certificate is the responsibility of the architect, reliance on the quantity surveyor's valuation is normal and thus, in practice, transfers liability for its accuracy.) In order to verify the contractor's application, it is sensible for the client's surveyor to arrange a meeting on site with the contractor's surveyor and, where necessary, make any adjustment to the contractor's statement. Where such an adjustment occurs, the client's surveyor is obliged to provide the contractor with a statement showing the revised amount of valuation, together with supporting detail to allow the contractor to identify and understand the changes that have been made.

Extent of measurement

The extent to which measurement will be necessary in making valuations for certificates will depend on the nature of the job and the stage it has reached. It may very often be possible to take the priced bill, identify the items that have been done, and build up a figure in that way. Some items will, of course, be only partly done, and in that case a proportion will have to be allocated. At a first valuation, for instance, there may be little beyond foundations, and to identify the appropriate items in the bill should not be difficult. If it is agreed that the foundations are two-thirds complete then the amount can be easily calculated. When, however, it comes to the superstructure, it may be necessary to take approximate measurements of such things as brickwork, floors and roofs. The surveyor should always bear in mind the value of the works being measured in determining the most appropriate approach. To avoid excessive labour in the measurement of relatively complex but low cost items for valuation purposes, it may be more reasonable to agree with the contractor's surveyor that 'half the total value of the plumbing works' is a fair assessment of the works carried out. It should be remembered that detailed measurement for valuation purposes will likely have no further bearing beyond the interim valuation in hand and is therefore of passing value only.

Figure 13.1 indicates the apportionment approach to the preparation of an interim valuation, reflecting both a necessary pragmatism and also a reasonable level of accuracy. In this example, concrete work has been collected into work between ground and first floor, first floor and second and second floor to roof. A tour of the site will allow a prompt but reasonable assessment.

The surveyor should consider several approaches within the calculation of the interim valuation. As indicated by the example, the usual method of valuation is to assess the total amount of work performed at each interim stage on a gross basis

Interim valuation 5; Date on site 20 March 2007						
Bill Ref	Brief description (Concrete works)	Total £	Previous %	Amount £	Present %	Amount £
18/3 B–G	Reinforced Concrete in Cols					
	Say 40% GF–1st	24 000	100	24 000	100	24 000
	Say 30% 1st–2nd	18 000	75	13 500	90	16 200
	Say 30% 2nd–Roof	18 000	Nil	Nil	10	1 800
18/4 A–F	Formwork to Cols					
	Say 40% GF–1st		100		100	
	Say 30% 1st–2nd		100		100	
	Say 30% 2nd–Roof		50		75	
18/5 A–H	Reinforcement in Cols					
	Say 40% GF–1st		100		100	
	Say 30% 1st–2nd		100		100	
	Say 30% 2nd–Roof		50		75	
	Total carried forward					

Fig. 13.1 Apportionment of works for valuation purposes.

(*not* as previous payment plus a little more). This approach negates any errors that may have occurred in a previous valuation, although inclusion of columns showing the previous assessment may be a useful guide. There are various further shortcuts that may be taken, for example, the annotation of a copy of the bills of quantities with inserted columns for amounts of work carried out. Information technology, of course, can assist greatly, either by use of an electronic copy of the bills, if available, or by the generation of a spreadsheet that could also be used in the assessment of amounts for fluctuations.

When dealing with housing, or other projects for which there are a large number of similar units, it should be possible from the bill of quantities to arrive at an approximate value of one house at various stages; for example:

- Brickwork up to damp-proof course
- Brickwork to first floor level, with joists on
- Brickwork to eaves
- Roof complete
- Plastering and glazing complete
- Doors hung
- Plumbing and fittings complete
- Decoration complete.

The value for different types of the same size of house will not vary sufficiently to make a difference for certificate purposes. The work done at any time can be valued by taking the number of houses that have reached each stage, and pricing for half and quarter stages. With such projects, if the contractor's surveyor intends to submit a detailed valuation in accordance with JCT 2005, details of the approach should be agreed with the client's surveyor at the outset to avoid disagreement and abortive work.

On projects without firm bills of quantities, the value of the work carried out will be assessed through measurement of work on site or, where appropriate, by proportion of the individual items on the tender summary. It may be convenient to invest some time at the commencement of the project to agree the value of particular work stages as a basis for interim valuations.

Towards the completion of the project, it is a good idea to check what is left to complete, as a safeguard against error in a cumulative total. For the last two or three valuations, the contract sum might be taken as a basis, and deduction made of all provisional sums and percentage additions and of work not yet done, the various accounts against the first of these being added together with percentage additions pro rata, adjustment being made for the approximate value of variations and price adjustment. Similarly, once identifiable sections of the works are almost complete, a similar approach can be adopted whereby, for example, with reference to Fig. 13.1 above, concrete on each floor can be calculated as complete less minor amounts of work not yet done.

Preliminaries items

Each valuation will take into account the pricing of the preliminary bill or preliminaries items. There may be a number of items separately priced, or there may be one total for sections or the whole in which case further analysis will be required. In addition, they may show an allocation between fixed and time-related costs in accordance with SMM7.

Each priced item should be considered and a fair proportion of each included. Single payments, time related costs and value related costs should be considered separately. Single payments occur where there is an identifiable item of work, for example a site notice board with a single cost payable on construction at the beginning of the project and a further payment on removal at completion. Time related items relate to items that attract costs on a regular basis during the works, for example, site management, the amount for which that is included in the preliminaries could be divided by the period of the contract to give a suitable monthly sum. Value related items could be valued on the basis of the proportion of the measured items completed by the contractor. Thus, if 25% of the measured works were completed, the same proportion of that element of the preliminaries would be incorporated accordingly. For example, the cost of electricity for the works, part of which will be linked to the amount of work carried out.

Some preliminaries items may be allocated to more than one or all of these categories, for example, the cost of telephones. A proportion of the telephone charges during the construction of the works will depend on works progress and will therefore be value related, part of the cost will be fixed (perhaps relating to connection fees) and some may be time related (standing charges). Similarly, the price for

Prelim item	Months												Total £
	1	2	3	4	5	6	7	8	9	10	11	12	
Site offices													
Electricity													
Water													
Telephones													
Temporary roads													
Insurances													

Fig. 13.2   Indicative example of a pro-forma for preliminaries assessment.

provision of offices, mess facilities and storage units could be split into delivery cost (single payment), weekly rent (time related) and removal cost (single payment), and valuation made accordingly.

If it is anticipated that the contract time will be exceeded, suitable reductions should be made on the time related items to relate payments on account more accurately to the work actually carried out. If an extension of time award has been made, this could affect the amount of the preliminaries allowance also.

A cashflow forecast of preliminaries costs relating to the duration of the works may be prepared for agreement at the first interim valuation. Once this is accepted it may be used for future valuations, not forgetting to monitor the progress of the works and adjust if necessary. A partial example of this is shown in Fig. 13.2; this would reflect the apportionment to single, time, and value related costs.

It should be noted that the use of either a time related or a cost related approach for the total preliminaries assessment at interim valuation stage is bound to be flawed and less accurate than the above method.

Nominated subcontractors and suppliers

The most significant change to the content of JCT 2005 SBC is the exclusion of provision for nominated subcontractors. The JCT consider that there has been little use of the nomination provisions in JCT 98 although have retained the provisions for listing named subcontractors (see 'Named subcontractors' below). Where the employer would like to use nomination it will now be necessary to use a different form of procurement (Lupton 2006).

Subcontractors

The contractor is allowed to appoint subcontractors to carry out portions of the works, provided written consent is obtained from the architect/contract administrator in accordance with JCT 2005, clause 3.7. This consent is required to permit the subletting and does not relate to the approval of the subcontractor selected by the contractor, although in practice, consent may be withheld until the details of the proposed subcontractor are known (Ndekugri and Rycroft 2000).

The payment of interim amounts to subcontractors is the responsibility of the contractor's surveyor. Applications for payment from subcontractors often form the basis of the main contractor's own application. Payments will be made to each subcontractor, generally in line with amounts certified to the main contractor. Reconciliation on a monthly basis will be required. The contractor's surveyor will analyse income received through certificate payments against costs of subcontractor payment and direct costs (see section 'Cost control and reporting' later in this chapter).

The client's surveyor should be aware that the interim accounts prepared by subcontractors should be considered as though the main contractor had produced them.

Named subcontractors

Named subcontractors are, as their title suggests, named by the architect/contract administrator. By providing a list of subcontractors and/or suppliers in the contract documents (at least three names must be provided for each portion of the works – as JCT 2005, clause 3.8.2), from whom specified work or materials should be obtained, the architect/contract administrator is able to restrict and therefore control possible sources of supply. Once a subcontractor has been so named (JCT 2005, clause 3.8), when finally selected by the contractor, they become subcontractors in exactly the same way as subcontractors chosen entirely at the discretion of the main contractor. No special requirements therefore arise regarding the client's surveyor's duties at interim valuation or final account stage.

In practice, named subcontractors may pose a problem to the contractor. The rates submitted by the contractor at tender stage for portions of the works covered by the named subcontractor provision are at the risk of the main contractor. It is therefore important that rates for the work, which is to be carried out by one of the named subcontractors, are adequate and that they have been obtained from one of the named subcontractors. Failure to conform in this regard, for example by using rates from an unnamed subcontractor with whom work is regularly sublet, is exposing the contractor to significant additional price risk. Additional names may be added to the list by either party if approval is obtained from the other (clause 3.8.3.1).

Unfixed materials

Besides the value of work done, most forms of contract allow payment to be made to the contractor for unfixed materials. The payment for such materials is an area of additional risk for the employer in that:

- At the time of payment, ownership of the materials may not be vested in the contractor but with subcontractors or suppliers. Therefore, in such situations,

payment by the employer does not transfer ownership from the contractor to whom it did not belong in the first place.
- The materials, being unfixed, may be more easily lost, damaged or stolen. This remains the risk of the contractor, in accordance with clause 2.24 unless, of course, insolvency of the contractor occurs.
- In the event of contractor insolvency, unfixed materials, irrespective of true ownership, may be easily removed from site.

The risks associated with unfixed materials are increased where the materials are stored off-site and this aspect is discussed further below.

In an attempt to protect the employer, JCT 2005 makes special provision regarding the payment for unfixed materials, both on-site and off-site. In consideration of the contract conditions, the surveyor should only include payment for unfixed materials stored on-site where:

- The materials are correctly stored. Storage on building sites may be less than ideal and wastage rates are generally considered to be too high. Payment for materials, which due to damage cannot be incorporated within the works, should not be allowed.
- The materials are delivered *reasonably, properly and not prematurely*. Materials delivered too far in advance of their incorporation within the works should not be paid for. There are several factors that should be considered in determining the reasonable timing of deliveries, including material availability, ownership of price risk (e.g. will the employer gain benefit from early purchase and delivery thus avoiding an imminent price increase) and works progress.
- The materials are required for the project. For example, materials being stored on a particular site for convenience of the contractor and eventual use on another site should not be allowed.
- Adequate insurance of the unfixed materials is provided by the contractor.

In practice, the surveyor will benefit by requesting a list of materials to be submitted by the valuation date, to be checked during the site visit. Where the contractor submits an application for payment, in accordance with the provisions in JCT 2005, this should be provided. The surveyor should bear in mind that consideration should be given for changes in material stock occurring between submission of the application and site visit. An item included as unfixed material in an application for payment prepared several days before the valuation date, may be incorporated within the works by the time of site verification. If this occurs, the value of the works will have increased, without any consideration in the payment application, and therefore, if any adjustment is made it should be an increase rather than decrease.

JCT 2005 states that the contractor must also be paid for unfixed materials stored off-site. This payment depends on adherence to the following conditions:

- Materials or goods or items pre-fabricated for inclusion, which should be in accordance with the contract, should be identified in a list, provided by the employer, that should be annexed to the contract bills. This provision ensures that: the contractor knows, in advance, the extent of the materials off-site that will be allowed and may make a tender allowance for costs in connection with any item excluded from the list.

- Proof that the property is vested in the contractor, thereby allowing transfer to ownership upon payment of the certificate, should be provided.
- If stated as a requirement in the contract particulars, a surety bond for the value of such materials should be given.
- The materials are to be set apart in the premises in which they are manufactured, assembled or stored and must be correctly labelled showing the name of the employer and the intended works.
- Proof of adequate insurance cover for the materials must be provided by the contractor, protecting the interests of the employer and the contractor in respect of the specified perils (clause 4.17).

The contract differentiates between uniquely identified 'listed items' (e.g. purpose made glazing units) and those that are not uniquely identified (e.g. possibly, sanitary ware).

Problems have arisen over retention of title of goods when certifying payments for materials on or off-site. In the case of *Dawber Williamson Roofing Ltd* v. *Humberside County Council* (1979) 14 BLR 70, it was held that title in a quantity of roofing slates had not passed to the main contractor at the time he was paid by the client. The subcontractor was not paid, and the main contractor subsequently went into liquidation. As title had not been passed to the main contractor, the subcontractor (plaintiff) was entitled to be reimbursed the value of the slates by the client. It is important therefore to ensure that such materials are, at the time they are paid for by the client, the lawful property of the contractor. JCT 2005 attempts to ensure that the value of any unfixed materials included within an interim certificate become the property of the employer. Despite the contractual provision in JCT 2005, which is intended to achieve greater security for the employer, there may be co-existent supplier agreements that retain the title of the goods with the supplier, until paid for, and thus prevent the contractor from transferring ownership.

Advance payment (JCT 2005, clause 4.8)

JCT 2005 makes provision for the employer to make an advance payment to the contractor, to be supported by an advance payment bond provided by the contractor (unless not required). The amount of the advance payment, and the timing of the payment, will be stated in the contract particulars. At interim valuation stage, where a prior advance payment has been made, it will be necessary to take into account the reimbursement provisions. The amounts and timing of these reimbursements will also be stated in the contract particulars.

Variations and claims

At interim valuation stage, the assessment of the value of some variations, provisional items or sections and claims may cause difficulty. For instance, where the foundations are 'All Provisional' it may not be possible to adequately remeasure the works prior to their completion on site and the contractor may need to be paid in total. In such cases, where the eventual construction varies substantially from the provisional measurement, the risk of under or over-payment is increased. Similarly, a major variation requiring significant re-measurement may be completed on site before such work can be fully remeasured. The surveyor will need to use varying methods of approximate estimating in order to make a fair allowance for such items.

Contractors' clams for loss and expense are discussed in Chapter 14 on Final Accounts. With regard to payment for these at interim stage, the surveyor should note that such amounts should be paid in full to the contractor once ascertained, without the deduction of retention. There may be some pressure on the client's surveyor to include payments on account prior to ascertainment; no such payment should be incorporated without the direction of the architect/contract administrator, although as later stated, in practice the surveyor's advice is usually sought in such matters.

Price adjustment

If the traditional method of adjusting fluctuations is in operation, that is, by the use of schedule 7; fluctuations option B (JCT 2005), it is valuable to start checking the records of price adjustment at an early stage, at any rate the labour portion, when a running total can be kept month by month for inclusion in the certificate valuation. Materials are rather more difficult to keep up to date than labour, owing to the time lag in rendering invoices. In either case, increased costs, whether in respect of labour or materials, should not be included in interim certificates unless supporting details have been made available.

As stated elsewhere, by far the most common method of adjusting fluctuations is by way of the price adjustment formulae schedule 7; fluctuations option (JCT 2005). With this method, reimbursement of increased cost is automatic in each interim certificate by the application of the formulae to the current indices relevant to the proportions of the work categories actually carried out. This is more fully discussed in Chapter 14. It is now usual for information technology to be used to assist in this process.

Retention

It is standard in building contracts to incorporate a provision whereby the client is entitled to retain part of the assessed value of the works at interim valuation stage. In JCT 2005 a maximum of 3% of the contract sum will be used unless a different percentage is stated in the contract particulars.

Retention benefits the client by providing a sizeable fund at the end of a contract that acts as an incentive to the contractor to complete the works and to make good defects where they occur. There are no perceived benefits to the contractor however, and a retention bond, in accordance with clause 4.19 (JCT 2005) may be used as an alternative, requiring a surety to be approved by the employer.

There are some components of the interim valuation that are not subject to retention and these are listed in clause 4.16.2 of the standard form. With reference to this document, these are outlined below:

- Reimbursement of fluctuations, unless calculated by the formula method as in schedule 7; fluctuations option C (JCT 2005)
- Remedial work in connection with damage for which insurance provision had been made under clause 6.7 option B and clause 6.7 option C (i.e. where employer takes out insurance)
- Reimbursement of loss and expense claims in accordance with clause 4.23 and also clause 3.24 relating to loss and expense concerning the discovery of antiquities
- Reimbursement of amounts paid by the contractor in connection with: fees and charges relating to local authorities and statutory undertakers clause 2.21; opening up and inspection of the works, provided the works are found to be in accordance with the contract clause 3.17; compliance with architect's instructions which infringe upon royalties and patent rights clause 2.23; taking out insurance in accordance with clause 6.5.1 (insurance against injury to persons or property); schedule 3, clause A.5.1 (additional premium in the provision of terrorism cover); schedule 3 clauses B.2.1.2 and C.3.1 (premiums paid to make good default by employer); clause 2.6.2 (additional premium relating to early use by the employer)

When work is substantially complete, part of the retention sum is released, the balance being held as security for making good of defects that may be found necessary within the rectification period. In compliance with the standard form, this retention release occurs in the first interim valuation following the issue by the architect/contract administrator of the practical completion certificate. Hence, this provides a clear incentive for the contractor to reach this stage. In the standard form, half is to be released. The remainder of the retention is released after the expiry of the rectification period or the issue of the certificate of making good, whichever occurs later. Again, this provides a clear incentive for the contractor, on this occasion, to promptly remedy any defects that may have occurred.

The rectification period, if nothing is stated in the contract particulars to the contract, is 6 months. The purpose of the rectification period is to provide a reasonable period during which defects may become apparent. Therefore, its length should be situation dependent rather than an automatic inclusion or default to 6 months.

Where part of the works is to be handed to the employer prior to contract completion, the release of retention for that section will be dealt with accordingly. This is incorporated within clauses 2.33, 2.34 and 2.35 of the standard form.

Clause 4.18 of JCT 2005 covers the rules pertaining to the treatment of retention money. The employer's interest in the retention is as a fiduciary trustee, whereby the employer holds the money on behalf of the contractor, although without obligation to invest. This is intended to provide some protection to the contractor against default by the client. The contractor can direct the employer to place the retention funds in a separately identified bank account. Thereafter, the employer should issue a confirmatory certificate to the architect/contract administrator, copied to the contractor.

Liquidated damages

To many clients, the completion of a construction contract by a particular date is vitally important and therefore most contracts incorporate a fixed date. Where the contractor fails to achieve this date, it is likely that the employer will suffer a loss. To compensate for such occurrences, most contracts and certainly JCT 2005 under clause 2.32, allow the employer to deduct an amount, i.e. liquidated damages, calculated on the basis of a rate stated in the contract particulars (e.g. £10000 per week or part thereof) from amounts due to the contractor. The delay may be calculated on the basis of the completion date stated in the contract particulars, or such extended period if an adjustment of completion date has been granted by the architect in accordance with clause 2.26 of JCT 2005.

The pre-ascertainment of the rate of damages is an important aspect of this provision. It must represent a reasonable pre-estimate of the loss that will be incurred by the client in the eventuality of a delay and must thereby provide all parties with a known amount at commencement of the contract. The benefits of this are clear since otherwise, actual damages would need to be calculated on each such occasion, a costly process and carrying with it additional risk to the contractor in that at the time of entering into the contract, the full contractual liability for delay is unknown. Once damages are fairly pre-ascertained, the amount paid to the employer in the event of a delay is limited to that stated in the contract particulars.

Although a delay in completion may have occurred, neither the surveyor nor the architect/contract administrator is entitled to deduct any liquidated and ascertained damages from the amount due to the contractor within either an interim or final certificate. In accordance with the contract, this entitlement falls to the employer. The architect/contract administrator must issue a non-completion certificate to the employer, advising of the contractor's failure to complete the works by the completion date, as a pre-requisite to any deduction. Although any action may lie with the employer and architect/contract administrator, it is good and pro-active practice for the surveyor to monitor any situation where liquidated damages may arise and advise the architect/contract administrator and employer where necessary. This advice should include a calculation of the deduction that may be applied.

The employer is entitled to waive the right to deduct liquidated damages or reduce the amount to be applied by notification to the contractor in writing. In practice, this may be done in consideration of: the contractor's withdrawal of a contractual claim thus avoiding a lengthy dispute; or a reduction in the actual loss to the employer relative to the amount pre-ascertained, although as stated above, the employer is under no such obligation.

Predetermined stage payments

Predetermined payments on account, based on stage payments, have been traditionally used on large multiple contracts such as housing; however, to use such a method on a complex building contract is much less feasible. Predetermined payments are not featured within the JCT 2005 contract; however, the process involved in stage payments may have some practical application within the preparation of an interim valuation, without any contractual basis.

Where stage payments are used, the contractor is paid an agreed amount when work reaches a certain stage of completion, for example when the substructure is complete £8 500 will be due, and so forth with additional payments upon the completion of each determined progress point such as external walls, upper floors and roof.

Use of stage payments allows both parties, contractors and clients, to know in advance the probable cash flow for the contract and to arrange their finances accordingly, although the completion of each stage is still dependent on contractor's progress. From the client's surveyor's point of view, a great deal of time is saved: the time taken to prepare an interim valuation is, where stage payments are used, a matter of hours not days, which it often is under the traditional system. For contractors the same saving of time is available, although for their requirements it is often necessary to know in more detail the value of work done for their own internal purposes and for paying subcontractors.

Previous certificates

A careful check should be made to ensure that the figure shown as already certified is correct, as a slip here may make a serious error in the valuation. The architect/contract administrator should confirm the amount of the previous payment, or the figure may be referred to as 'previous valuations', the architect/contract administrator being asked to verify before certifying.

Specimen forms

A specimen valuation is set out on the standard RICS valuation form in Fig. 13.3. This form is similar in design to the corresponding RIBA certificate form in Fig. 13.4 and gives all the information necessary to enable the architect to complete the certificate.

Care should be taken not to refer to the valuation as the certificate or to the surveyor as certifying. The surveyor only recommends, and it is for the architect/contract administrator to certify, who may take into account other matters than those within the surveyor's sphere, such as defective work.

Certificate/valuation papers

The surveyor's copy of each statement should be kept together with all the papers relating to one valuation. The attached papers will include such things as lists of unfixed materials and interim statements of price adjustment. Similarly, related computerised records should be retained within a suitable folder or file system with adequate back-up storage.

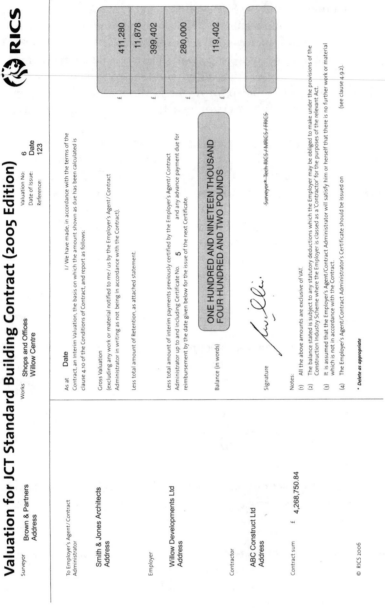

Valuation for JCT Standard Building Contract (2005 Edition)

Surveyor   Brown & Partners
           Address

Works   Shops and Offices
        Willow Centre

Valuation No: 6
Date of Issue: Date
Reference: 123

To Employer's Agent/Contract
Administrator

Smith & Jones Architects
Address

As at   Date   I / We have made, in accordance with the terms of the
Contract, an Interim Valuation, the basis on which the amount shown as due has been calculated is
clause 4.10 of the Conditions of Contract, and report as follows:

Gross Valuation
(excluding any work or material notified to me / us by the Employer's Agent / Contract
Administrator in writing as not being in accordance with the Contract).

£   411,280

Less total amount of Retention, as attached statement:

£   11,878

£   399,402

Employer

Willow Developments Ltd
Address

Less total amount of interim payments previously certified by the Employer's Agent / Contract
Administrator up to and including Certificate No.   5   and any advance payment due for
reimbursement by the date given below for the issue of the next Certificate.

£   280,000

£   119,402

Balance (in words)

**ONE HUNDRED AND NINETEEN THOUSAND
FOUR HUNDRED AND TWO POUNDS**

Contractor

ABC Construct Ltd
Address

Signature   Surveyor Tech RICS / MRICS / FRICS

Notes:
(1)   All the above amounts are exclusive of VAT.
(2)   The balance stated is subject to any statutory deductions which the Employer may be obliged to make under the provisions of the
      Construction Industry Scheme where the Employer is classed as a 'Contractor for the purposes of the relevant Act.
(3)   It is assumed that the Employer's Agent/Contract Administrator will satisfy him or herself that there is no further work or material
      which is not in accordance with the Contract.
(4)   The Employer's Agent/Contract Administrator's Certificate should be issued on   (see clause 4.9.2).

Contract sum   £   4,268,750.84

* *Delete as appropriate*

© RICS 2006

Fig. 13.3   RICS Valuation Form, with permission © RICS.

Interim Certificate

SBC

Issued by: address:	Smith and Jones Architects Address	
Employer: address:	Willow Developments Ltd Address	Job reference: 456
		Certificate no: 6
Contractor: address:	ABC Construct Ltd Address	Date of valuation: Date
		Date of issue: Date
Works: situated at:	Shops and Offices Willow Centre	Final date for payment: Date
Contract dated:	Date	

This Interim Certificate is issued under the terms of the above-mentioned Contract.

		£	
Gross Valuation	. . . .	£	411,280.00
Less Retention as detailed on the attached Statement of Retention	£	11,878.00
Sub-total	£	399,402.00
Less reimbursement of advance payment	£	-
Sub-total	£	399,402.00
Less total amount previously certified	£	280,000.00
Net amount for payment	**£**	119,402.00

*All amounts are exclusive of VAT.
The Employer shall in addition
pay the amount of VAT properly
chargeable.*

I/We hereby certify that the **amount due** to the Contractor from the Employer is (in words)

One hundred and nineteen thousand four hundred and two pounds

To be signed by or for
the issuer named
above

Signed _____ J. Smith,

This is not a Tax Invoice.

Distribution	☐ Employer	☐ Contractor	☐ Quantity Surveyor	☐ File copy

F501A   for SBC   RIBA CONTRACT ADMINISTRATION FORMS   © RIBA Publishing 2006

Fig. 13.4   RIBA Certificate, with permission © RIBA.

Valuation on insolvency

In the event of the contractor becoming insolvent it is prudent for the client's surveyor to make a valuation of the work executed, by taking the necessary measurements of the work up to the stage at which work ceases or is continued by another contractor. The valuation will include unfixed materials and plant, which the architect/contract administrator will be responsible for seeing are not removed from the site (see JCT 2005, clauses 2.24 and 8.5). The purpose of such a valuation is that all parties may be aware of the financial position, and some estimate of money outstanding to the insolvent contractor can be made. It is important to note that a major change occurring in JCT 2005 is the need for the employer to action determination which can be done immediately. In the previous standard form, JCT 98, determination occurred at contractor insolvency, without the need for any action.

Insolvency of the contractor is likely to severely disrupt the works and result in costs to the client. It is a complex area of involvement, which increases risk to the professional advisor. There may be several actions that the surveyor is able to take in such situations and further discussion on this aspect is contained in Chapter 15.

Cost control and reporting

It is vital to the client and, of course, the design team, that costs are effectively managed throughout the construction of a project. Clients generally desire the final cost of a project to be no more than the contract sum; for some clients this may be their most dominant concern throughout the project. It is the role of the client's surveyor to try to manage these costs by a process of monitoring design and site developments and advising the client and members of the design team of their likely impacts and remedies.

The client's surveyor, if working closely with the members of the design team, should be aware at the earliest opportunity of proposed variations to the contract, including drawing amendments. Advance knowledge of proposed changes enables a full evaluation in terms of cost, quality and programming implications to be carried out in advance of their issue.

Although the initial estimate of variations to the contract is likely to be of a budgetary nature based on approximate measurements and notional rates or merely lump sums, it is important that such estimates be progressively updated as more detailed information becomes available in the form of firm measurements, quotations or daywork records. It is also necessary for the surveyor to review all correspondence and meeting minutes issued on the project in order to identify the potential cost implications of the issues contained therein. Similarly, it is also beneficial if the client's surveyor is aware of what is actually happening on site. Occasionally, changes may occur which are undocumented but for which the client may be liable.

Regular financial reports will be required to advise the client of the anticipated outturn costs. These are commonly produced at monthly intervals. As mentioned earlier, the report will be tailored to meet specific client requirements. Certain clients will only require a simple summary statement of the current financial position (see Fig. 13.5), others will require a detailed report identifying the cost implication of each instruction (whether issued or anticipated) and the reason for it. On a complex project, this may result in a lengthy document. In addition to the advice given in the regular financial statement, it may be necessary to provide the client and members of the design team with cost advice more promptly if a major issue arises.

The regular report will identify adjustments to the contract sum in respect of the following:

- Issued instructions
- Adjustment of provisional sums
- Remeasurement of provisional work
- Dayworks
- Increased costs (if applicable)
- Provision for claims and anticipated future changes.

An updated cash flow (see later in this chapter) may also be included with the report to identify for the client the actual current level of expenditure relative to that previously anticipated. This will allow the adjustment of the future budgetary provision. Supporting calculations should be filed with each report for future reference when it comes to reviewing and updating the costs.

Where significant cost increases occur, it is good practice to prepare an outline of possible remedies to such budgetary excess, before advising the client. Almost inevitably this information will be required and advance preparation for this request will be helpful.

The contractor's surveyor also submits regular financial reports to his senior managers and directors. These will show a different emphasis to reports prepared for the benefit of the client. The contractor will carry out frequent cost/value reconciliations which are likely to have three broad cost elements:

- *Actual value* The real value of the works in accordance with the quantity of work performed valued at the rates estimated at the time of tender, contained in the contract documents.
- *Actual cost* This is the real cost of carrying out the works on site, irrespective of the tender estimate, upon which the contract sum is based. Predominantly, this will be work carried out by subcontractors.
- *External valuation* This is the valuation that is agreed by the client's surveyor and is included in the interim certificate for payment by the client. In a perfect world, this amount, subject to retention, should be equal to the 'actual value' whenever an interim valuation is carried out. In practice however, the nature of the interim valuation process makes this a difficult objective to achieve.

The relationship of these three elements of cost, value and external valuation is of key importance to both the contractor and the client. Figure 13.6 considers three project situations that may occur at interim valuation and financial report stage:

SHOPS AND OFFICES DEVELOPMENT

WILLOW CENTRE

COST REPORT NO. 6

Date

SUMMARY	£	£
Contract Sum		4,268,751
Less Contingencies		120,000
		4,148,751
Adjustments for:		
Instructions Issued – Section 1	62,138	
Named subcontractors – Section 2	(8,619)	
Provisional sum Expenditure – Section 3	(4,603)	
Anticipated Variations – Section 4	30,190	
Ascertained Claims	-	
	79,106	79,106
Anticipated Final Account	£	4,227,857
Current Approved Sum		4,268,751
Balance of Contingencies	£	40,894

Notes: Costs exclude VAT, Professional fees and direct client costs

Fig. 13.5 Financial statement.

	Project A	Project B	Project C
Actual value	£15 000	£16 000	£14 000
Actual cost	£14 000	£15 000	£15 000
External valuation	£16 000	£14 000	£16 000

Fig. 13.6   Cost reconciliation.

- *Project A*   The contractor is in a profit making situation since the cost of performing the works is less than the value of the works contained in the contract sum. Also, in the short term relating to the interim stage, the contractor's cash flow is positive. This project is therefore in a relatively harmonious position, albeit that the client is paying in excess of the true value of the works. This additional payment is in effect a 'hidden borrowing' by the contractor.
- *Project B*   As with Project A, the contractor is in a profit-making situation. However, in the short term, due to the shortfall between actual cost incurred and payment received, the contractor's cash flow is negative. This may prove to be a serious situation for the contractor, leading ultimately to liquidity problems.
- *Project C*   The contractor is in an overall loss making situation since the cost of performing the works is more than the value of the works contained in the contract sum. This position may be disguised in the short term due to the excessive external valuation, which, as for Project A, contains some 'hidden borrowing'. Unfortunately, whilst the contractor may be avoiding a short-term cash flow problem, if this cost/value balance continues in the long term, the contractor will suffer a loss on the project. In such a situation, in addition to the obvious problems to the contractor, others may also suffer negative consequences. The contractor will be motivated toward attempting to find legitimate opportunities to retrieve the loss from other parties including the client, the subcontractors and suppliers, the consultants and possibly employees.

Cash flow

A forecast of cashflow is normally requested by clients and provided by the surveyor. This should be prepared in association with the contractor since it will be greatly influenced by the intended programme of works. Software is available which will assist in the preparation of a cashflow forecast based on criteria specific to the project and the 'S' curve of expenditure shown in Fig. 13.7.

Cashflow forecasts are useful for two reasons: they may be used as a basis upon which to arrange project finance, and they can assist in monitoring the progress of the works. With regard to the latter, it may be found useful to have on file a note of the net amount of the contractor's work (excluding provisional sums), so that an eye can be kept on the proportion of this kept in each valuation. The value of certificates can be graphically represented to enable comparison with the anticipated cash flow. There may be reasons, such as site conditions or inclement weather, that result in a shortfall against that anticipated, but any serious departure from

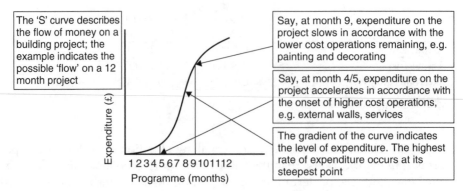

The 'S' curve describes the flow of money on a building project; the example indicates the possible 'flow' on a 12 month project

Say, at month 9, expenditure on the project slows in accordance with the lower cost operations remaining, e.g. painting and decorating

Say, at month 4/5, expenditure on the project accelerates in accordance with the onset of higher cost operations, e.g. external walls, services

The gradient of the curve indicates the level of expenditure. The highest rate of expenditure occurs at its steepest point

Expenditure (£)

1 2 3 4 5 6 7 8 9 10 11 12
Programme (months)

Fig. 13.7　Indicative 'S' curve.

regularity should be looked into as it may be an early indication of delays and difficulties being experienced by the contractor.

Discussion topic

Outline and discuss the rationale for, and the impact of, the major changes that have occurred with the issue of JCT 2005 Standard Building Contract with regard to subcontractors.

Please note, this example may be read in conjunction with an equivalent task in Chapter 14 (Final Accounts).

Introduction

To begin with, some general notes about JCT 2005. There has been an attempt to improve the style and layout of the contract with the intention of making it more readable and accessible than JCT 98. JCT 2005 has achieved the objective of improved style and layout and hopefully practitioners will agree once they become familiar with its new structure. One particular change to the layout of the contract which practitioners will find of benefit is the incorporation of several key supplements within a single document. This includes the provisions for sectional completion, fluctuations and the contractor's design portion which are incorporated within the main body of the text and supported by a series of dedicated schedules.

In terms of content, it may appear that little has changed, however, whilst being relatively few in number, there are some major changes, albeit in some cases merely reflecting common practice. The situation relating to subcontractors represents one area of significant change.

The demise of the nominated subcontractor

To those who first entered the profession in say, the 1960s and 1970s, it is perhaps hard to contemplate that the nominated subcontractor is, in terms of JCT recognition, no more. However, in reality, the process of nomination has for many years been increasingly out of favour with clients and practitioners. It seems that it is now rarely used, if ever, and therefore this major omission from the standard form of contract will perhaps not come as a surprise.

The process of nominating subcontractors was expedited through the use of prime cost sums which allowed an architect (and hence the client, via the architect) the opportunity of selecting a particular firm to carry out certain items of work (i.e. a nominated subcontractor) or a firm to manufacture and supply materials (i.e. a nominated supplier). This opportunity to nominate a particular organisation was used where the work was of a specialist nature and where the architect desired greater control of the subcontractor choice and thereby quality. It was also used to overcome known shortages. For example, in periods of strong market activity the lead-in time for, say, structural steelwork may have been such that if no order was placed for steel until a main contractor had been appointed, delays would be almost certain. The ordering of the steelwork by the employer from a preferred subcontractor prior to the appointment of the main contractor, and thereafter nomination of the same subcontracting organisation, resolved this problem. Steelwork manufacture could in effect begin before the main contractor was known.

Although advantages from nomination were clear, the procedure imposed significant interference upon the freedom of the contractor in carrying out the works. The position of the nominated subcontractor, and to a certain extent the nominated supplier, was a special one in terms of the contract. This placed certain obligations upon the employer, for example:

- At interim valuation stage, any subcontractor application for payment needed to be included in the valuation as a separately identifiable amount. Payment of this was via the main contractor, which included an allowance for the contractor's profit and overheads. This process was monitored closely by the quantity surveyor and where the contractor failed to pay the nominated subcontractor in accordance with the provisions and timescales set out in the contract, direct payment of the nominated subcontractor by the employer could be made and subsequently deducted from payments due to the main contractor.
- The most difficult aspect of nomination, as far as risk to the employer was concerned, related to contractual provisions concerning any default by the subcontractor. Delays on the part of the nominated subcontractor and nominated supplier were considered to be 'relevant events' for which the contractor could claim an extension of time. Thus, where the subcontractor was underperforming, there was considerably less incentive for the main contractor to mitigate any negative effects. This in turn could frequently lead to the client suffering financial loss.

In addition to the problems that could be created by the above obligations, the complexity of contractual provisions surrounding the use of nomination, clearly evidenced by the amount of attention given to the subject in any relevant textbook, created substantial additional risk to the employer.

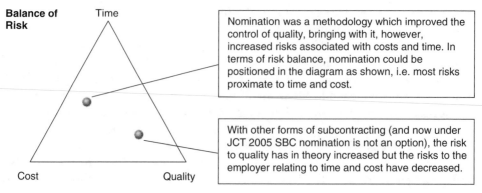

Fig. 13.8   Changed perceptions of risk balance.

The rationale for the removal of the nomination procedures from the standard form of contract can be viewed and summarised from the perspective of risk management (Fig. 13.8).

Named subcontractors

The decision to remove nomination from the standard form does therefore leave the employer with the problem of control relating to specialist areas of the contract. In an attempt to provide for this, JCT 2005 has retained the listing provisions, previously set out in clause 19.3 (JCT 98) but now contained within clause 3.8. This procedure is now established practice, and allows the client to name subcontractors by listing a minimum of three firms in the contract documents from which one will be selected by the contractor to carry out an identified part of the works. Since further names may be added by either party prior to commission, the contractor could use an alternative, provided the employer has no reasonable objections. Therefore, listing does not provide the employer with as much control of subcontractor selection as may be desired.

If the employer requires more control, it will be necessary to adopt an alternative form of procurement, for example, construction management, or to use a different form of contract, for example the 'Major Project Form' which accommodates the pre-selection of specialists the contractor will be responsible for, or the intermediate form of contract (JCT 2005 Intermediate Building Contract) which continues with a method to allow the appointment of a named subcontractor. It is also possible to specify a single subcontractor in the bill of quantities, however, this action brings with it similar problems to nomination, only less well defined and tested (Lupton 2006).

Simplification of payment process

In terms of payment, absence of the nominated subcontractor simplifies matters considerably since the provisions for monitoring their payment throughout the contract were complex. In fact, as stated, the whole matter of subcontracting is much simplified since the term domestic subcontractor is also now redundant in that all subcontractors are treated within the Conditions of Contract in the same way. The JCT have also issued a standard building subcontract and a standard building sub-contract with subcontractors design for optional use with JCT 2005. Irrespective of their use, sub-letting arrangements need to be in accordance with clauses 3.7–3.9. These have been reworded since JCT 98 but are substantially the same.

References and bibliography

Association of Consultant Architects Limited. *ACA Standard Form of Contract for Project Partnering*. 2000.

Chappell D. *The JCT Design and Build Contract 2005*. Blackwell Publishing. 2007.

Davison J. *The 2005 What's New? A comparison with previous forms*. RICS Business Services Ltd. 2006.

Hackett M., Robinson I. and Statham G. *The Aqua Group Guide to Procurement, Tendering, and Contract Administration*. Blackwell Publishing. 2006.

Joint Contracts Tribunal Limited. *Standard Form of Building Contract 2005 Edition. With Quantities*. Sweet and Maxwell. 2005.

Joint Contracts Tribunal Limited. *Intermediate Building Contract 2005 Edition*. Sweet and Maxwell. 2005.

Lupton S. *JCT Standard Forms of Building Contract 2005 Editions: Part 1*. The International Construction Law Review. 2006.

Ndekugri I. and Rycroft M. *The JCT 98 Building Contract, Law and Administration*. Butterworth Heinemann. 2000.

14 Final Accounts

Introduction

The majority of construction projects result in a final cost that is different to that agreed by the client and contractor at commencement of the construction works. The calculation and agreement of this final construction cost, the final account, is usually of the utmost importance to both the employer and contractor. Therefore, parties to the contract need to ensure that the final account incorporates a fair valuation of the works carried out. This chapter deals with the principles of measuring for variation accounts and the practical implications of contract conditions covering the calculation and agreement of the final account. Within lump sum contract arrangements the price agreed by the client at commencement will usually require adjustment for several matters, including:

- Variations
- Provisional measurements
- Provisional sums
- Fluctuations
- Claims.

These are considered below. In addition to contractual provisions, there are other external factors that are likely to influence the project environment, the contract administration and the preparation of the final account. The degree to which this applies may depend on a range of factors including:

- *Status of documentation at formation of the contract*
  The quality of the contract documentation will have a bearing on the proximity of the final project cost with that of the agreed contract sum. For example, a thorough pre-contract design process allied to accurate contract bills is likely to result in fewer post-contract changes than projects with incomplete designs and less accurate contract bills.
- *Skills of the contract administration and cost management team*
  An experienced contract administration and cost management team will be more able to maintain construction costs within the client budget.
- *Market forces*
  In times of recession, contractors and subcontractors may be operating within particularly small profit margins. In such economic conditions, the tendency toward contractual claims may be more acute.

- *Client and contractor attitude*
  An adversarial contract environment, a frequent occurrence in the construction industry, is likely to be both a consequence of and a contributor to cost variations. In recent years, the construction industry has recognised this problem and has encouraged a change of attitude, for example via the promotion of partnering (see Chapter 10 on Procurement).
- *Accuracy of contingency allowance*
  It is accepted practice to provide an allowance for unknown circumstances that may result in an increase to the construction cost established at tender stage. This allowance is known as the contingency and is usually provided as a lump sum addition to the contract. A well considered and realistic contingency allowance will assist in maintaining the final cost within the client's budget.
- *Resources*
  Limited post-contract services, sometimes a consequence of clients wishing to reduce fees, is likely to reduce the extent and quality of post-contract control, resulting in a greater risk of cost variation.

Variations

General procedures

Once a contract has been concluded, its terms cannot be changed unless the contract itself contains provision for variation, or the parties make a further valid agreement for alteration. The standard form contracts contain extensive machinery for variation, but the only variations thereby permitted are those that fall clearly within the contractual terms. If the desired change is not covered by those terms it can only be effected properly by fresh agreement. In this connection care should be exercised to ensure that the new agreement is itself a valid contract and, in particular, that it is supported by consideration given by both parties. When making visits for interim certificate valuations or site meetings, the surveyor should keep an eye on the variations in the contract that have arisen. It is often valuable to have seen the work in the course of construction.

The responsibility for issuing instructions is with the architect/contract administrator. The surveyor requires such instructions from the architect/contract administrator as an authority to incorporate the value of any resultant variation within the final account. Instructions from the architect/contract administrator may be issued for many reasons; however the most common is to amend the design in some way. Although JCT 2005, clause 3.12.1 states that 'All instructions issued by the architect/contract administrator shall be in writing', it is not necessary for the architect/contract administrator to issue an instruction using a standard form. It is, however, both usual and good practice, and an example of an instruction from the architect/contract administrator using JCT 2005 is shown in Fig. 14.1. Under clause 3.12 provision is made for dealing with instructions other than in writing and this is discussed later in this chapter.

The contract may also provide (e.g. JCT 2002, clause 2.14.3) that errors in the bill of quantities shall be treated as a variation and adjusted. Although no specific

			Architect's Instruction
Issued by: Address:	Smith and Jones Architects Address		

Employer: Address:	Willow Developments Ltd Address	Job reference:	456	
		Instruction no:	10	
		Draft no:	.	
Contractor: Address:	ABC Construct Ltd Address	Created date:	Date	
		Sheet:	1	of 1
Works: Situated at:	Shops and Offices Willow Centre			
Contract dated:	Date			

Under the terms of the above-mentioned Contract, I/we issue the following instructions:

	Office use: Approximate costs	
	£ omit	£ add
Construct inspection chamber and Install drainage pipework in Accordance with drawings 456/12B, 13C and 14D		
	ON ARCHITECT AND QS COPIES ONLY	

To be signed by or for the issuer named above

Signed _____ T. Smith _____   Sub-total

Amount of Contract Sum	£	
±Approximate value of previous issued Instructions	£	
Sub-total	£	
±Approximate value of this Instruction	£	
Approximate adjusted total	£	

Distribution			
☑ Contractor | 2 |	☐ Quantity Surveyor	☐ Clerk of Works	☐
☑ Employer | 1 |	☐ Structural Engineer	☐ Planning Supervisor	☐ Other
	☐ M&E Consultant	☐	☑ File | 1 |

Fig. 14.1   Architect's instruction. With permission © RIBA.

instruction will be required for doing this, the architect/contract administrator who has to certify the final account upon completion, and the employer, should be told of any substantial items, even where fault may lie with the client's surveyor.

Drawing revisions

Many instructions from the architect/contract administrator will be accompanied by revised drawings. These should be dated on receipt and used to update a database of drawings which should be carefully maintained, as during the pre-contract stage. It is good practice to indicate that these drawings are post-contract, likewise drawings which are superseded by new drawing issues should be recorded and identified as such. This procedure will reduce the risk of using redundant information in the preparation of the final account.

There is some possibility that drawings will continue to be issued during the tendering period, i.e. once the contract bills have been completed and issued to the tendering contractors. If the changes contained in these revisions are relatively minor, it will be adequate for the client's surveyor to measure and price approximately and advise accordingly. However, if the changes are significant, it may be necessary to prepare addendum contract bills (see Chapter 12) and issue them as a further document to the tendering contractors; thus the cost implications of the revisions will be included in the contractor's tenders.

Surveyors should also note the practice of design consultants to include on their drawings a schedule of revisions, indicating an outline of the changes contained in each updated drawing issue. This may be very helpful in identifying variations to design; however, it should not be relied on solely since some drawing revisions may not be adequately scheduled. Each new drawing issue should be carefully examined to identify any amendment to the design.

A separate file for instructions from the architect/contract administrator should be kept with other variation related correspondence. Such information may be found to be explanatory of the instructions and indicate the designer's intent, so helping when it comes to measurement.

Procedure for measurement and evaluation

Oral instructions

Instructions that have yet to be formally authorised by the architect/contract administrator are occasionally contentious and care needs to be taken to ensure that they are valid in terms of the contract. Whilst the contract states that all instructions must be in writing, in practice they are not. There are occasionally situations of convenience or perceived necessity whereby an instruction may be given orally. With regard to instructions other than in writing, the contract (clause 3.12) states that:

- The instruction shall be of no immediate effect.
- The instruction must be confirmed in writing by the contractor within seven days of issue and, if not dissented to by the architect/contract administrator

following a further seven days from receipt, will take effect. This requirement on the part of the contractor is relaxed where the architect confirms an oral instruction within the seven-day confirmation period.

A contractor wishing to ensure payment for a variation arising from an oral instruction should therefore follow the contract accordingly. Failure to follow the confirmation procedure may result in non-payment, although the architect/contract administrator may confirm any such instruction before issue of the final certificate.

Measurement

The evaluation of variations will usually include two distinct operations: measurement, either from revised drawings or from site, and valuation. The procedures for both measurement and costing are outlined below.

Irrespective of who does the work, it is likely that the measurement of variations would be more easily carried out in the office although there are situations where it may be useful or essential to physically see the work carried out; for example, where provisional quantities form part of the contract and require site measurement, such as replastering in refurbishment projects. There may be several approaches to site measurement, depending upon circumstance. For instance, traditionally site measurements may be taken using a dimension book that will be 'worked up' later in office conditions. Alternative approaches may include the use of schedules of items that can be neatly pre-prepared in the office and used as a site tick sheet, or drawings that can be accurately dimensioned or annotated during a site visit. Site measurement can be a cold, dirty and dangerous task and any initiative that improves the process should be tried! Where measurement occurs on site, it is not essential that the omissions be set down as the additions are measured. It is more pragmatic while on site to measure the additions, leaving the omissions to be evaluated in the office.

As a general principle, adjustment will be made by measuring the item as built and omitting the corresponding measurements from the original dimensions. However, there will be occasions when it may be easier to adjust a contract item by either 'add' or 'omit' only. For example, if all emulsion on walls is amended from two coats to three coats, an 'add' item of the contract quantity at the price for the additional coat would be a more efficient approach.

It is very important to keep omissions and additions distinct, and it is suggested that the words 'omit' or 'add' should always be written at the top of every page and at every change from omission to addition. This advice may sound pedantic, but the scale of error, should inversion between 'omit' and 'add' occur, may be very large. When the measurement, calculation and billing process are separated and performed by various staff within an organisation it is easy to see how this type of error could result. Each item of variation should be headed with a brief description and instruction number. In accounts for public authorities, the references for instructions from the architect/contract administrator may well be required by the auditors.

Grouping of items within the final account

Before any abstracting or bill of remeasurements is started, the surveyor should decide on the suitable subdivision into terms that will be adopted in the account. Since items may or may not correspond with the instructions from the architect/contract administrator they may be arranged in a different order: an instruction may be subdivided or several grouped together, if their subject matter suits. The arrangement of variations within the final account may reflect other objectives, for example, where fluctuations require the categorisation of work or where capital allowances may be sought on some aspects of the completed building.

Quite probably, the adjustment of foundations will be the first variation for which measurements are taken. If there is a large difference between the contract provision for foundations and the as built foundations, there may be some urgency, for remeasurement and provisional agreement. In a situation of under-provision, the contractor will require to be paid for the additional amount of work and the employer will need to accommodate the revised cost in the budget provision. Conversely, where the contract allowance is in excess of the as built foundations, the consultant will need to ensure that over-payment does not occur and the only way to do this accurately is to carry out a remeasure. At this stage of the contract it may not be known what other variations there will be; however the foundation adjustment can be regarded as an item with which other variations will not interfere to any great extent. In the event that there is a minor change in plan for which an instruction from the architect/contract administrator is issued, any subsequent change to the foundations may be dealt with as a variation within a variation. Unless there is any special reason for distinguishing, the lesser variation will be absorbed in the greater. When it eventually comes to adjusting for the change in plan, this will be done for the superstructure only.

It might, however, happen that the complete value of the change in plan is separately required. For example, in the rebuilding of a fire-damaged building, for which reimbursement from insurers applied, it might be that a change in plan was being made at the client's request and additional expense. In that case the foundation adjustment would have to be subdivided to give the separate costs required. If the variation is a completely additional room, then the foundation measurements for that room can easily be kept separate from those for the foundations generally, which is preferable as it gives more relative values.

As the list of variations develops, the surveyor will be able to decide on how to group them. For instance, there may be one instruction from the architect/contract administrator for increasing the size of storage tanks, another for the omission of a drinking water point and a third for the addition of three lavatory basins. Each of these will be measured as a separate item, but the surveyor may, for convenience, decide to group these together as 'variations on plumbing'. But if the client has ordered the three lavatory basins and does not know about the storage tanks or drinking water point (which are changes in the ideas of the architect/contract administrator), it may be advisable to have the value of the extra for basins separately available for reporting. It is good practice to highlight the 'reason for variation' in cost reports to ensure that the client is able to appreciate why costs have changed.

It will be convenient to group the very small items under the heading of 'Sundries', preferably in such a way that the value of each can be traced.

The role of the clerk of works

If there is a clerk of works, the client's surveyor should ensure that arrangements are made for records of hidden work to be kept in the form required. This will assist when examining accounts submitted by the contractor or otherwise prepared. Clerks of works vary, of course, in their ability and experience and it would be unwise to assume that the clerk of works knows exactly what is required without any guidance. Also, other duties may prevent the clerk of works from carrying out remeasurement work as and when it is necessary. The depths of foundations, position of steps in foundation bottoms, thickness of hardcore and special fittings in drainage (i.e. items of work which will become hidden) are all examples of items that the clerk of works may be asked to note and record. If these records are carefully kept and agreed at the time with the foreman, the client's and contractor's surveyors should have no difficulties from lack of knowledge.

The clerk of works may also play an important role in verifying materials, labour and plant records which relate to variations for which dayworks are to be utilised in their evaluation (see later in this chapter).

Where employed, a clerk of works may issue instructions on behalf of the architect/contract administrator. Such instructions will not take effect unless the architect confirms them in writing within two working days of issue.

Pricing variations

Pre-costed variations

The preparation and agreement of a final account can be a time consuming and expensive process for both parties, so much so that in many cases it has been found that the measurement period often exceeds the contract period. An attempt to limit this problem is contained within JCT 2005. In addition, where a major and complex variation occurs, there may be considerable cost uncertainty at the time of issuing the required instruction to carry out the works. In an attempt to overcome these problems, the principle of pre-costing variations is incorporated within JCT 2005 by way of clause 5.3 and schedule 2.

Under these conditions of contract, the architect/contract administrator has the power, if thought appropriate, to seek a firm price quotation from the contractor for variations before confirming the order. Notification of the intended variation is sent to the contractor, who is then required to price the instruction, breaking the cost down into the cost of the work and the cost of any concomitant prolongation and disruption. The price is then submitted for scrutiny by the quantity surveyor, who passes recommendations to the employer who, if the price is acceptable, provides the contractor with an acceptance in writing. This is followed by the issue of a confirmed acceptance from the architect/contract administrator. This procedure is subject to a strict timetable, and provision is made for adequate information to be supplied by both the architect/contract administrator and the contractor. If an acceptable price cannot be achieved using this procedure, then the traditional methods of agreeing the cost are adopted.

Under the terms of clause 5.3 and schedule 2, this procedure is subject to the agreement of the contractor who is required, in addition, to provide a method statement.

Pre-costing of variations may save a great deal of time and provide cost certainty with regard to the variation affected. However, there are drawbacks, not least of which is that pre-costed variations would be expected to be more expensive than those arrived at by traditional means:

- The pre-costing of variations lacks the element of competition provided in the traditional approach wherein tendered contract rates form the basis for pricing variations
- The costing of prolongation and disturbance, both totally unknown factors at the time of pricing, has to be something of a gamble, and contractors must err on the pessimistic side in order to safeguard their position
- The time frame is such that the contractor may have inadequate time to prepare (thus encouraging mistakes) and the client little time to consider (thus encouraging oversight)
- In the event that the employer does not accept a schedule 2 quotation that has been prepared on a reasonable basis, a fair and reasonable amount will be added to the contract sum for the abortive work in connection with its preparation.

Where a schedule 2 quotation is used, fluctuations if applicable will not be applied.

The pricing of measured work

The methods of pricing variations are described in the contract conditions (Clause 5.6 to 5.10, 'Valuation Rules', in JCT 2005) and follow a logical and hierarchical sequence in terms of their choice of application:

- Where applicable, the rates in the contract bills (or relevant schedule of rates) should be used. Clearly this will apply to omission; however, additions may vary in terms of not only specification but also the nature of the work execution. For example, it is reasonable to expect that the rate for paint to soffits 2.5 m above floor level will be less than the same painting specification at a height of 5 m, which will involve additional labour. Likewise, there may be a significant change in the quantity of work resulting in less or greater efficiency.
- Where contract bill rates are not applicable, they may be adjusted to take into account the revisions in working conditions and quantity. In the 'paint to soffits' example above, the new rate for valuation of the variation may be adjusted to

reflect the additional labour involved. It may be possible to use other bills of quantities or schedule items of work in this agreement. For example, we may have an alternative painting specification, at both soffit heights, showing a price differential of say, £0.20/m² reflecting the additional labour involved at the greater height. It may therefore be reasonable to use this price differential in establishing a revised rate for the new item of work.

- Where variations do not resemble any contract bills item in terms of either specification or work context, a 'fair valuation' should be carried out. This may involve returning to the first principles of estimating whereby the rate for the item of work is considered in terms of material, labour and plant costs incorporating an addition for overheads and profit. This process may be avoided or reduced where the new item of work is relatively common and/or where there is available evidence of similar work, perhaps from a recent similar project.

Dayworks

Certain variations, which it may not be reasonable or possible to value on a measurement basis, may be charged on a prime cost basis. Daywork sheets, 'referred to as vouchers in JCT 2005, clause 5.7, will be rendered for these by the contractor; they set out the hours of labour of each named operative, a list of materials and details of plant used. It is the duty of the architect/contract administrator to verify these, not later than the end of the week following that in which the work was carried out. If there is a clerk of works, it will usually be one of his/her duties to verify the dayworks voucher on behalf of the architect/contract administrator. The clerk of works' signature is not in any way authority for a variation, nor does it signify that the item is to be valued on a daywork basis instead of by measurement. When there is no clerk of works the architect/contract administrator will be expected to sign the sheets. Neither the architect/contract administrator nor employer's surveyor, without being continuously on site, which is unlikely, can correctly guarantee that the time and material are correct, but, if these appear unreasonable for the work involved, they can make enquiry to satisfy themselves.

JCT 2005, clause 5.7 provides for pricing daywork as a percentage addition on the prime cost. The definition of prime cost is laid down in *Definition of Prime Cost of Daywork Carried Out Under a Building Contract* (RICS/Construction Confederation), and the percentages required by the contractor are to be inserted in a space to be provided in the contract documents. These tendered percentage allowances will be used in the calculation of any dayworks arising during the project. (See Fig. 14.2.)

For work within the province of some specialist trades, there may be different definitions of prime cost agreed, and these must be taken into account in the preparation of subcontracts.

The provision made for daywork should be taken into account when considering the amount of any provisional sum for contingencies.

Generally, the attitude of the client's and the contractor's surveyors towards dayworks is very different. As far as the client is concerned, payment for work on the basis of cost reimbursement, which is what dayworks is, brings several

ABC CONSTRUCT LIMITED	
JOB No. 2436	DAYWORK No. 120
CONTRACT: WILLOW CENTRE	AUTHORISATION
W/E: DATE	AI No.
	CVI No. 102
	OTHER

DESCRIPTION OF WORK

Break out floor adjacent manhole. Insert 2 no. 100D 30 degree bends to form swan-neck. Install 100D vertical pipe within manhole with r/eye and rest bend. Adapt manhole benching, insert branch bend and Make good benching, brickwork and floor slab.

LABOUR

Name	Trade	M	T	W	T	F	S	S	Total	Rate	Add %	£
C Rose	B/layer			5					5	8.53	70	72.50
											Total Labour £	72.50

MATERIALS / PLANT / SPECIAL CHARGES

	No	Unit	Rate	Add %	£
Materials					
100D uPVC drain	1.5	m	5.35	10	8.83
...30 degree bends	2	No	10.86	10	23.89
...rest bend	1	No	12.18	10	13.40
...r/eye	1	No	15.34	10	16.87
...coupler	2	No	2.97	10	6.53
Site mixed concrete	1	Item	5.00	10	5.50
Cement and sand	1	Item	2.50	10	2.75
Plant					
Kango	5	Hrs	1.10	10	6.05
Angle grinder	5	Hrs	0.55	10	3.03
Task lighting	5	Hrs	0.20	10	1.10
Transformer and leads	5	Hrs	0.20	10	1.10
			Total materials etc £		89.05
			Total for sheet £		**161.55**

Signed by ABC CONSTRUCT LIMITED	Signed by Client's Representative
Date	Date

Fig. 14.2 Daywork sheet.

disadvantages. With dayworks, the contractor will be reimbursed the full cost of all labour, materials and plant used, plus a percentage addition for overheads and profit. Therefore, the contractor has no incentive to complete the works in an efficient manner; in fact, as the cost of the variation work increases, so does the contractor's overheads and profit. For this reason, from the client's perspective, the use of dayworks will be resisted. However, they are likely to be to the advantage of the contractor. As with cost reimbursement contracts, dayworks can be expected to result in higher costs than with the methods of measurement and valuing previously outlined.

In practice, on most projects, there are likely to be situations for which dayworks are the only fair method of valuation, i.e. where work cannot be properly measured; for example, work to remedy a design error, which has been discovered post-construction, as shown in Fig. 14.2.

With reference to the above example of a completed dayworks voucher (Fig. 14.2), please note that the signatures affirm that the work recorded thus has been carried out, not that the work shall be valued on a dayworks basis, nor if priced, that the rates are correct in accordance with the contract. When calculating the amounts for inclusion into the final account, these should include the tendered allowance for overheads and profit, which may be a different amount for each of the three categories, materials, labour and plant.

Overtime working

Though there is nothing to prevent a contractor's employee from working overtime, subject to trade union control, this is normally entirely a matter for the contractor's organisation. No extra cost of overtime can be charged without a specific order. Where, therefore, overtime is charged on a daywork sheet, it will be corrected to the standard time rates unless there is some such special order.

It may be that owing to the urgency of the job, a general order is given for overtime to be worked, the extra cost to be charged as an extra to the contract. Or the order may be a limited one with the object of expediting some particular piece of work. When an operative paid, say, £10.00 per hour works an hour per day extra at time-and-a-quarter rate, i.e. £12.50, the quarter hour (£2.50) will be chargeable in such cases. As a matter of convenience, on the pay-sheet, if the normal day is 8 hours @ £10.00 and 1 hour @ £12.50, it will be $9^1/_4$ hours @ £10.00. The quarter hour is not 'working time' at all, and is therefore sometimes called 'non-productive overtime': the extra cost of payment for overtime work over normal payment. Any charges that are to be based on working time, such as daywork (where overtime is not chargeable as extra), must exclude the quarter hour. Where the extra cost of overtime is chargeable, the data will be collected from the contractor's pay sheets and verified if necessary from the individual operative's time sheets.

Provisional measurements

It often happens that such work as cutting away and making good after engineers is covered by provisional quantities of items such as holes through walls and floors, or making good of plaster and floor finishings. The original bill may have been taken from a schedule supplied by the engineers, and the need for remeasurement on site must not be overlooked. These items should be distinguished in the contract bills by the label of 'provisional' where applicable to single items, or where an entire section is provisional, for example, foundations, the entire section would be marked as 'Substructures; ALL PROVISIONAL'. Such labelling would denote the need for the remeasurement of those items affected. It does sometimes happen that the provisional quantities reasonably represent the work carried out and can therefore be left without adjustment, but this should not be done merely to avoid what is certainly a rather laborious job. Non-technical auditors are apt to frown upon such procedure. One of the few things they can do to check a technical account is to go through the original bill and see that all provisional items have been dealt with. An appendix to the variation account, showing how this has been done, can be of help to an auditor.

The main reason for the incorporation of provisional quantities within the contract bills is:

- To make a reasonable provision for work which will be required but for which exact detail is unavailable
- To obtain contract rates for provisionally measured items, which can be used in subsequent remeasurement.

The amount of work thus included should be a reasonable representation of the work involved and not a wild guess.

The valuation of remeasured approximate quantities may be based on the associated bill rates. However, where the approximate quantity allowance in the bills is not a reasonably accurate forecast, adjustment will be necessary to accommodate the cost differential caused by the revised quantity. (See example in Fig. 14.3.)

It should be noted that 'provisional quantities' are not to be confused with 'provisional sums' which are discussed below. In terms of the contract, which in this matter follows the General Rules of the Standard Method of Measurement (SMM7), provisional quantities shall be used:

'Where work can be described and given in items in accordance with these rules but the quantity of work required cannot be accurately determined, an estimate of the quantity shall be given and identified as an approximate quantity [i.e. being marked 'Provisional' as discussed above].'

Provisional sums

Provisional sums are provided at tender stage for items of work for which there is little information or for work to be executed by a statutory body, e.g. electricity supply. Their inclusion provides a tender allowance for known work that cannot be properly measured or valued until later in the project. In terms of the contract, as with provisional quantities discussed above following SMM7, provisional sums

Scenario: The refurbishment of 65 houses for a local authority
Treatment of existing plasterwork:
The existing plasterwork is of variable condition throughout the properties and therefore it is not cost effective to remove and replace all plaster throughout the scheme, nor is it possible to prepare a detailed schedule of the location and extent of plaster to be hacked off and replaced. Not only is the latter unforeseeable at design stage, it is practically impossible due to access to individual properties, the effects of damage arising from other refurbishment works (e.g. chasing for new wiring) and concealment due to wall coverings, furniture, etc.

Therefore, a sensible approach in pre-contract measurement is to approximate the amount of replastering per house (marking such quantities as '(Provisional)' and to later remeasure the exact amount of work carried out. The approximation may be based on an inspection of say, two to three properties, or previous experience.

Extract from notional final account

BQ Ref		Quantity	Unit	Rate	£	
	House type: C: 23 Giles Street					
	Adjustment of provisional quantities					
	Omit					
18/3 B	Hack off/key/bond/2 ct plaster (Provisional)	28	m²	6.25	175	00
					175	00
	Add					
18/3 B	Hack off/key/bond/2 ct plaster	47	m²	6.25	293	75
					293	75

Fig. 14.3 Example of provisional quantities.

are used 'where work cannot be described . . . it shall be given as a Provisional Sum and identified as for either "defined" or "undefined" work as appropriate'. Examples of these are provided below.

Defined work

Defined work relates to work that is not completely designed but for which particular information is available, which with reference to SMM7 incorporates:

- The nature and construction of the work
- A statement of how and where the work is fixed to the building and what other work will be fixed thereto

- A quantity or quantities which indicate the scope and extent of the work
- Any specific limitations and the like identified in Section A35 [of SMM7].

For instance, for an ornamental steel canopy at the entrance to a hotel: the nature and construction are given in the description of the work; the type and extent of work are clearly stated; the location of the canopy is given and the method of fixing shall be by 'bolting to steel columns'.

Undefined work

Undefined work is identified as necessary at the time of tender; however, in addition to the lack of full design information upon which to measure, the particular details outlined above for defined work are unavailable. For instance, it may be known that landscaping work for the hotel project is required, however no further detail is available. A provisional sum calculated on the basis of a previous project is included in the contract bills.

The important distinction to be made between defined and undefined provisional sums relates to programming, planning and pricing of related preliminaries. In the case of the steel entrance canopy, the contractor is required to allow for such in the tender submission. Provided no changes to these details occur, no adjustment will be made to the preliminaries, nor will the contractor be able to make reference to programme or planning matters in connection therewith. However, with regard to 'undefined work' due to the 'inferior' information available at tender stage, no such allowance is made. When the design of the related work is finalised and priced, the contractor will be entitled to include costs relating to preliminaries items and will also give consideration to planning and programming matters.

Fluctuations

Fluctuations are an allowance for building cost inflation that may or may not be reimbursed to the contractor, subject to the provisions of the contract. The amount of this cost factor depends on levels of inflation existing during the contract period. At times, this may be negligible; however at certain times, for instance in the 1970s, building cost inflation may be at a very high level.

It is important to note that building cost inflation should be considered apart from tender price levels; it is quite possible for both to be moving in opposite directions at the same time. For example, as indicated by Fig. 14.4, between 1990(i) and 1991(i), tender price levels fell by 15%, whilst building cost inflation increased by approximately 8%.

When such circumstances prevail, it is quite possible for an unfortunate client to suffer a relatively high level of poor value due to an untimely commencement date and, with the benefit of hindsight, misjudged contract option. Considering the tender and building cost indices above, a client that opted for a fluctuating contract commencing in 1990(i) would be considerably worse off than had the project been delayed to 1991(i) and let on a firm price basis.

Date	Tender index	General BCI
1990 (i)	137	128
1991 (i)	116	138
% change	$(137-116)/137 * 100 = -15.33\%$	$(138-128)/128 * 100 = +7.81\%$

Fig. 14.4   BCIS indices 1990 (i) and 1991 (i).

Fluctuations, also referred to incorrectly as 'increased costs' and 'escalations' (since both terms suggest a single direction of price movement), can be calculated by either a traditional method or by a formula method. JCT 2005, clause 4.21 outlines the choice of method to be used by reference to schedule 7, options A, B and C. Where neither option B (labour and materials cost and tax fluctuations) or option C (formula adjustment) are selected and stated in the contract particulars, option A (contribution levy and fluctuations) applies.

Option A – firm price contracts

In situations where the client wishes to distribute the risk of building cost inflation to the contractor, option A is used. The choice of this contract option will require consideration of several factors including contract duration, market conditions and levels of present and predicted building cost inflation. Although such a contract choice will transfer the main burden of inflation risk to the contractor, option A does provide for the reimbursement of some inflation costs to the contractor where caused by statutory matters such as changes relating to contributions and taxes on labour. These are identified in JCT 2005. In addition, clause A12 of schedule 7 provides for a percentage to be inserted in the contract particulars by the contractor, to be applied to all fluctuations to allow for cost inflation relating to some preliminaries items, overheads and profit.

Option B – the traditional method

Labour

The traditional way of adjusting fluctuations in the cost of labour and materials, although today rarely encountered in practice, is by way of a price adjustment clause (JCT 2005, option B). Under such a clause any fluctuation in the officially agreed rates of wages, or variation in the market price of materials, is adjusted. JCT 2005, clause B13 of schedule 7, provides for a percentage to be inserted in the contract particulars by the contractor, to be applied to all fluctuations to allow for cost inflation relating to some preliminaries items, overheads and profit.

The checking of wages adjustment should be fairly straightforward on an examination of the contractor's pay-sheets. The rates of wages are officially published by the construction industry joint council or other wage-fixing body so there should be no doubt as to the proper amount of increases or decreases, or the dates on which they came into effect. To these increases will be added allowances for increases arising from any incentive scheme and any productivity agreement and for holiday payments as set out in the contract. These increases will apply to workpeople both on and off-site and to persons employed on-site other than workpeople.

Care must be taken that there is no overlapping with the rates charged for daywork when dealing with price variations. If daywork has been priced at actual rates (as required by JCT 2005), say with labour 10p per hour above basic rates, the number of hours so charged in daywork must be deducted from the total on which price adjustment is being made. In this way the contractor gets, for the hours charged in daywork, a percentage on the difference in cost, whereas adjustments under the price variation clause are strictly net differences. If, of course, the contract provides for daywork to be valued at basic rates, the point does not arise.

Materials

The adjustment of materials prices is more difficult. The contractor will produce invoices for those materials from which the quantities and costs can be abstracted and the value will be set against the value of corresponding quantities at the basic prices. Prices must be strictly comparable. If the basic rate for eaves gutters is for 2 m lengths, an invoice for 1 m lengths cannot be set against that rate. The 1 m length rate corresponding to the 2 m length basic rate must be ascertained. There is also the difficulty of materials bought in small quantities, perhaps by the foreman from the local ironmonger, when again the price paid is not comparable with the basic rate. JCT 2005, option B, says 'if the market price . . . increases or decreases' and these are significant words. When in doubt the applicability of the contract wording must be considered.

As for labour, reference must be made to the rates charged in daywork for materials, and adjustment made, if necessary, on the totals being dealt with for price adjustment. Invoices should be requested for *all* materials appearing on the basic list. Claims for materials not appearing on the basic list are to be excluded from the calculation. The surveyor is responsible for seeing that fluctuations in either direction are adjusted. This is another case where non-technical auditors would be apt to worry if all items did not appear in the account.

The surveyor should also see that the quantities of the main materials on which price adjustment is made bear a reasonable relation to the corresponding items in the bill of quantities and variation account. An approximation, for instance, can be made of the amount of cement required for the concrete and brickwork, and any serious discrepancy should be investigated.

Option C – formula adjustment

The more common method of price adjustment in building contracts is by way of formulae to calculate the adjustment.

Unlike the traditional method, in which a calculation of actual amounts of increases and decreases are made, the formula method uses indices to calculate the amount of reimbursement. As such, on a particular project, this calculation will therefore be technically inaccurate since it is a generalised approach for use across the construction sector. However, the use of the formula method is generally more popular than the traditional method for several reasons:

- The results are likely to be more predictable, particularly in unstable economic conditions
- The protection against price fluctuations is more comprehensive. With the traditional method there is likely to be a considerable shortfall in overall recovery on the contract.
- The formula method greatly simplifies the administration of price fluctuation provisions
- The simplified process facilitates prompt payment of fluctuations in interim valuations
- The scope for dispute is greatly reduced
- Contractors can quote competitively on current prices with the confidence that reimbursement will be in terms of current prices throughout the contract
- The formula method can benefit from the application and use of information technology.

The JCT Formula Rules 2005 explain the formulae and provide information and assistance to those using them. The formulae are of two kinds: the building formula and specialist engineering installations formulae.

The building formula uses standard composite indices (each covering labour materials and plant) for similar or associated items of work, which have been grouped into work categories. For example, the breakdown of concrete work categories reflects the differing material input and weighting of the index calculations. Despite the general heading of 'concrete' each of the items is subject to different cost influences – e.g. labour balance, cement, aggregates, fuel, timber, steel – and therefore justifies a different category. Rationalisation within the process can be seen by consideration of the work group for 'concrete: in situ'. The weighting of the two main material components, cement and aggregate, will differ for $1:2:4$ concrete and $1:3:6$ concrete; however no differentiation is made in the calculation of fluctuations by the formula method. In reality, an increase in the cost of cement will have a greater impact on $1:2:4$ than $1:3:6$ concrete; however this detail is sacrificed for ease of application (i.e. to limit the work categories to a manageable number).

The formula is applied to each valuation, which will need to be separated into the appropriate work categories. In practice this is done at the commencement of the contract by annotating the contract bills. As stated previously, the use of IT can assist in this process.

There are alternative applications of the formula available. Each of the work categories indices may be applied separately (Section 2; Part I of the rules). This provides the most sensitive possible application of the formula. Alternatively, the work categories may be grouped together to form work groups (Section 2; Part II of the rules). Clearly, as explained above, the fewer work groups used, the less sensitive will be the indices to changes. It must also be practicable to analyse the tender and the value of work carried out in each valuation period into the selected work groups. This entails less work in separating the value of work carried out in every valuation.

These alternative uses are described in detail in the Formula Rules 2005 mentioned above, which also give notes on the application of the formula at pre-contract, interim valuation and final account stages, with sample forms and worked examples.

The specialist engineering installations formulae (Section 2; Part III of the rules) cover electrical installations, heating, ventilating and air-conditioning installations, lift installations, structural steelwork installations, and catering equipment installations. These formulae use separate standard indices for labour and for materials, the respective weightings of which are to be given in the tender documents, except for lift installations where the weightings are standardised. In each case the formula is expressed in algebraic terms and has been devised in conjunction with the appropriate trade association. It is intended that these specialist formulae will normally be applied to valuations at monthly intervals.

Completing the account

In terms of the contract, time limits for all parties to adhere to are provided:

- Clause 4.5.1: 'not later than 6 months after the issue by the architect/contract administrator of the practical completion certificate the Contractor shall provide the . . . Quantity Surveyor with all documents necessary for the purposes of the adjustment of the contract sum'
- Clause 30.6.1.2: 'not later than 3 months after receipt . . . by the Quantity Surveyor' of the information from the contractor, the client's surveyor should prepare the final account and submit it to the contractor, via the architect/contract administrator.

These timescales will be affected by the completeness of the information provided by the contractor and the communication process between the parties.

As stated, whilst it is the client's surveyor's role to prepare the final account, the information required to do so will be provided by the contractor. Normally, subject to the considerations relating to variations discussed above, this will effectively include the preparation of the variation account.

In practice, the agreement of the final account will be helped by a good and sensible working relationship between the surveyors involved. The contractor, having examined the account, is fairly certain to have some criticism. Unless the criticisms are of a minor nature, which can be settled by correspondence, an appointment will be arranged for the contractor's surveyor to call at the client's surveyor's office and go through the points. The new contract provisions relating to variations where schedule 2 quotations are fully applied should limit the extent of any disagreement at final account stage.

Audit

It is frequently assumed that the main role or function of an audit, whether it is of a company balance sheet, profit and loss account or the final account of a construction project, is to detect errors or more importantly fraud. This is an incorrect assumption and forms only subsidiary objectives. An audit of a final account, or any account, involves the examination of the account and the supporting documentation and more importantly the designated procedures involved. This enables an auditor to report that the account has been prepared to provide a true and fair view of the account. The auditor will:

- Compare the final account with contract bills
- Examine the records available
- Discuss aspects with relevant staff
- Examine the procedures used
- Prepare a report on the findings.

The auditor will use the skill and diligence that is normally expected and the process will involve an examination of all transactions involved. A technical audit, perhaps using the skills and knowledge of a quantity surveyor, may sometimes precede the more usual audit process.

The extent of the examination of the final account will depend on the auditor's experience and assessment of the internal controls that have been adopted. The audit will set objectives and incorporate what is known as the audit plan. The process is not intended to repeat that which has already been carried out by the quantity surveyor. It is chiefly concerned:

- With the processes used
- That the processes have not been departed from without good reason
- That the accounts are free from error
- That the final account is a fair and true record
- That sound accounting principles have been adopted and used throughout.

The following are the essential features of the audit plan:

- Critical examination and review of the system used for the preparation of the final account and the methods of internal control adopted
- Critical examination of the final account in order that a report can be made to the client as to whether the accounts are a true and fair record
- Ensure that the accounts have been prepared on sound accounting principles and in accordance with professionally accepted procedures.

The auditor will give particular attention to those types of transactions that could offer particular facilities for fraud. The amount of checking that will need to take place will depend on the quality and reliability of the system that is used. A good system usually relies on the collusion of two or more individuals and this provides some safeguard on which the auditor can base judgements.

The majority of errors that are discovered will be due to miscalculation, carelessness or ignorance. However, on occasions, what may appear to be nothing more than a simple clerical error, may ultimately be found to be fraudulent manipulation. Errors may be the result of careless arithmetic, although the use of information technology has now greatly reduced these occurring in practice. An auditor can never guarantee the discovery of all fraud, since ingenious schemes to avoid possible detection may have been introduced. There are many examples of these in the business world and they are frequently reported in the technical press (Smith 1992). The auditor must, however, be able to show that reasonable skill and care have been exercised in designing the audit trail. Areas that appear dubious will be checked and spot checks made generally where it is considered necessary.

The culmination of the auditor's work results in a written report being sent to the client indicating the opinion regarding the reliability of the outcomes and processes used. It will also offer suggestions of how these might be improved. The auditor's report will comment on the following.

The tender

The auditor should make sure that any relevant standing orders have been complied with and that the tenders were received and opened in the prescribed manner. Where the lowest price has not been accepted, the auditor will look for careful documentation as to why this was not so. The auditor will need to be satisfied that appropriate arithmetical and technical checks were performed on the contractor's price and that errors were corrected within the agreed rules. Good tendering practices should be in evidence, with some account of modern methods of contractor selection being considered and justified.

Interim payments

It is usual for clients to maintain a contracts payment register to record all of the payments made in respect of a project. It will be closed on the agreement and payment of the final account. The auditor will need to scrutinise the payments made to ensure that they are in accordance with the contract and the certificates from the architect/contract administrator.

Variations

The auditor will need to establish that the correct protocols were applied when issuing and authorising variations as described in the contract documents. The various conditions of contract lay down precise rules of instruction and valuation. These should have been followed unless there are good and documented reasons to the contrary.

The final account

The settlement of the final account is a lengthy and complex process and frequently extends beyond the period stated in the contract. The auditor's final task is to ensure that the amount of the final payment added to the sums that have already been paid equals the final account for the project. In order to reduce any possible delays in making the payment to the contractor, the auditor should examine the account as speedily as possible. Contractual claims may remain to be agreed for some time and the auditor may need to include provision for the auditing of these at some later date.

Where the client has been partially responsible for either the purchase of construction materials or the execution of a specialist portion of the works, the auditor will need to be satisfied that these have not been included within the final account.

Liquidated damages

Liquidated damages become due where the contractor fails to complete the works on time, unless there are agreed reasons that are beyond the control of the contractor. It is a rare occasion when an auditor will need to verify such an amount since extensions of time are frequently granted to cover for such a delay. However, the auditor should be satisfied that such a waiver was only granted for adequate and carefully documented reasons.

Fees

An auditor will also be involved in verifying the fees paid to the consultants. Commonly these will be by agreement between the client and the consultants, frequently following a fee bidding process. They are usually influenced by the tender sum and the additional work involved during the progress of the works.

Timing and resources

It is not intended that final account preparation should commence only upon the completion of the construction of a project. Whilst this may be the case in some situations, it is neither correct nor good practice. If we assume a project duration of 18 months, there will be ample opportunity for final account measurement during the course of the works. For example, at completion of the substructures, say after month 2/3 of the programme, it will be in the interest of all parties to remeasure and if possible finalise the account for this element of the works. At this time, events are fresh in the mind and uncertainties relating to payment and final project cost can be reduced by prompt agreement. Continuation of such action throughout the course of the works will reduce the time required for completion of the final account at the end of the contract period. Again, this is advantageous to all parties. In the case of the client's internal financial arrangements it may be very important to achieve an early completion. A further factor that should be borne in mind relates to the turnover of surveying personnel. In the event that final account measurement is left until the end of the project, it is quite possible that either (or both) the contractor or project consultant no longer employs the surveyors responsible for the earlier stages of a project. In such situations, measurement by surveyors without the benefit of direct involvement during the works will probably result. This is much more difficult, time consuming and therefore inefficient. As stated previously, where the schedule 2 quotation process is applied, this will be of great assistance.

In recognition of the factors outlined above, it may be beneficial to maintain a 'running' final account for which measurements and agreements occur throughout the project. Unfortunately, this philosophy may go awry, for example, due to lack of available resources or where disputes arise. This is regrettable and will almost certainly lead to delays in the completion of final accounts, some of which are legendary in their duration, despite the obligations stated in the contract. Clients are entitled to be critical of our practices in these situations; it is often stated that construction is the only sector of business where the client is not aware of the final cost until several months – or years! – after product purchase.

Despite the view above that it is expedient to prepare aspects of the final account throughout the progress of the works, it is important to add a note of caution to this advice. It is not advisable to start too soon on measurement of variations when future developments, which cannot be foreseen, might affect the surveyor's work. One might, for instance, measure a number of adjustments of foundations or drains, only to find later that the whole of one of these sections must be remeasured complete.

Discussion topic

Consider and discuss some of the challenges that can face the QS practitioner in managing costs relating to variations and outline the approaches to valuation accommodated within the contract provisions. Highlight, where relevant, the changes that have occurred with the issue of JCT 2005.

Challenges facing the quantity surveyor in the cost management of variations

One of the biggest challenges facing the quantity surveyor is the effective management of costs relating to variations. This is applicable to both the contractor's surveyor and the employer's surveyor and relates to: the process of interim valuations and the submission of regular cost reports, and the completion of the final account.

With regard to action at interim stage, frequently there is considerable pressure to prepare an accurate cost estimate soon after issue of an architect's instruction and provide prompt advice to the client. Likewise, there is likely to be similar pressure to calculate an amount for inclusion in the interim payment to the contractor which will also be based on an estimate of the final cost of the variation. Although eventually these two estimates will be one and the same, at interim stage, the consultant quantity surveyor will be inclined to:

- Report a figure to the client that will reflect a worst case scenario resulting in a higher cost estimate than eventuality
- Show caution in assessing the basis for payment to the contractor resulting in a lower cost estimate than eventuality.

It could perhaps be argued that this inconsistency is unprofessional, however, it would be a natural response in order to: reduce the chances of future criticisms and possibly claims from the client as a consequence of an underestimate; reduce the risks relating to overpayment to the contractor.

Conversely, to the contractor's surveyor, these standpoints are likely to be reversed: amounts claimed at interim valuation stage will seek to enhance cash flow whilst in reporting final agreed variation costs to the management team, conservatism would seem a wise approach. These two positions reflect the differing perspectives of both quantity surveyors' viewpoints and indicate an understandable risk-averse attitude.

With regard to the agreement of the final project costs, although the contract does set out timescales for the completion of the final account, and these are discussed within, in the past, the length of time to complete such accounts has on occasion been legendary. The reasons for this are several, including: the volume and complexity of some variations are such that final agreement isn't possible at point of instruction issue; shortage of quantity surveying resources; changes in personnel with the loss of staff involved during the course of a project prior to agreement of the final account. The consequences of this delay can be problematic in terms of closing the account and understanding total financial liability under the contract. It also reflects badly upon the quantity surveying profession.

Not only can the evaluation of some variations be a technically demanding aspect of the quantity surveyor's role, but it can also be very time consuming, expensive and risky. The case study below, whilst exceptional, is by no means a rare occurrence and is intended to give an impression of the type of position the quantity surveying practitioner may face.

Case study

Project: A five storey refurbishment project for a residential developer with a contract sum of £15 000 000 with the contractor appointed using the JCT 2005 form of contract.

In month 2 of the project, the architect, responding to directions from the client, issues an instruction significantly amending the design of the basement level to introduce a high quality gymnasium and swimming pool. Six months earlier the consultant quantity surveyor had given an estimate of £1 000 000 for this work, based upon outline sketch plans and elevations and the input of potential specialist subcontractors, however, at that time it was decided not to proceed. Subsequently, market research has shown that the inclusion of high class leisure

(continued)

facilities in the residential complex will significantly increase saleability and that selling prices will be well above any costs incurred, resulting in an increase in the developer's profit.

The architect's instruction has an immediate consequence upon the contractor: certain constructed foundation works require alteration due to the revised location of ground beams and delays occur due to incomplete structural design. This is an urgent problem to the quantity surveying team on both sides and will require the prompt injection of additional resources to resolve.

In terms of preparing an interim valuation: there are clearly additional works to consider (involving specialist subcontractors and suppliers); almost certainly a request for an extension of time and a contractual claim for loss and expense; dayworks in connection with changes to work already constructed. In addition, the consultant quantity surveyor needs to report with accurate cost advice to the client, and likewise, the contractor's quantity surveyor to his/her management team. And, of course, not only are there cost management issues to consider, but the procurement of specialist subcontractors and possibly designers – in very tight timescales.

Unfortunately, situations of this type are often impossible to manage, within the timescales allowed, without some compromise in the short term and this can carry additional professional risk. Whatever actions are taken, irrespective of the skill and experience of the practitioners involved, this additional uncertainty should be made clear to the client. With regard to the example, hopefully the quantity surveyor would have been invited to provide new cost advice prior to issue of the instruction showing that the £1 000 000 was no longer applicable. If this opportunity hadn't been provided, it would be important to immediately advise the client of the significantly changed circumstances.

Contract provision

Although the real world complexities outlined above and underlined by the difficulties arising in the case study example will demand some rationalisation in a short-term approach, the evaluation of variations should eventually adhere to the procedures set out in the Conditions of Contract.

As stated within the main text, the evaluation of variations will usually require two distinct operations: measurement and valuation. Practical guidelines on the methods and approaches to measurement are detailed within, as are methods of valuation, which are described in the Contract Conditions (clause 5.6 to 5.10, 'Valuation Rules' in JCT 2005).

The main change that has occurred with the issue of JCT 2005 is the removal of the short-lived provisions which *allowed* the contractor to submit a price statement. In JCT 98 there was recognition and formalisation of what had become generalised practice, whereby the contractor's surveyor evaluated the cost relating to variations and submitted a price statement to the client's surveyor for approval. Under clause 13.4.1.2 (JCT 98), 'Alternative A: Contractor's price statement', the contractor was allowed, upon receipt of an instruction, to submit a price for carrying out the work known as the price statement. The process was prescribed, in detail, in the contract. The provision merely reflected the contractor's primary role in the process and didn't interfere with the contractual standing of the architect's instruction. Although the contract specified some situations where the 'Alternative A' arrangement should not apply, it was generally intended to be the manner in which variations were to be priced. Under the same clause, 'Alternative B' provided a default methodology to be used where Alternative A was unsuccessful. With 'Alternative B' the contractor played a more passive role whilst the client's surveyor carried out the valuation works. This method of valuation was in the manner which was previously accepted procedure but not generally practised.

In addition to these two alternatives, JCT 98 also provided for the '13A quotation' process whereby, upon request from the architect, the contractor was *required* to submit a price, including an assessment relating to extension of time and loss and expense. JCT 2005 has retained this provision now described as the 'schedule 2 quotation' under the terms described in clause 5.3 and the separate schedule 2 conditions.

In effect, the difference between the two previous approaches to valuation ('price statement' and '13A quotation' (now the 'schedule 2 quotation') is clear. With the price statement approach, the contractor could elect to submit a price, without a request being made by the architect, whereas with the 13A quotation (schedule 2 quotation JCT 2005) the contractor was/is obliged to submit a price and if requested, shouldn't carry out the variation until agreement of the price is reached. If the 'schedule 2 quotation' approach isn't used, the client's quantity surveyor will now follow the traditional approach (previously established in JCT 98 as 'alternative B'). Please see within for a full description of the valuation rules appertaining to this method.

The pre-costing of variations facilitated by application of the 'schedule 2 quotation' methodology may save a great deal of time and provide cost certainty with regard to the variation affected. However, there are drawbacks. With reference to the case study above, if prior agreement had been required, the contractor would have needed to accept substantial extra risk and this would have been translated into additional costs to the client. Other disadvantages are described within.

References and bibliography

Hackett M., Robinson I. and Statham G. *The Aqua Group Guide to Procurement, Tendering, and Contract Administration.* Blackwell Publishing. 2006.

Joint Contracts Tribunal Limited. *Standard Form of Building Contract 2005 Edition.* Sweet and Maxwell. 2005.

Joint Contracts Tribunal Formula rules 2005; Sweet and Maxwell Ltd. 2006.

McGuinness J. *The Law and Management of Building Subcontracts.* Blackwell Publishing. 2007.

Ndekugri I. and Rycroft M. *The JCT 98 Building Contract, Law and Administration.* Butterworth Heinemann. 2000.

Smith T. *Accounting for Growth.* Business Books. 1992.

15 Insolvency

Introduction

Insolvency is a generic term that covers both individuals and companies. Bankruptcy applies specifically to individuals and liquidation to companies. The law relating to bankruptcy is governed by the Bankruptcy Act 1914 and that of insolvency is contained in the Insolvency Act 1986 and relevant sections of the Companies Act 1985. Unlike death or illness, the effect of these do not by themselves have the effect of terminating a contract.

The construction industry is responsible for more than its fair share of bankruptcies and liquidations. Limited liability companies become insolvent and then go into liquidation. The annual report form the Department of Trade and Industry (DTI) is often greeted with a mixture of sensationalism and 'as expected' as far as the construction industry is concerned. Builders and contractors always head the list.

A company becomes insolvent when the value of everything that it owns comes to less than the value of its debts. An example of a statement of a contractor's affairs, following liquidation, is illustrated in Fig. 15.1. A company may enter voluntary liquidation or, more commonly, have been forced in this direction by a single creditor. The single creditor is usually a financier, such as a bank, who refuses to extend a loan or chooses to call in a debt. Even a profitable company may suffer in this way because of a shortage of cash to pay its bills. Hence, it suffers a cash-flow crisis.

In times of a recession or slump, the rate of the collapse of companies in the UK exceeds over 1000 per week (1990s recession), many of which are in the construction industry. For companies to weather the business cycle it is essential that they are careful with whom they do business, employ sound financial disciplines and verify and monitor companies. In times of recession, new as well as long-established companies cease to trade and more companies begin to pay their bills late. This causes other firms to fail. A way out of the vicious circle has been to introduce legislation to improve payment performance. During the recession of the early 1990s, insolvencies in the construction industry more than doubled from an already high base. But it is not just in times of recession that insolvency occurs. It is a common feature throughout the business cycle.

The kind of firm that goes into bankruptcy or liquidation ranges from the one-person business, subcontractors, suppliers, general contractors through to, on occasion, the national contractor. Employers and consultants are also not immune from

Harry Parker (Builders) Ltd
Statement of Affairs 08-03-07

	Book value		Estimated to realise
Assets			
Land and buildings		330 000	
Less Norwich Union Group	48 000		
Less National Westminster Bank	223 860	− 271 860	58 140
Goodwill	5 700		0
Plant and equipment – owned	195 000		45 000
Value subject to HP	40 180		
Less outstanding HP	37 550		2 630
Motor vehicles	36 500		27 250
Stock – Chair frame department	13 250		13 250
– Builder's yard	90 000		35 000
Work in progress	372 000		185 000
Debtors Chair frame dept.	16 500		16 500
Trade contract	274 650		190 000
VAT	3 000		3 000
Cash ABU plc	13 500		
Rogers Plc	4 000		17 500
Available assets			593 270
Liabilities			
Preferential creditors			
PAYE	55 200		
NHI	47 750		
Holiday pay	19 250		122 200
Amount available to unsecured creditors			471 070
Unsecured creditors			
Redundancy pay	23 800		
Trade and expenses	685 550		
Sale contractors	395 500		1 104 850
Deficiency regarding creditors			− 633 780
Share capital			
20 000 Ordinary shares of £1.00 each fully paid			20 000
Total deficiency			− 653 780

Fig. 15.1 Statement of affairs.

these statistics. When this happens a knock-on effect is created and other firms tee-tering on the edge often go out of business as a result. The reasons for construction company failures include the following:

- A recognition that the construction industry is a risky business and often the risks involved are not fully evaluated
- Construction projects are used as economic regulators, with price fluctuations depending considerably on the state of the market

- Competitive tendering has been cited to explain the high incidence of contractor and subcontractor failure, although this has not been verified
- The quality of management and financial expertise, which has traditionally relied on the self-made man image without proper training, is frequently lower than in other industries
- Neither the best of British brains nor the aristocracy, not necessarily the same, have involved themselves in the construction industry
- Insolvency has a knock-on effect on other, often smaller, firms, who are more vulnerable.

Avoiding the possibility of contractor insolvency in the first place is the preferred route to be followed, although this can never be completely assured. The Code of Practice for the Selection of Main Contractors identifies the following precautions that should be taken prior to inviting a firm to submit a tender for a construction project:

- Check the firm's financial standing and record
- Identify the firm's general experience and reputation for the type of project being considered
- Evidence recent experience of building over similar contract periods
- Examine the adequacy of management capability
- Establish whether the firm has the capacity to carry out the proposed project.

The general reasons for any company failures are shown in Fig. 15.2, indicating the owner's viewpoint and those of the creditors. The perception of these two different groups regarding the financial failure of a company is markedly different in many respects. This may be a recognition that firms do not have clear perceptions about themselves.

The role of the quantity surveyor

It is important, for all concerned, that quantity surveyors carry out their work to the highest professional standards and integrity. In this context they should only work for bona fide employers. They should ensure that procurement practices are fair and reasonable and that risks are allocated to those parties who are best able to control them. Only financially stable construction firms should be recommended and where necessary performance bonds used. Interim payments should always be in accordance with the contract provisions, remembering also that these represent the lifeblood of the industry. Quantity surveyors should, as far as possible, seek to protect employers from undue loss and expense resulting from additional payments and extensions of time. One of the important roles of the quantity surveyor is financial propriety between all parties involved in the construction project.

Prior to a contractor going into liquidation, there are often signs that the firm is in some sort of financial difficulties. These may include:

	Owner's opinion %	Creditor's opinion %
ıs depression	68	29
ıncı... ،ent management	28	59
Insufficient capital	48	33
Domestic or personal factors	35	28
Bad debt losses	30	18
Competition	38	9
Decline in assets values	32	6
Dishonesty and fraud	0	34
Excessive overhead expense	24	9

Fig. 15.2 Reasons for company failures.

- Complaints from subcontractors and suppliers about the non-payment of their accounts
- Pressure to maximise interim payments
- A reduction in the progress of the works due to a reduced labour force or a lack of supply of material deliveries
- Changes in members of staff
- Rumours, which may be unfounded, and which may be especially damaging to a firm which has known difficulties.

It may be necessary to make photographic records of the works at the time of the insolvency.

Scenario

Liquidation is rarely a total surprise. For example, a contractor had cash-flow problems in paying subcontractors, especially a national supplier of ready-mixed concrete. Cash would eventually be received from profitable contracts currently being completed by the contractor. But rumours began to emerge about the firm and especially its financial standing. The contractor, at the same time, was restructuring its operations and introduced a new policy, for office-based staff, that company cars would no longer be provided. This appeared to endorse the rumour and the contractor's shares began to slide. Rumours began to spread, both from the supplier and by staff in the firm and eventually by others in the industry. There was a belief that there was no smoke without fire! Contracts, where the firm had submitted the lowest tender, were surprisingly awarded to other firms. This fact then became public knowledge. Suppliers of materials and goods became cautious, exacerbating the cash-flow problem. The lack of new orders meant redundancies, increasing the contractor's problems and credibility. On fixed price contracts, it is arguable that more profit is made at the beginning of the project than towards the end. It is a belief that new contracts tend to subsidise those nearing completion. New contracts would therefore have helped the contractor's position. News eventually broke that the firm had suffered major losses on a project overseas. That was the final straw. The bank called in its loans, which the firm was unable to meet. The firm went into receivership on the basis of rumour and predicted loss. Without the rumour it may have survived. But who knows?

The role of the liquidator

The commonly held view about receivers and their advisors is that of an undertaker employed to perform the last rites for an ailing construction firm. However, the true role adopted by the receiver is to preserve and salvage any or all the parts of the firm. This is based on the fact that where insolvency occurs all those involved are losers; there are no winners. The types and values of assets which can be recovered vary considerably depending on whether the company is a contractor or developer, manufacturer or even a consultant. Contractors have become wise enough to separate the company's activities into groups such as plant hire, house building, general contracting, etc., so that if hard times occur then the whole business might not be lost. When insolvency becomes a possibility then speed becomes of the essence. The whole assets of the contractor must be maintained and not traded against favours elsewhere.

In the construction industry when receivers are appointed, a firm of specialist quantity surveyors arrives at the same time. For each contract there are usually three choices available:

- Completion of the project. This is desirable and the more advanced a project is towards completion, the better is the likelihood of this occurring
- Abandon it. This is the least desirable option, because there are usually assets still tied up in it
- Sell the project to a third party.

In each of the above cases, the quantity surveyor calculates the amounts of work done, the materials on site, what is owed to subcontractors, suppliers and other creditors, amounts outstanding in retention, etc. In about 95% of receiverships, projects are either completed or sold to other firms.

For example, a contractor could be 60% of the way through a £10 m project, but of the £6 m work completed, the contractor might only have received £5 m owing to work being completed since the last interim valuation and outstanding retention sums. Coupled with this there is a good possibility that future profits may exist in the remaining work that needs to be carried out. However, life is never down to simple arithmetic and other factors will need to be taken into account, such as the amounts owing to the different creditors, the lowering of site morale and motivation, and the fact that the project might have been financially front-end loaded.

The law

The machinery to wind up a company may be set in motion either by the debtor themselves or by the creditors. The courts make the receiving order and require the debtor to submit to the official receiver or trustee in bankruptcy a statement of affairs showing assets and liabilities (see Fig. 15.1). A creditor's meeting is usually held to decide whether to accept any suggested scheme or whether to declare the firm as insolvent.

The official receiver will eventually, on completion of the liquidation, turn all of the assets into cash and then divide this amongst the creditors, paying them at a rate of so many pence per pound that is owed. Certain creditors may have preference, e.g. those who hold preference shares and the Inland Revenue. Most creditors rank equally. Insolvency is broadly concerned with a firm or company's inability to pay its debts. Within the Insolvency Act 1986 various situations can arise:

- *Voluntary liquidation*   This can be the result of either the share-holders or creditors making a resolution which is then accepted by the company.
- *Compulsory liquidation*   This occurs because a company refuses to cease trading and is the result of a court order for a company to be wound up.

Creditors

Creditors may be secured, i.e. those who hold a mortgage, charge or lien upon the property of the debtor, or they may be unsecured. Unsecured creditors may be ordinary trade creditors, preferential creditors or deferred creditors. Preferential debts include the costs of insolvency proceedings. The proceeds of liquidation are distributed in accordance with the following hierarchy:

- Fixed charge holders
- Liquidators' fees and expenses
- Preferential creditors such as the Inland Revenue, Customs and Excise, National Insurance contributions, pensions and employees' pay
- Floating charge holders
- Unsecured creditors
- Shareholders.

Determination of contract (contractor insolvency)

All of the standard forms of construction contract incorporate provisions aimed at regulating the events of a contractor's insolvency. It usually results in one of two courses of action being taken. Either the contractor's employment is terminated and a new contract probably made with a new contractor, or alternative ways are examined to enable the same contractor to complete the works. On those occasions where the personal skill of the contractor is the essence of the contract, completion may only be acceptable under those terms. This situation is likely to be unusual in the construction industry and will usually only be applicable to specialist firms.

It should be recognised that insolvency does not by itself have the effect of terminating the contract, since the liquidator will normally have the power, after obtaining leave, to carry on the business of the debtor as may be necessary for the beneficial winding up of the company.

A liquidator has the power within twelve months of appointment to disclaim unprofitable contracts which are still uncompleted at the commencement of the insolvency. The liquidator may decide to disclaim, when perhaps most of the more financially fruitful work has been completed, or when some retention, such as for example, sectional completion, has been released.

Where the work is reasonably far advanced, a liquidator is not likely to disclaim it if this can be avoided. Where a relatively small amount of work needs to be completed up to completion, then this will enable outstanding retention monies that are due to be released. When a liquidator does disclaim a contract at this stage, the employer is likely to be in the most favourable position, since the retention sums can be used as a buffer when inviting other firms to tender for the outstanding work that needs to be completed.

Where the construction work is at a relatively early stage on a project when insolvency occurs, it is the employer who will want to ensure that the contractor completes the contract as originally envisaged. The employer, under these circumstances, is likely to have to pay higher prices to have the work completed by another firm, especially where the original contract was awarded through some form of competitive tendering. However, there is much less inducement on the part of the liquidator to want to carry on with such a project, especially if the rates and prices in the contract documents are economically unfavourable.

Where an employer is faced with poor performance from a contractor who is continuing with the project, there is still the right within the contract to determine the contractor's employment; for example, where the progress of the works is not being maintained.

Provision in the forms of contract

Clause 27 of the JCT 98 form of contract, clause 8 of JCT 2005 and clause 63 of the ICE Conditions of Contract make provisions for insolvency. Similar clauses are found in the other forms of contract. The contractor is to inform the employer if it is intended to make a composition or arrangement with creditors, or the contractor becomes insolvent. More usually where the contractor is a company, the employer must be informed if one of the following occurs:

- The contractor makes a proposal for a voluntary arrangement for a composition of debts or scheme of arrangement in accordance with the Companies Act 1989 and the Insolvency Act 1986
- A provisional liquidator is appointed
- A winding-up order is made

- A resolution is passed for voluntary winding-up, other than in the case of amalgamation or reconstruction
- Under the Insolvency Act 1986 an administrator or administrative receiver is appointed.

The forms of contract allow the employer in the above situations to:

- Ensure that the site materials, the site and the works are adequately protected, including offsetting any of the costs involved against monies due to the contractor
- Employ and pay others to carry out and complete the works
- Make good defects
- Use all temporary buildings, plant, tools, equipment and site materials
- Purchase materials and goods that are necessary to complete the works
- Have any benefits for the agreement for the supply of materials or goods assigned from the contractor
- Make payments to suppliers or subcontractors for work already executed and deduct these amounts from any monies due to the contractor
- Sell any temporary buildings, plant, tools and equipment (for the benefit of the contractor), if the contractor does not remove these when requested
- Withhold the release of any retention monies
- Calculate the costs involved of the contractor's insolvency and show these as a debt to the contractor.

Where the temporary buildings, plant, tools, equipment and site materials are not owned by the contractor, the consent of the owner will usually be required. The costs for these and the use of the contractor's facilities, described above, must be paid for by the employer. However, such costs will be offset against any payments that are made to complete the works. The legal title to the goods and materials must be clear (*Dawber Williamson Roofing* v. *Humberside County Council* (1979)).

Factors to consider at insolvency

Until the contractor's legal insolvency becomes a fact, the employer is in a rather frustrating position. The employer must continue to honour the contract, otherwise the contractor may pursue a claim for breach of contract. Any interference on the part of the employer may also exacerbate the unfortunate situation.

The following are some of the principal actions that an employer should take immediately following the insolvency of a contractor.

Secure the site

The construction site should be secured as quickly and as expeditiously as possible. This is to prevent unauthorised entry or vandalism. It is also to avoid the possibility of materials or equipment being removed by the contractor or subcontractors. Although this would be in breach of the contract, it is clearly more beneficial to the employer to retain these rather than to have to go through the courts for their return when their whereabouts may have become unknown or even unrecoverable.

Materials

Materials on site for which the employer has paid, belong to the employer unless their title is defective. Furthermore they cannot be removed from site without the employer's written permission (JCT 2005 clause 2.2.4). Similar conditions exist in all the major forms of contract. Materials on site that have not been paid for can be used by the employer to complete the works. No payment is made directly, since they will help to reduce the total costs of achieving completion. They will therefore reduce the financial indebtedness of the contractor. Materials off site for which the employer has paid are also in the ownership of the employer, providing that the provisions of clause 2.2.5 (JCT 2005) have been complied with. A list of the materials on site should be quickly made with a note of which of these have been paid for by the employer.

Plant

The contractor's directly owned plant and other temporary structures on site can be used freely to secure completion. Hired plant and that which belongs to subcontractors cannot be used without express permission and payment. However, no plant passes directly to the ownership of the employer. Hired plant that is temporarily supporting the structure will obviously be retained by the employer, but at the usual hire rates.

Retention

Retention by the employer is one of the main buffers the employer has to face any possible loss. Any balance as a result of underpayment to the contractor or for work completed since the last interim payment, will be used for similar effect.

There has in recent years been a considerable debate about retention monies. How should they be retained? Who should hold or own them? It is essential to maintain the status quo for the benefit of the employer.

Other matters

It will be prudent to ensure that insurances remain effective at all times. If a bond holder is involved, then these must be kept informed of what is happening and any progress that is being achieved.

Completion of the contract

An early meeting of everyone concerned should be arranged to consider the best ways of completing the works that are outstanding. It is usual for the liquidator to attend this meeting, but where this is not possible, then the liquidator should be kept fully informed of all the decisions that are made.

If the liquidator decides to disclaim the contract there are several different ways that can be used to appoint a replacement contractor. It will be necessary, at a later date, to show the liquidator and the bond holder that due care was taken to complete the project as economically as was originally intended. The bulk of the work involved in drawing up a new contract will be performed by the quantity surveyor.

Where the amount of work carried out from the original contract is small, then the contract bills may be used to invite new contractors to tender for the work. Adjustments for the work completed will, of course, need to be made. If the contract is in its early days, then it is likely that priced documents will be available from the previously unsuccessful firms. Where the work is significantly advanced, then new documents or addendum bills of quantities will need to be prepared. The time delay involved in this may be unacceptable to the employer and alternative methods may need to be adopted. A new contract may, for example, be negotiated with a single contractor, perhaps on the basis of a single percentage addition to the original contract. Again adjustments for the work completed will need to be made. The main advantage here is that the new contractor can start work on site immediately, with a minimum of delay, if this is possible. A third alternative is to pay for the remainder of the work on a cost plus basis, which may not be favourable to the employer. It is likely that the original contract was the most competitive and any new arrangement will therefore be more expensive.

With each of these methods, an allowance in the form of a provisional sum needs to be made to cover any remedial works or making good defects that have been left by the insolvent contractor. Whilst some of these defects may be apparent, others may not show for some time. Where the work is piecemeal and almost at practical completion, it may be better to employ a contractor who is suitably able to carry out such work; for example, if the outstanding work includes macadam hard standings then it may be more appropriate to employ a firm who specialises in this kind of work.

An alternative method of securing completion is to award the new contract to a contractor for a fee. The selected contractor would then complete the works on exactly the same basis as the original contractor, using the contract rates and prices and accepting total responsibility for the project including its defects. The negotiated fee might be based on a percentage or lump sum to cover the higher costs involved in completing the project and the responsibility, for example, for defects.

A further suggestion is to offer the contract to a new firm on exactly the same basis as that for the original contract, using the same rates and prices. This may appear, at first, to be an unattractive option for three reasons, which no contractor will accept. Firstly, the insolvent contractor was probably selected on a lowest price tender that no other firms could match, hence the prices could be uneconomical for other firms. Secondly, costs will need to be expended on rectifying the partially completed works. Thirdly, the new contractor will be responsible for the first contractor's work, including the making good of any defects. However, the attractiveness of this proposal lies in the fact that retention sums are outstanding and these may provide more incentive than the three disadvantages combined.

The employer's loss

The main sources of loss to the employer are the probable delays in completion and the potential additional costs that might be incurred. If the original contractor had not become insolvent then the loss due to delays would normally have been recovered by way of liquidated damages. The employer is still entitled to show this as a loss, using the basis of calculation shown in the appendix to the form of contract. However, in practice, because of the contractor's insolvency, there may be insufficient funds to make this payment.

A second source of loss involves the additional payments that the employer may need to make to complete the project. This will include the temporary protection and extra site security, which might include erecting additional hoardings, fencing, locking sheds and a twenty-four hour guard. Additional insurances are likely to be required, especially to cover the period prior to appointing the new contractor. Additional fees will also be necessary for the services involved in placing and managing the new contract.

Expenditure involved

It is necessary to calculate a hypothetical final account, which would have been the amount payable to the original contractor had this firm completed the works. For most of the items involved this entails a translation of the actual final account by using the rates and prices from the original contract. A careful note must be made of dayworks, which may otherwise, in the original contract, have been described as measured works. Claims for loss and expense that might have arisen may be a matter for professional judgement. In the case of fluctuations, had the original contract been on a fixed price basis then no account of these need be considered for the original contractor. The additional fluctuations due to the later completion can be genuinely deducted as some compensation against liquidated damages. Remedial work charged by the new contractor for the unsatisfactory work of the original contractor will be excluded from the hypothetical final account. Examples of typical calculations are shown in Figs 15.3 and 15.4.

Termination of contract (employer insolvency)

Whilst a greater number of contractors become insolvent than employers, nevertheless the insolvency of employers is something that does occur from time to time during the execution of a contract. Clause 28.3 (JCT 98), for example, includes the same provisions as clause 27.3 relating to the contractor (JCT 2005 clauses 8.1 and 8.12). When insolvency of the employer arises, the employer must inform the contractor in writing of its occurrence. The following are the contractual consequences of insolvency of the employer:

Contract 1		
Amount of original contract		1 356 000
Agreed additions for variations, prime cost		
sums, dayworks, fluctuations, claims[1]		82 800
Amount of the final account had the original		
contractor completed the works		1 438 800
Amount of completion contract	838 200	
Agreed addition for second contractor only	63 050	
Amount paid to the original contractor[2]	495 000	
Additional professional fees	12 500	1 408 750
Debt due to the original contractor		£30 050

[1] This includes all variations priced at the original contractor's rates and prices.
[2] The total amount of the first contractor's certificate, plus work done since then, was estimated to be £570 000. The original contractor actually receives in total £525 050.

Fig. 15.3 Examples of financial summary.

Contract 2

Amount of original contract		2 292 000
Agreed additions (as above)		111 150
Debt due to the original contractor		2 403 150
Amount certified to original contractor	240 100	
Amount of completion contract	2 163 050	
Agreed lump sum fee	225 000	
Additional professional fees	36 000	2 664 150
Debt payable by the trustee in bankruptcy[1]		£261 000

[1] If there was a bond for 10% of the contract sum, then the employer should receive £229 200 towards the above debt. If the final dividend of five pence in the pound was paid to creditors at some later date, the employer would receive a further £1590 (£31 800 @ 5p). The ultimate employer's loss is therefore £30 210.

Fig. 15.4 Examples of financial summary.

- The contractor can remove from the site all temporary buildings, plant, tools, equipment, goods and materials. This must be done reasonably and safely to prevent injury, death or damage occurring.
- Subcontractors should do the same.
- Within 28 days of determination of the employment, the employer must pay the contractor the retention that has already been deducted, prior to determination.
- The contractor prepares an account setting out the following:
  - The total value of work properly executed up to the date of determination, calculated in accordance with the contract provisions.
  - Any sum in respect of direct loss and expense incurred either before or after determination.
  - The reasonable costs associated with the removal of temporary works, plant and materials.
  - Any direct loss or damage caused to the contractor by the determination.
  - The cost of materials or goods properly ordered for the works which the contractor has already paid. These items become the property of the employer.

Construction contracts involve giving credit to the employer by the contractor in respect of work carried out and goods and materials supplied to the site prior to the issue of an interim certificate. The payment of the certificate, which can take place up to 14 days after certification, is not made in full but is subject to retention. Also a contractor has no lien on the finished work and the employer's insolvency may place the contractor in some real difficulty until matters are finally resolved. This difficulty can result in the contractor's own insolvency. A prudent contractor will, prior to signing a contract, wish to establish whether an employer is able to keep their side of the bargain.

Under JCT 2005 the employer's interest in retention monies is fiduciary as trustee on behalf of the contractor. This is to protect the contractor in the event of an employer's insolvency. It is now strongly recommended that the full provisions of the contract are invoked. A separate bank account should be used, jointly in the names of the employer and the contractor, for the depositing of retention monies.

The employment of the quantity surveyor and architect may not necessarily end at the employer's insolvency. However, if they are to continue with their work they will require some assurances from the liquidator that they will be paid for their services in full.

Insolvency of the quantity surveyor or architect

Insolvency of architects or quantity surveyors is fortunately an unusual occurrence, but in common with other businesses nevertheless does occur. Where an architect's or surveyor's practice becomes insolvent then the employer must within 21 days nominate a successor (articles 3 and 4, JCT 2005). The contractor can raise reasonable objections and if these are upheld then the employer must nominate other firms. Under renomination, the firms appointed cannot over-rule anything that has previously been agreed and accepted. These include matters relating to certificates, opinions, decisions, approvals or instructions. The new firms can make changes, as necessary, but these will always constitute a variation under the terms of the contract.

Performance bonds

A performance bond is a written undertaking by a third party, given on behalf of the contractor to an employer, wherein a surety accepts responsibility to ensure the due completion of the contractual works. Local and central government have frequently requested a contractor to take out a bond for the due completion of the work. This is often in the order of 10% of the contract sum. Main contractors may also require performance bonds to be provided by their own subcontractors as a form of guarantee or insurance.

Bond holders are frequently banks or insurance companies. They are responsible for paying the amounts involved, up to the bond limit, should the contractor default, as in the case of insolvency. However, they have no control over the way in which the work is carried out or the project completed.

When an employer declares that a contractor has failed to perform the works adequately, the surety has normally three courses of action to follow:

- Pay damages up to the full value of the bond
- Engage another contractor to complete the work
- Make such arrangements that the contractor is able to finish the works.

Other bonds may be used which guarantee performance of the contractor. In these instances the surety will satisfy themselves that the bid tendered is responsible, practical and complete. The charge for the bond is included within the contractor's tender and is thus ultimately borne by the employer as a form of insurance.

Wherever possible standard forms of bond should be used where each party is then clearly aware of its implications. The ICE Conditions of Contract provide a standard form of bond.

Discussion topic

The data shown in Fig. 15.5 is an industry analysis of bankruptcies in England and Wales between 1994 and 2004. Making reference to this and other sources of data and information, what do these figures suggest about the construction industry?

The data shown in Fig. 15.5 has been taken from the *Construction Statistics Annual 2005*, published by the then Department of Trade and Industry (DTI). They are based on the definition of the construction industry as given in the revised 2003 Standard Industrial Classification. It represents insolvencies and bankruptcies of construction firms. Bankruptcies include employees, unemployed, directors of companies as well as individuals whose occupation is unknown. Company insolvencies include

| | Bankruptcies | | Company liquidations | |
| | Construction self-employed | | Construction companies | |
Year	Number	Percentage of total	Number	Percentage of total
1994	3362	13.1	2401	14.4
1995	2783	12.7	1844	12.7
1996	2713	12.4	1610	12.0
1997	2182	11.0	1419	11.3
1998	1919	9.8	1325	10.0
1999	1911	8.8	1529	10.7
2000	1741	8.1	1474	10.3
2001	1783	7.6	1509	10.1
2002	1637	6.7	1840	11.3
2003	1781	6.4	1728	12.2
2004	1658	4.6	1653	13.6

Fig. 15.5   Insolvencies and bankruptcies of construction firms in England and Wales (*Source*: Department of Trade and Industry).

partnerships. Liquidations of companies include both those which are compulsory following court orders and creditors' voluntary liquidations when debtor companies wind-up their affairs after agreeing terms with their creditors.

In 1992 bankruptcies in all industries were running at 32 106 or over 6 000 each week. In the construction industry in that year, there were 4 692 bankruptcies, just short of 100 per week. This was the year when, in recent times, the number of bankruptcies reached their all-time high (Ashworth and Harvey 1997). This was also a peak year in respect of business failures in terms of both bankruptcies and liquidations generally.

The first and obvious indication from this data is the fall in both the number and percentage of bankruptcies and liquidations amongst construction firms, which is a welcome sign for an industry that, unfortunately, usually tops the lists when compared with other industries. However, the data that are published in this respect need to be informed by both the type of industry and its relative size. These two aspects are sometimes easily forgotten or ignored, but are nevertheless of particular importance.

Bankruptcies in the construction industry fell markedly from 13.1% to 4.6% over the period 1994–2004. Historically, these had been at a much higher level throughout the 1980s with an average figure equivalent to about 17% each year.

Contrastingly, company insolvencies at 13.6%, whilst relatively static, are marginally higher than the average of 11.69% for the period. Insolvencies fell up to 2001 but then began to rise again in consecutive years up to 2004. In the early 1990s, insolvencies in the construction industry were running at a much higher level of about 15%.

Whilst the number of individuals or firms who went into bankruptcy or liquidation during this period is counted, there is no indication of the relative size of these or the amount of debt that they had accumulated. It must be assumed, however, with the large number of firms or organisations involved that some comparisons can be made.

The trend data from this information on bankruptcies in the construction industry is also very encouraging, since this has shown a reduction in number since 1994, and even going all the way back to 1988. But the trend for liquidations is not so encouraging, since, whilst these continued to fall to 2001, they have begun to rise again. The trend is firmly and surprisingly on the increase, taking into account the current excellent state of the construction industry. It must therefore not be assumed that firms only go into liquidation in times of recession, although notably the figures do then increase.

Other data (www.dtistats.net.insolv) on bankruptcies and liquidations indicates that by comparison the total number of all bankruptcies between 1992 and 2002 declined over this period of time for most of the other industries in the UK. In 1992, for example, there were 32 106 recorded cases of bankruptcy. By 2002 these had fallen to 24 292, representing a fall of 24%.

The growth in the numbers of small firms, which peaked in the early 1990s (Ashworth and Harvey 1997), and their subsequent consolidation might also have had some bearing on these figures. In 1992, at the beginning of the last recession in the construction industry, the figures for bankruptcies peaked at 4 692. The current data shows a reduction from that date by almost 65% in terms of the number of individuals and self-employed.

Construction industry cases in 1992 represented 14.61% of all bankruptcies. The construction sector also represented by far the greatest proportion when compared against the other industries that are listed. Hotels and catering were the next recorded highest at 7.37%, roughly half of that recorded for the construction industry. By 2002, when the latest set of figures were available, recorded bankruptcies had fallen to 24 292, of which 1637 were in the construction industry (6.37%). Bankruptcies had fallen across all industries but, even with this dramatic fall, the construction industry still accounted for the largest proportion of cases.

Bankruptcies, amongst individuals and the self-employed in the construction industry, have fallen considerably since the middle of the 1990s. It is hoped that improved education and training may have played their part in this and the acceptance of better practices encouraged by organisations, such as *Constructing Excellence* and its predecessor, *Rethinking Construction*. This, hopefully, might be partly true in the absence of any detailed analysis or correctly interpreted correlation. However, one reason for the sharp decline in the number of bankruptcies is the more stable and buoyant economy that has been created. The construction industry has over the last ten years been without its boom and slump characteristics that had plagued it in the latter half of the twentieth century.

By 2004 the total number of bankruptcies generally had fallen over the ten-year period by almost a quarter (24.34%) to 24 292. The construction sector by comparison had fallen much faster by 65%. The construction industry in 2004 was experiencing a period of sustained growth and activity. Its single, biggest problem was in recruiting sufficient staff at the levels currently required and gaps in skills were evident as the industry began to rely more upon foreign workers, especially from Eastern European countries. The industry had become profitable, margins were higher and these provided a cushion against, for example, problems of late payment. This would lead one to believe that the proposed working practices outlined earlier by Latham (1994) and Egan (1998) had made some contribution towards these figures.

However, the quarterly figures for corporate and individual insolvencies for the final quarter of 2005 were not so encouraging. The figures show that these rose when compared to the same period in 2004. The figures show the total number of company liquidations rose by 14.2% when compared to the same period in 2004. Compulsory liquidations rose by 35.7% and creditor voluntary liquidations by 1% over this same period. The figures also show that receiverships fell by 16.3%, but administrations continued to increase with a 47.3% rise on the same period a year earlier. The number of companies entering administration has risen for the eighth successive quarter (source: DTI 2005).

The figures also show that the total number of individual insolvencies has increased by 46% on the same period last year. Bankruptcies have risen by 30.9% and Individual Voluntary Arrangements (IVAs), an alternative to bankruptcy, have risen by 95% when compared to the same figures for the previous year. These are figures that relate to all industries, but within them construction industry analysts have identified similar figures for the construction industry.

As a health warning, there is always a danger of looking at these figures in isolation from other data such as bank lending, company, profit, return, cash flow and turnover. For companies to weather the business climate it is essential that they are careful with whom they do business, employ sound financial disciplines and verify and monitor companies they deal with. In times of recession new, as well as established companies fail as more companies do not pay their bills on time.

References and bibliography

Ashworth A. and Harvey R.C. *The Construction Industry of Great Britain*. Butterworth-Heinemann. 1997.

CIB. *Code of Practice for the Selection of Main Contractors*. Thomas Telford. 1997.

Department of Trade and Industry. *Construction Statistics Annual 2005*. HMSO. 2005.

Egan J. *Rethinking Construction*. Department of Industry and the Regions. 1998.

Latham M. *Constructing the Team*. HMSO. 1994.

Rajak S. and Davis P. *Insolvency a Business by Business Guide*. Butterworth Tolley. 2001.

Websites

www.dtistats.net.insolv

www.dunnandbradstreet.org

16 Contractual Disputes

Introduction

From time to time quantity surveyors find themselves involved in contractual disputes either in litigation in the courts, in arbitration or in alternative dispute resolution cases (ADR) cases. Their involvement is sometimes as witnesses of fact: that is, someone who was actually there at the time as project surveyor or manager. However, more often they are involved as expert witnesses, adjudicators, arbitrators themselves or as neutrals or mediators in ADR cases. Each of these situations is considered in this chapter, as alternatives to litigation in the courts, under the following headings:

- Litigation
- Arbitration
- Adjudication
- Alternative dispute resolution
- Expert witness
- Lay advocacy.

Disputes are a common feature of the construction industry. They occur daily and fortunately many are solved amicably between the parties involved, without the need to resort to one of the above scenarios. However, the number of reported legal cases suggests a certain litigious nature driving the construction industry. The costs associated with resolving these differences of opinion or interpretation are often high and damage the image of the industry. In the light of this, those responsible for drafting forms of contract for use by the industry have increasingly included provision for the parties to undertake some process or processes of ADR before proceeding further. Thus, both the Partnering Contract (PPC 2000) and Standard Building Contract (JCT 2005) include specific provision for mediation.

Why disputes arise

The construction industry is a risky business. It does not build many prototypes, with each different project being individual in so many respects. Even apparently identical building and civil engineering projects that have been constructed on different sites create their own special circumstances, are subject to the vagaries of different site and weather conditions, use labour that may have different trade practices and result in costs that are different. Even the identical project constructed on an adjacent site by a different contractor will have different problems and their associated different costs of construction. Disputes can therefore arise, even on projects that have the best intentions. Even when every possibility of disagreement has been potentially eliminated, problems can still occur – such is human nature. Some of the main areas for possible disputes occurring are:

General

- Adversarial nature of construction contracts
- Poor communication between the parties concerned
- Proliferation of forms of contract and warranties
- Fragmentation in the industry
- Tendering policies and procedures.

Employers

- Poor briefing
- Changes and variation requirements
- Changes to standard conditions of contract
- Interference in the contractual duties of the contract administrator
- Late payments.

Consultants

- Design inadequacies
- Lack of appropriate competence and experience
- Late and incomplete information
- Lack of coordination
- Unclear delegation of responsibilities.

Contractors

- Inadequate site management
- Poor planning and programming
- Poor standards of workmanship
- Disputes with subcontractors
- Delayed payments to subcontractors
- Co-ordination of subcontractors.

Subcontractors

- Mismatch of subcontract conditions with main contract
- Failure to follow and adopt agreed procedures
- Poor standards of workmanship.

Manufacturers and suppliers

- Failure to define performance or purpose
- Failure of performance.

Litigation

Litigation is a dispute procedure which takes place in the courts. It involves third parties who are trained in the law, usually solicitors and barristers, and a judge who is appointed by the courts. This method of solving disputes is often expensive and can be a very lengthy process before the matter is finally resolved, sometimes taking years to arrive at a decision. The process is often extended, upon appeal, to higher courts involving additional expense and time. Also, since a case needs to be properly prepared prior to the trial, a considerable amount of time can elapse between the commencement of the proceedings and the trial.

A typical action is started by the issuing of a writ. This places the matter on the official record. A copy of the writ must be served on the defendant, either by delivering it personally or by other means such as through the offices of a solicitor. The general rule is that the defendants must be made aware of the proceedings against them. The speed of a hearing in most cases depends on the following:

- Availability of competent legal advisers to handle the case, i.e. its preparation and presentation
- Expeditious preparation of the case by the parties concerned
- Availability of courts and judges to hear the case.

The amount of money involved in the case will determine whether it is heard in the County Court or High Court. Where the matter is largely of a technical nature the case may be referred in the first instance to the Technology and Construction Court, formerly the Official Referee's Court. A circuit judge whose court is used to hearing commercial cases usually presides over these cases, and hence handles most of the commercial and construction disputes. Under these circumstances a full hearing does not normally take place, but points of principle are established. The outcome of this hearing will determine whether the case then proceeds towards a full trial.

Under some circumstances, the plaintiff may apply to the court for a judgment on the claim (or the defendant for a judgment on the counterclaim), on the ground that there is no sufficient defence. Provided that the court is satisfied that the defendant (or plaintiff) has no defence that warrants a full trial of the issues involved, judgment will be given, together with the costs involved.

Every fact in a dispute that is necessary to establish a claim must be proved to the judge by admissible evidence whether oral, documentary or of other kind. Oral evidence must normally be given from memory by a person who heard or saw what took place. Hearsay evidence is not normally permissible.

In a civil action the facts in the dispute must be proved on a balance of probabilities. This is unlike a criminal case where proof beyond reasonable doubt is required. The burden of proof usually lies on the party asserting the fact.

Arbitration

Disputes between parties to a contract are traditionally heard in the courts, but in building contracts the chosen method has more often been arbitration. Arbitration to resolve disputes in building contracts comes about following the agreement of the parties, either when the dispute arises, or more often as a term of the original contract. For instance, the JCT forms of contract all provide that if a dispute arises between the parties to the contract then either party can call for arbitration. When such clauses exist in contracts the courts, if asked, will generally rule that arbitration, having been the chosen path of the parties, is the proper forum for the dispute to be heard and will stay any legal action taken in breach of the arbitration agreement under section 9 of the Arbitration Act 1996.

The JCT publish arbitration rules for use with arbitration agreements referred to in the JCT contracts. They set out rules concerning interlocutory (intermediate) matters, conduct of arbitrations and various types of procedures: without hearing (documents only), full procedure with hearing, and short procedure with hearing, each containing strict timetables.

Arbitration and the law relating to it are subjects on their own. The traditional advantages of arbitration over the courts are four-fold:

- Arbitration proceedings are quicker than the courts
- Arbitration is cheaper than litigating in the courts
- The parties get a 'judge' of their choosing, a person knowledgeable about the subject matter in dispute, but with no knowledge of the actual case, rather than a judge imposed on them
- Unlike proceedings in the courts, arbitration proceedings are confidential.

With regard to the first of these traditional advantages, provided the proceedings are kept simple then arbitration can still prove quicker than proceeding through the courts. However, the modern tendency is to involve lawyers at all stages and to complicate disputes. This has eroded this advantage, and there is now often very little difference between the time it takes to get a dispute settled either in arbitration or in the courts. Equally, arbitration is no longer the cheaper option that it used to be. Legal costs, and the fact that the parties have to pay for the 'judge and courtroom' in arbitration proceedings instead of having them provided at the taxpayer's expense, have eroded this traditional advantage as well, although under section 65 of the Arbitration Act 1996 the arbitrator may limit costs to a specified amount unless the parties agree not to let this happen.

However, the third and fourth advantages, choice of judge and confidentiality, still exist and are considered by many to be of overriding importance. Hence the continued popularity of arbitration references, although statutory adjudications under the Housing Grants, Construction and Regeneration Act 1996 appear to have resulted in fewer references.

The duty of arbitrators is to ascertain the substance of the dispute, to give directions as to proceeding, and to hear the parties as quickly as possible and make their decision known by way of an award. Arbitrators hear both sides of an argument, decide which they prefer and award accordingly. They cannot decide that they do not like either argument and substitute their own solution. This would lead to an accusation of misconduct (see below).

An arbitrator may be named in a contract, although this is rare. It is usually thought better to wait until the dispute arises and then choose an appropriate person. The choosing will be by the parties. Each side exchanges suitable names and usually an acceptable choice emerges. If the parties are unable to agree then a presidential appointment will be sought from a body such as the Chartered Institute of Arbitrators, the Royal Institute of British Architects or the Royal Institution of Chartered Surveyors. If a quantity surveyor is thought the most appropriate choice then the President of the RICS would be the most likely to be approached. When appointed, an arbitrator calls the parties together in a preliminary meeting. At this meeting the nature and extent of the dispute are made known to the arbitrator and an order is sought for directions, fixing a timetable for submission of the pleadings (i.e. points of claim, points of defence, etc.). The directions normally end by fixing a date and venue for a hearing.

As the date for the hearing approaches, the parties will keep the arbitrator informed of the progress they are making in working their way through the timetable. If a compromise is achieved and the matter is settled, they will inform the arbitrator immediately. At the hearing each party puts their case and then calls their witnesses, first the witnesses of the facts and then the expert witnesses. All witnesses normally give evidence under oath and are examined and cross-examined by the parties' advocates.

All arbitrators are bound by the terms of the Arbitration Act 1996. Their awards, once made, are final and binding. If an award is not honoured, then the aggrieved party can call on the courts to implement the award. The only exception to this is if the arbitrator is found to be guilty of misconduct, or the arbitrator is wrong at law. If either of these events occurs a party can apply to the courts for leave to have the arbitrator's award referred back for reconsideration or, in extreme cases, to seek the removal of the arbitrator.

Being wrong at law is self-explanatory. The arbitrator may be a lawyer but more likely does not hold that qualification. There is therefore no disgrace in getting the law wrong. If in doubt, an arbitrator has the power under the Act to seek legal assistance and often this is good advice.

Misconduct is more difficult to describe. It is not misconduct in the usual understanding of the word. It can be defined in this context as a failure to conduct the reference in the manner expressly or impliedly prescribed by the submission, or to behave in a way that would be regarded by the court as contrary to public policy. Failure to answer all the questions asked and hearing one party without the other party being present are examples of misconduct that the courts would consider and direct as they thought fit.

Quantity surveyors are well suited to act as arbitrators in construction disputes, as the dispute frequently involves measurement, costs and loss and/or expense and interpretation of documents. These are all matters falling within the expertise of quantity surveyors. When the matters in dispute concern quality of workmanship or design faults then arbitration is best left to architects or engineers. Equally, when matters of law are the prime consideration then the arbitrator should be a lawyer or at least have a law qualification.

Adjudication

Statutory adjudication

Sections 108 to 113 of the Housing Grants, Construction and Regeneration Act 1996 impose statutory adjudication in most written construction contracts (there are some exceptions, e.g. oil and gas contracts and private homes). This scheme has revolutionised dispute resolution for construction contracts (which include contracts between employers and construction professionals) because if one party to the contract wishes a dispute to be heard he can call for adjudication, and the adjudicator, when appointed, has 28 days to make his award unless this period is extended by mutual agreement of the parties. The award will be binding and must be paid forthwith and will stand unless overturned in a later arbitration or court case. Many cases in the Technology and Construction Court now deal with the enforcement of adjudication awards and the judges are determined to uphold them unless it can be shown that there is no contract in fact or some really major breach of natural justice has occurred.

Obviously in the limited time available it is difficult to resist a claim which may have been copiously compiled (usually by the contractor or subcontractor) before the adjudication is demanded, and quantity surveyors must look sharp to analyse it and prepare a response to be put before the adjudicator. The statutory scheme must either be replicated in the construction contract or it will be imposed. It has therefore largely supplanted other adjudication provisions found in standard forms of contract.

One of the benefits of adjudication is that it can often lead to a settlement without the matter going any further. This is because a party that has lost in an adjudication will think very carefully before proceeding with very expensive litigation or arbitration. They might well lose again, with the additional penalty of paying the other side's costs. Current statistics suggest that few adjudications lead on to full arbitration or litigation.

Alternative dispute resolution

Alternative dispute resolution, or ADR as it is generally known, originated in the USA and was adopted in the UK in the 1980s. It is now practised worldwide. ADR provides a means of resolving disputes without resorting to arbitration or the courts. In that respect it is nothing new. Quantity surveyors and contractors have over the years traditionally settled disputes by negotiation. With ADR the process is somewhat more formal. The advantages of ADR can be summarised as follows:

- Private: confidentiality is retained
- Quick: a matter of days rather than weeks, months or even years
- Economic: legal and other costs resulting from lengthy litigation are avoided.

However, none of these advantages will be achieved unless one vital ingredient is present. There must be goodwill on both sides to settle the matter on a commercial rather than a litigious basis. If this goodwill does not exist, then the parties have no option but to resort to arbitration or the courts, without wasting further time and resources.

ADR can take a variety of forms, and of these the first two are the most commonly met:

- Mini-trial
- Mediation
- Mutual fact-finding
- Mutual expert
- Private judging
- Dispute resolution boards.

Mini-trial

Each of the parties is represented, generally but not necessarily by a lawyer. The lawyer makes a short presentation of their client's case to a tribunal. This presentation will have been preceded by limited disclosure of documentation, or discovery as it is known. The purpose of this limited discovery is to ensure that each party is aware of the opposite side's case and is not taken by surprise by the presentation.

The tribunal usually takes the form of a senior managerial representative of each party and an independent advisor referred to as a neutral. It is important to the success of the mini-trial that each of the parties' representatives should not have been directly involved in the project. There will then be less emotional involvement than with someone who has lived with the dispute and has difficulty in taking a detached view. It is also essential that both parties' representatives have full authority to settle the matter. There is nothing worse than arriving at what appears to be consensus, for one party then to disclose that they can only agree it subject to authorisation from a chairman, board, council, chief officer or some other third party.

The neutral has to be someone with a knowledge of the industry but no knowledge or interest in the dispute. In this respect the same requirements as that of a good arbitrator are required. However, unlike an arbitrator, whose task it is to hear the arguments and decide which is preferable, a neutral becomes much more involved, listening, suggesting and giving advice on matters of fact and sometimes on law as well.

After the initial presentation, experts and witnesses of fact may be called, following which the managers enter into negotiation with a view to coming to a consensus. The length of these negotiations will depend very much on the complexity of the matters in dispute. They will be assisted by the neutral, who may if the parties remain deadlocked, give a non-binding opinion. This may lead to a settlement after further negotiation.

Once a negotiated settlement is reached, the neutral will there and then draft the heads of a statement of agreement, which the parties will each initial. This will then be followed by a formal agreement ending the matter.

Mediation

This is a less formal method of proceeding than that described for a mini-trial; however it is more commonly used and is becoming more popular than the mini-trial. The parties, with assistance from their experts or lawyers, will select a neutral whose background will reflect the matters in dispute. A preliminary meeting will be arranged by the neutral to discover the substance of the dispute and to decide how best to proceed.

At a subsequent meeting the parties will make formal presentation in a joint session. This is then followed by a series of private meetings, or 'caucuses' as they are termed, between the neutral and each of the parties on their own. The neutral moves from one caucus to the next, reporting, with agreement, the views of each party in turn. This should lead to the neutral's being able to suggest a formula for agreement, which in turn may lead to a settlement. Such agreement is terminated in the same way as in a mini-trial.

Mutual fact-finding

Resolution of a dispute by mutual fact-finding is an informal procedure. The parties, possibly at a different level from those closely involved, take a pragmatic and commercial approach to settling the dispute with or without the assistance of a mutual expert.

Dispute resolution boards

The appointment of dispute resolution boards under specific forms of contract (e.g. dispute adjudication board under the FIDIC conditions of contract) is becoming more common. These boards comprise individuals appointed by the parties to the contract, the boards' role being to adjudicate on disputes referred to them by either party.

Conclusion

To summarise, ADR provides a dispute resolution mechanism that concentrates on resolving disputes by consensual, rather than adjudicative, methods. To quote from the Centre for Dispute Resolution's introduction to ADR, 'These techniques are not soft options, but rather involve a change of emphasis and a different challenge. The parties cooperate in the formulation of a procedure and result over which they have control.' The main features are easily captioned under:

- *Consensus* – a joint objective to find the business solution
- *Continuity* – a desire to find a solution in the context of an ongoing business relationship
- *Control* – the ability to tailor a solution that is geared towards a business result rather than a result governed by the rule of law, which may be too restrictive or largely inappropriate
- *Confidentiality* – avoiding harmful washing of dirty linen in public.

Quantity surveyors have a role to play in ADR. They are obvious candidates for the post of neutral in construction disputes, provided they have good negotiating skills and have undergone the necessary training.

Expert witness

Quantity surveyors, particularly those more experienced, may be briefed to give expert witness in a variety of circumstances:

- Litigation in the High Court or in the County Court
- In arbitration
- Before a tribunal.

While proceedings in court tend to be more formal than in arbitration, the same rules apply.

Expert evidence is evidence of opinion to assist with the technical assertions of each party. The expert must be seen at all times to be independent and have no financial interest in the outcome. Expert witnesses are there to assist the court, the arbitrator or the tribunal and must not be seen to be solely advocating their client's case. To emphasise this, two quotations from court judgments are relevant.

In the case of *Whitehouse* v. *Jordan* (1981) 1 WLR 247, Lord Wilberforce warned:

'Whilst some degree of consultation between experts and legal advisers is entirely proper it is necessary that expert evidence presented to the court should be and should be seen to be the independent product of the expert and uninfluenced as to form or content by the exigencies of litigation. To the extent that it is not, the evidence is likely to be not only incorrect but self-defeating.'

The second quotation comes from the judgment of Sir Patrick Garland in the case of *Warwick University* v. *Sir Robert McAlpine and Others* (1988) 42 BLR:

'It appeared to me that some (but by no means all) of the experts in this case tended to enter into the arena in order to advocate their client's case. This led to perfectly proper cross-examination on the basis: "You have assembled and advanced explanations which you consider most likely to assist your client's case". It is to be much regretted that this had to be so. In their closing speeches counsel felt it necessary to challenge not only the reliability but the credibility of experts with unadorned attacks on their veracity. This simply should not happen where the court is called upon to decide complex scientific and technical matters.'

An expert opinion must be based on facts, and these facts have to be proved unless they can be agreed. The process of agreement is achieved by meetings of the experts. These are frequently ordered by a judge or arbitrator to agree facts and figures wherever possible, and to narrow the issues in dispute. These meetings are usually described as being 'without prejudice'. That is, nothing discussed can be used in evidence. Such meetings are followed by a joint statement of what has been agreed, and of those matters that are still outstanding. In fact, such meetings can often lead to a settlement of the dispute without troubling the tribunal further.

The Civil Procedure Rules, which now govern procedure in the courts following the far-reaching reforms of Lord Woolf, strictly regulate the procedure relating to expert evidence and should be studied by all involved in the process of adducing such evidence.

Acting as an expert witness is a time-consuming and therefore an expensive exercise. There needs to be careful reading of all relevant matters, some research where appropriate and a studiously prepared proof of evidence. As has been stated above, such proofs of evidence have to be seen as being in the sole authorship of the expert. While they may listen to advice, the final document, on which they stand to be cross-examined, is theirs and theirs alone.

Lay advocacy

Advocacy in the High Court and in the County Courts is restricted to barristers and solicitors. There is no such requirement when it comes to arbitration or appearing before a lay tribunal. Anyone can advocate their own or their client's case.

It sometimes happens that when a matter is strictly technical, such as a dispute on measurement or computation of a final account, an arbitrator may direct that the two quantity surveyors should appear before him or her to present their client's case. When this happens, quantity surveyors are acting as lay advocates and are dealing with matters very much within their own expertise. Occasionally it is suggested that a quantity surveyor might act as lay advocate on the whole case, not just that part of it within his or her own discipline.

Advocacy is a skilled art that is not easily learned. It involves meticulous preparation, the ability to ask the right questions in the right way, and the ability to think on one's feet and respond quickly to answers from witnesses or points made by one's opponent. For this reason quantity surveyors should consider very carefully whether or not to accept an invitation to act as advocate, and unless they are quite confident that they are able to, should politely decline.

Claims

It is evident, from society in general, that as individuals we are becoming more claims conscious. Firms of lawyers are now touting their services, often on a contingency fee basis, although this is still more the exception than the rule in construction cases.

Contractual claims arise where contractors assess that they are entitled to additional payments over and above that paid within the general terms and conditions of the contract for payment of work done. For example, the contractors may seek reimbursement for some alleged loss that has been suffered, for reasons beyond their control. Claims may arise for several different reasons, such as:

• Extensions of time
• Changes to the nature of the project
• Disruption to the regular progress of the works by the client or designer
• Variations to the contract.

Claims arise in respect of additional payments that cannot be recouped in the normal way, through measurement and valuation. They are based on the assumption that the works, or part of the works, are considerably different from or executed under different conditions than those envisaged at the time of tender. The differences may have revised the contractor's intended and preferred method of working, and this in turn may have altered or influenced the costs involved. The rates inserted by the contractor in the contract bills do not now represent fair recompense for the work that has been executed.

Where a standard form of contract is used, attempts may be made by contractors to invoke some of the compensatory provisions of the contract, in order to secure further payment to cover the losses involved. As with many issues in life, contractual claims are rarely the fault of one side only. If the claim cannot be resolved, then one of the methods of dispute resolution referred to earlier may need to be invoked. Special care, therefore, needs to be properly exercised in the conduct of the negotiations since they may have an effect upon the outcome of any subsequent legal proceedings. Claims usually reflect an actual loss and expense to a contractor. Once the contract is made, the contractor is responsible for carrying out and completing the works in the prescribed time and for the agreed sum. Entitlement to additional payments occurs if a breach of contract is committed by the employer or a party for whom the employer is responsible, or if circumstances arise that are dealt with in express terms of the contract providing for additional payment.

Contractual claims – loss and expense

Contractual claims are claims that have a direct reference to conditions of contract. When the contract is signed by the two parties, the contractor and the employer (or promoter), this results in a formal agreement to carry out and complete the works in accordance with the information supplied through the drawings, specification and contract bills. Where the works constructed are of a different character or executed under different conditions, then it is obvious that different costs will be involved. Some of these additional costs may be recouped under the terms of the contract, through, for example, remeasurement and revaluation of the works, using the appropriate rules from the contract or express terms of the contract allowing the recovery of loss and expense or cost. Other additional costs that an experienced contractor had not allowed for within the tender may need to be recovered in a different way. It should be noted that the costs incurred by contractors because of their own mistakes in pricing or whilst executing the works cannot form the basis of a successful claim.

Contractual claims – extensions of time

The two factors, time and money, are often inextricably linked. Hence a claim for loss and expense may include some element relating to the cost to the contractor of an extension to his preliminaries costs and the like. However, most forms of contract also allow for certain situations where the contractor has been delayed and may claim an extension of time but for which they will not receive any additional payment. Proving an entitlement to an extension of time in such cases may be of significance to the contractor as it will free them, importantly, from liability to pay the client liquidated and ascertained damages.

Ex-gratia payments

Ex-gratia payments are claims not based on the terms or conditions of contract. The carrying out of the works has nevertheless resulted in some loss and expense to the contractor. The contractor has completed the project on time, to the required standards and conditions and at the price agreed. Perhaps, due to a variety of different reasons, and at no fault of the contractor, a loss has been sustained that cannot relate to the express or implied terms of the contract. On rare occasions, a sympathetic employer may be prepared to make a discretionary payment to the contractor. Such payments are made out of grace and kindness. They may be made because of a long standing relationship and trust between employer and contractor, or because of outstanding service and satisfaction provided by the contractor. Nevertheless such payments are rare.

The quantity surveyor

The details of claims submitted by the contractor are usually investigated by quantity surveyors, since claims invariably have a financial consequence. A report will subsequently be made to the architect, engineer or other lead consultant. The report should summarise the arguments involved and set out the possible financial effect of each claim. Quantity surveyors frequently end up negotiating with contractors over such issues, in an attempt to solve the financial problems and agree wherever possible an amicable solution. This is preferable to lengthy legal disputes described above.

The quantity surveyor may well be requested to negotiate with the contractor on specific issues relating to a claim before there is resort to adjudication, arbitration or litigation. To guard against commitment, correspondence in the period before action is often marked 'without prejudice'. This will usually preclude reference to such correspondence in subsequent litigation but it is not, as is sometimes mistakenly believed, a safeguard against any form of binding obligation. In fact should an offer made in such correspondence be accepted, a binding agreement will usually result. This may be so even if the matter agreed relates only to a detail and the negotiations as a whole eventually collapse.

The decision on the outcome of a claim usually rests with the architect, or engineer on civil engineering works. This is except in such matters of valuation as the parties to the contract have entrusted to the quantity surveyor. The quantity surveyor, in a preliminary consideration of them, should remember the principles that must guide the architect or engineer in a decision. The following thoughts are suggested as a guide to a decision on claims:

- What did the parties contemplate on the point at the time of signing the contract? If there is specific reference to it, what does it mean?
- Can any wording of the contract, though not specifically mentioning it, be *reasonably* applied to the point? In other words, if the parties had known of the point at the time of signing the contract, would they have reckoned that it was fairly covered by the wording?
- If the parties did not contemplate the particular matter, what would they have agreed if they had?
- If the claim is based on the contract, does it so alter it as to make its scope and nature different from what was contemplated by the parties signing it? Or is it such an extension of the contract as would be beyond the contemplation of the parties at the time of signing it? In either case the question arises whether the matter should not be treated as a separate contract, and a fair valuation made irrespective of any contract conditions.
- The value of the claim in monetary terms should not affect a decision on the principle.
- If the claim is very small, however, whichever party is concerned might be persuaded to waive it, or it may be eliminated by a little 'give and take'.

In particular, any action of the client that may have been a contributory cause should be given due importance. If the claim is based on unanticipated misfortune, consideration of what the parties would have done, if they had anticipated the possibility, will often indicate whether it would be reasonable to ask the client to meet the claim to a greater or lesser extent.

The architect or engineer may make a recommendation to the client. If the contractor does not accept the client's offer then it may be necessary to invoke a third party to help resolve the difficulty, although if the architect or engineer is the certifier under the contract he must make a decision on the validity of the claim and certify accordingly.

Forms of contract

Within the industry there are a great number of standard and model forms of contract, the choice of which depends upon the procurement route adopted by the parties. All contain mechanisms for identifying and addressing situations where the contractor may be entitled to additional payment over and above that due to him in the normal course of valuation of the works. The processes within most forms will be similar to those outlined in the two examples illustrated below.

The Standard Form of Building Contract, JCT 2005, seeks to clarify the contractual relationship between the employer and the contractor. As far as possible, ambiguities have been eliminated, but some nevertheless remain. If such forms or conditions of contract were not available, then the uncertainty between the two parties would be even greater. This could have the likely effect of increasing tender sums. Under the present conditions of contract, the contractual risks involved are shared between the employer and the contractor. Claims may arise most commonly under clause 4 and these are known as loss and expense claims. They may also arise due to a breach of contract. The contractor must make a written application to the architect, in the first place, stating that a direct loss and expense has occurred or is likely to occur in the execution of the project. The contractor must further state that any reimbursement under the terms of the contract is unlikely to be sufficient (clause 4.23). This information should be given to the architect as soon as possible in order to allow time to plan for other contingencies.

Under clause 53 of the ICE Conditions it is the contractor's responsibility to inform the engineer within 28 days of an event that a claim may arise. The contractor must keep the necessary records in order to reasonably support any claim that may be subsequently made. Without necessarily accepting the employer's liability that a claim may exist, the engineer can instruct the contractor to keep and maintain such records. Where required the engineer is then able to inspect such records and have copies supplied where these are appropriate.

As soon as it is reasonably possible, the contractor should provide a written interim account providing full details of the particular claim and the basis upon which it is made. This should be amended and updated when necessary or when required. If the contractor fails to comply with this procedure, then this might prejudice the further investigation of the claim and any subsequent payments by the employer to the contractor.

Contractors are entitled to have such amounts included in the payment of interim certificates. However, in practice a large majority of claims are often not agreed until the completion of the contract. In these circumstances the contractor is entitled to receive part of the claim included in an interim certificate, if it has been accepted in principle. Additional monies in the form of interest payments may also become due.

Contractors

Many contractors have well organised systems for dealing with claims on construction projects and the recovery of monies that are rightly due under the terms of the contract. They are likely to maintain good records of most events, but particularly those where difficulties have occurred in the execution of the work. However, some of the difficulties may be due to the manner in which the contractor has sought to carry out the work and they thus remain the entire responsibility of the contractor.

Claims that are notified or submitted late will inevitably create problems in their approval. In these circumstances the architect or engineer might not have the opportunity to check the details of the contractor's submission, and suitable records might not have been maintained. Such occurrences will not be looked on favourably by either the architect or engineer and the employer.

The contractor must prepare a report on why a particular aspect of the work has cost more than expected, substantiate this with appropriate calculations and support it with reference to instructions, drawings, details, specifications, letters, etc. The contractor must also be able to show that an experienced contractor could not have foreseen the difficulties that occurred. They will also need to satisfy themselves that the work was carried out in an efficient, effective and economic manner.

Example

The construction of a major new business park on a green-field site requires a large earthmoving contract. The quantities of excavation and its subsequent disposal have been included in contract bills and priced by the contractor. During construction, due to variations to the contract and the unforeseen nature of some of the ground conditions, the quantities of excavated materials increase by 25% by volume, all of which needs to be removed from the site.

The contractor's pre-tender report indicates that a variety of tips at different locations and distances from the site will be used for the disposal of the excavated materials. The contractor's tendering notes indicated that the tips nearer the site would be filled first. In this case they result in lower haulage costs, and, in this example, also have lower tipping charges. The disposal of the excavated material therefore includes two separate elements:

- haul charges to the tip
- tipping charges.

In the contract bills, the rate used by the contractor for disposal of excavated materials represents an average calculated rate. This is based on the average haul distances and average tipping charges using the weighted quantities in each tip (Fig. 16.1). The actual rate for the disposal of the excavated materials can be calculated as shown in Fig. 16.1.

To continue to apply the contract bill rates is unfair. The rates no longer reflect the work to be carried out and the contractor's method of working. Other questions will also need to be asked. Is the type of material being excavated similar to that described in the contract bills? Does this material bulk at the same rate? Is it more difficult to handle? Does it necessitate the same type of mechanical plant? Some other factors that the contractor may also consider include:

Contract bills
Excavated materials for disposal in the contract bills 1 000 000 m³
Contract bill rate is based on:

Tip A 500 000 m³	distance 1 km	Tip charge £0.10 per m³
Tip B 300 000 m³	distance 2 km	Tip charge £0.20 per m³
Tip C 200 000 m³	distance 4 km	Tip charge £0.30 per m³

$$\text{Average distance} = \frac{500\,000 + 600\,000 + 800\,000}{1\,000\,000} \qquad = 1.9\,\text{km per m}^3$$

$$\text{Average tip charge} = \frac{50\,000 + 60\,000 + 60\,000}{1\,000\,000} \qquad = £0.17\,\text{per m}^3$$

Final account
Actual quantities in the final account

Tip A 500 000 m³	distance 1 km	Tip charge £0.10 per m³
Tip B 300 000 m³	distance 2 km	Tip charge £0.20 per m³
Tip C 300 000 m³	distance 4 km	Tip charge £0.30 per m³
Tip D 150 000 m³	distance 6 km	Tip charge £0.50 per m³

$$\text{Average distance} = \frac{500\,000 + 600\,000 + 1\,200\,000 + 900\,000}{1\,250\,000} \qquad = 2.56\,\text{km per m}^3$$

$$\text{Average tip charge} = \frac{50\,000 + 60\,000 + 90\,000 + 75\,000}{1\,250\,000} \qquad = £0.22\,\text{per m}^3$$

Note: bulking of the soil has been ignored.

Fig. 16.1 Earthmoving claim.

- The increase in the amount of excavated materials, on this scale, may also have other repercussions, such as an extension of the contract time.
- The method of carrying out the works might also now be different from that originally envisaged by the contractor.
- Different types of mechanical excavators may have been more efficient than the plant originally selected to do the work.
- The mechanical plant on site may no longer be the most appropriate to do the job. This is especially so where cut and fill excavations are considered, where motorised scrapers may need to be substituted for excavators and lorries.
- The contractor may also be involved in hiring additional plant at higher charges and employing workpeople at overtime rates, in order to keep the project on schedule.

The preparation of the claim includes two aspects:

- A report outlining the reasons why additional payments should be made to the contractor.
- An analysis showing how the additional costs have been calculated.

The client's quantity surveyor

The quantity surveyor will recognise that most building and civil engineering contractors desire to carry out the works to the complete satisfaction of the employer. However, their main reason for being in business is to make a financial profit from the project.

The client's advisors are very much in control of the project and can approve or disapprove of the contractor's methods of working. They can issue instructions to the contractor under the terms of the contract, order additional works, approve or nominate subcontractors, etc. They will also have had some input into appointing the contractor, before the contract was awarded. A contractor may have to request information from the architect or engineer. This may be a positive step on the part of the contractor in alleviating a claim at some later stage. The architect or engineer must respond to such requests within a reasonable time.

After receiving a contractual claim from the contractor the following should be considered:

- Is the contractor's claim reasonable?
- What are the costs involved?
- What clauses in the conditions of contract are relevant?
- What basis is the contractor using to justify a claim?

The quantity surveyor may be tempted to dismiss a contractor's claim out-of-hand. It is well recognised that claims generally, and contractors' are no exception, may be inflated on the assumption that it will be contested and a reduced amount negotiated. The contractor may not be entirely faultless in the matter giving rise to their claim, and some of their losses could be of their own making. The agreed amount of the claim is therefore always likely to be lower than that originally requested or calculated by the contractor.

So the instinct at first may be to reject the claim, lest any acceptance might appear to reflect badly on those appointed by the client or promoter. It may imply that they have not been carrying out their duties under the contract correctly. Where this is the case, the employer might have some redress against the architect or engineer for professional negligence. Where the matter cannot be resolved, by the parties involved, then other ways of dealing with the problem, as described above, may need to be adopted.

The changing face of disputes

We have seen in this chapter how, over the years, a number of approaches to the resolution of construction disputes have been developed, each seeming appropriate to differing types and scales of dispute. We have also seen how there is an on-going shift away from litigation and arbitration towards adjudication and ADR as acceptable forms of dispute resolution. Only recently has the relationship between arbitration and litigation seen a reversal, with litigation taking the place of the former as the default (preferred) option after adjudication, as decreed by JCT 2005.

At present, neither adjudication nor any of the ADR techniques produce decisions which are considered legally binding by the courts, but who is to say that this too will not change. As we have seen, the courts are already giving increasing credence to the decisions of adjudicators. Perhaps before too long legislation will be produced which does indeed give this status to the decisions of adjudicators, such that a greater than ever proportion of matters can be decided conclusively at a grass roots level, which must be best for time and cost.

As regards the role of the quantity surveyor in all of the above, this chapter has been based largely upon the state of affairs (certain procurement routes engendering certain types of disputes) that are still largely prevalent within the construction industry at the time of writing. However, things are changing at an increasing pace and, as newer practices increase in currency, such as that of partnering, the adversarial undertone suggested at the start of this chapter may be changing too. Partnering, for example, calls for an open-book, trusting relationship between the parties. This seeks to break down the traditional, *us and them*, stances of both the client's and contractor's quantity surveyors. Perhaps in ten years time, or less, we shall be writing simply of the quantity surveyor, disputes of the kind described above being a thing of the past.

Discussion topic

A new client, with no previous association with the construction industry, is concerned by current press coverage of the effects on time and money of disputes on certain high profile construction projects. He requests a meeting to discuss strategies (a) for the avoidance of disputes and, (b) should they arise, their effective resolution. Prepare a set of notes as a brief for your meeting, addressing each of the above issues.

(a) Causes of dispute and their avoidance

If indeed your client is new to commissioning construction works you must explain the potential for dispute. This is a very real possibility at a number of points in time unless matters are approached in a certain way. This will vary depending upon the type and scale of their proposed works and can be affected by the procurement route chosen.

A major feature of the construction process is that, whatever the project, all the parties are involved in a unique operation or set of operations, each offering the potential for problems to arise and thus, at worst, some form of dispute. It cannot be guaranteed even when two buildings of identical design are built side by side on the same site, that they will exhibit exactly the same pace of construction, experience exactly the same ground conditions, or be affected by the exactly the same weather, for example. In most cases, one is concerned with a single new-build or refurbishment project, quite unique in the problems it may present. Problems usually give rise to delay or extra cost, both of which can give rise to dispute.

You may refer your client to a list of the number and type of interested parties who might become involved, such as that presented within this chapter. Who is actually involved, the risks which each are to bear and thus the effects of these will vary, as noted above, depending upon the size and complexity of the project and, from the client's perspective, the procurement route adopted.

Other chapters in this book have examined the importance of calculating and appropriately apportioning risk. It has been explained elsewhere how, as the procurement methods change, from traditional competitive tendering with priced bills of quantities to, say, design and build, the related responsibilities for design quality, cost and time will shift from the employer over to the contractor. As responsibility shifts, so too does the risk factor in each. Explain the responsibilities of the parties involved, depending upon the procurement method chosen. These will usually directly determine the risk they are carrying which will, in turn, determine the effect upon them of any problems that might occur, and thus any likelihood of their instigating formal dispute proceedings at some stage.

Common to the successful dispute-free progress of any procurement path is the need for timely, effective communication of information forming the contract, whether this be a particular contract term, a drawn detail or an item in the specification. Get this right as the client, and encourage all others involved in the process to do likewise, and one goes a long way towards avoiding disputes in the first place. The most embarrassing scenario for the client must surely be where they themselves, or their immediate representatives, architects or other consultants, have been late, inaccurate or unclear in the issue of information. This presents the clearest case for a contractual claim from the contractor and perhaps, subsequently, a dispute. It has been found, from research conducted by the Building Research Establishment, that up to 90% of all variations, which can be the route of claims for extra time or money, are caused by the client or their representatives, born of modifying or firming up details which might have been finalised at some pre-tender stage.

(b) Which dispute resolution technique to use

Ideally, contentious issues will be avoided as much as possible, or can be resolved through timely discussions between the parties, but there will be those which are not settled and for which a more formal and systematic approach must be adopted.

Where matters do call for a formal process of dispute resolution, the recommended course of action will depend upon the procurement route involved, as this will usually have pre-determined the form of contract in operation. Notwithstanding the common law right in all situations to take one's opposite number to litigation, the form of contract, for its part, will state clearly the options available and the route to be taken in seeking to resolve any dispute. Thus, for example, the JCT family of contracts have traditionally, until JCT 05, steered the parties through adjudication, then arbitration and then to litigation, and in that order. Newer standard forms, as they are known, sometimes offer a variation to this, for example, the *Engineering and Construction Contract* (ECC) encourages an attempt at mediation or conciliation before moving onto more serious territory.

Most recently, in JCT 05, as with the PPC 2000 (partnering) contract, mediation has been introduced to the more traditional contracts of long-standing. Also, in recognition of the actual costs and timescales involved, an emphasis has begun to be placed on litigation rather than the hitherto automatic recourse to arbitration.

In advising your client, you may find it helpful to summarise the key features of the differing dispute resolution techniques in a similar manner to that shown in Table 16.1, highlighting those which are in fact available to them, depending upon their chosen procurement route/form of contract. To some extent, one trades cost

Table 16.1 Principal features of techniques for dispute resolution

Technique	Cost	Time	Quality
ADR – that is; mediation, conciliation, etc.	Minimal – basic expenses and small fee perhaps, as for registered mediator	Quickest process – dependent only on the parties finding earliest opportunity to meet and discuss the issues	A private decision, not binding on either party. Only of lasting value if the problem is fully resolved and not returned to by either party
Adjudication	Cost of preparation should be small. The adjudicator's fee is shared between the parties	An on-site process which should be commenced as soon as possible upon notification of any dispute. The adjudicator has a limited time to decide, causing minimum disruption to the progress of the project	A private decision, once again not regarded as binding in law, but decisions of adjudicators are increasingly being upheld by the courts, suggesting that a subsequent challenge will often prove futile and to be avoided perhaps?
Arbitration	Costs to be paid for arbitrator's time and premises for hearing. Also, the parties may employ legal counsel etc. at additional cost	May be protracted over many months, as the arbitrator may give the parties a number of months to produce their respective evidence. The arbitrator may then take some time over his/her decision	The arbitrator's decision will be private, as above, but expert and binding unless challenged in the courts on a matter of legal process, not fact – quite rare
Litigation	Court and judge provided free, but costs of specialist legal and technical advice, if used, and of barristers in court may make costs very high	Can take place within six months or less, due to growing availability of technical and construction courts	The court's decision will be reported publicly, expert and binding and can only be challenged in a higher court (usually at great expense to the appellant). Only challenged on matters of legal process, not fact.

and speed against the quality or certainty of any decision. You must point out that, despite the increasing support given to adjudicators' decisions upon appeal, they are not as yet automatically binding on the parties, as are those of the arbitrator or judge. You should inform your client of the procedural requirements of each of the three main techniques: adjudication, arbitration and litigation. Although notices may not need to be given by the client in person, there are certain timescales etc. involved of which they will wish to be advised, as these may affect their own cash-flows and other arrangements.

The client should also be advised that the differing dispute resolution techniques carry with them varying degrees of privacy, ranging from total confidentiality in the case of on-site mediation, through to total disclosure in the case of litigation, where results are in the public domain (via Law Reports and the like). Both parties may seek to avoid publicity, for the sake of their reputation, and for this reason alone may attempt resolution of their dispute through ADR.

It may be that you are faced with a situation in which none of the existing standard forms meets your client's requirements or, indeed, they insist in any case upon your designing a contract (a bespoke contract) for a particular project. This may afford the opportunity to stipulate a specific dispute resolution technique according to the value of contract and/or the likely value of any dispute. This might seem appropriate, as a low value dispute may not warrant engaging in lengthy and/or costly proceedings. However, the client should be advised that only decisions arising out of arbitration or litigation will be binding. In all cases they should be advised that, under the Construction Act 2000, you must include a provision whereby either party may take a dispute to adjudication should they wish.

Additional reading

A very full and detailed account and discussion of the requirements and procedures involved in each of the dispute resolution procedures can be found in Part E of *Tolley's Guide to Construction Contracts* (Lewis *et al.*; updated annually).

References and bibliography

Appleby G. *Contract Law.* Sweet & Maxwell. 2001.
Barrett F.R. *Cost Value Reconciliation.* Chartered Institute of Building. 1992.
Bartlett A. *Emden's Construction Law.* Butterworth. (Annual.)
Brown H. and Marriott A. *ADR Principles and Practice.* Sweet & Maxwell. 1999.
Chappell D. *Parris's Standard Form of Building Contract: JCT 98.* Blackwell Science. 2001.
Chappell D. *Powell-Smith & Sims' Building Contract Claims.* Blackwell Publishing. 2004.
Chappell D., Marshall D., Powell-Smith V. and Cavender S. *Building Contract Dictionary.* Blackwell Science. 2001.
Darbyshire P. *Eddey and Darbyshire on the English Legal System.* Sweet and Maxwell. 2001.
Delves S. *Tolley's Guide to Construction Contracts.* Butterworth. 2005.
Elliot R.F. *Building Contract Disputes: Practice and Precedents.* Sweet and Maxwell. (Annual.)
Furmston M.P. *Cheshire, Fifoot and Furmston's Law of Contract.* Oxford University Press. 2006.
Furmston M.P. *Powell-Smith and Furmston's Building Contract.* Blackwell Publishing. 2006.

Harris B., Planterose R. and Tecks J. *The Arbitration Act 1996.* Blackwell Publishing. 2007.

Hibberd P. and Newman P. *ADR and Adjudication in Construction Contracts.* Blackwell Science. 1999.

Lewis S. *et al. Tolley's Guide to Construction Contracts.* Tolley. (Annual updates.)

Ndekurgri I. and O'Gorman C. *Construction Law and Contractual Procedures.* Butterworths. 2001.

Povey P.J., Wakefield R. and Danaher K.F. *Walker-Smith on the Standard Form of Building Contract.* Tolley. (Annual.)

Reynolds M.P. *The Expert Witness in Construction Disputes.* Blackwell Science. 2001.

Riches J. and Dancaster C. *Construction Adjudication.* Blackwell Publishing. 2004.

Rutherford L. *Oxborn's Concise Law Dictionary.* Sweet and Maxwell. 2001.

Scriven J., Pritchard N. and Delmon J. *A Contractual Guide to Major Construction Projects.* Sweet and Maxwell. 1999.

Sykes J. *Construction Claims.* Sweet and Maxwell. 1999.

Turner R. *Arbitration Awards.* Blackwell Publishing. 2005.

17 Project Management

Introduction

Increasingly, clients are adopting a project culture in all aspects of their business. Project management is not new or specific to construction contracts.

The organisation and management of construction projects has existed in practice since buildings were first constructed. The process long ago was much simpler but, as knowledge increased and societies became more complex, so the principles and procedures in management evolved. In some countries, notably in the USA, the management of construction works began to emerge as a separate and identifiable professional discipline some years ago alongside architecture and engineering. Because of the differences in the way the construction industry is structured in the UK, the professions have not developed in the same way, or to the same extent. It is now, however, being accepted by more and more clients that, to succeed in construction, someone needs to take the responsibility for the overall management of the construction project. The development of the standing of the project manager is shown further by contractual recognition, for example, within the new engineering contract (NEC) family of contracts. Project management is a very different function from either design or construction management and requires other, different qualities which are not necessarily inherent in the more traditional disciplines.

There has in recent years been a considerable interest amongst quantity surveyors in project management of one sort or another. This has been evidenced by the increased number of postgraduate courses, textbooks and other publications on the subject, and of practitioners seeking to specialise in this type of work. Some quantity surveyors anticipating the possible threat to their traditional role have seen project management as a source of work for the future. Others suggest that the financial expertise of quantity surveyors makes them ideally suited to such a role.

To many quantity surveyors, much of the project management service has been provided in the past as part of their extended traditional role, usually without acknowledgement or recognition. The degree to which this occurred was largely dependent on the relative level of skills and experience of the project surveyor, architects and engineers. However, it is important to distinguish between this extended quantity surveying involvement and formal project management. The quantity surveyor must take a substantial step in order to provide the comprehensive project management service required by clients.

As part of their agenda for change initiative, the Royal Institution of Chartered Surveyors has recognised the specialist activity of project management and its importance to the surveying professions by the establishment of a separate Project Management Faculty. In addition to the RICS, professional recognition of the project management function can be obtained via the Association of Project Managers, an organisation which, with regard to qualification, is more interested in professional competence than academic achievement and to which many quantity surveyors belong.

Justifying project management by adding value

The success of the design team in achieving the client's objectives is greatly influenced by their ability to recognise each other's activities and to integrate them to the full. However, it is the nature of most professionals to see the project objectives in terms of their own discipline and to operate, to a large extent, from within their specialist perspective. Project management provides the important management function of bringing the project team together and may be defined as:

> 'The overall planning, coordination and control of a project from inception to completion, aimed at meeting a client's requirements in order to produce a functionally and financially viable project that will be completed on time within authorised cost and to the required quality standards' (CIOB 2002)

This definition reveals the essence of project management as that of managing and leading the project team toward the successful completion of the client's project objectives. In execution, the role demands an array of management and technical sills with particular emphasis on the important aspect of human resource management.

Within most projects, there are several key aspects that are of importance to the client, and each will fall within the remit of the project management function. These include the fundamental considerations of time, cost and quality and their interrelationship, and within this framework, the management of procurement, risk and value, each of which is considered in detail elsewhere in this book. To justify the appointment of a project manager, it is important to be able to demonstrate the added value that the function brings to clients in respect to these key considerations.

There is some disagreement within the industry about the value of the service provided by dedicated project management, particularly from those who have previously performed this role, for example:

> 'It will be seen that in the orthodox process of managing building there is a tendency for middlemen of various sorts to come between the exponents. They emerge in the first place because they can offer some rationalisation of the cost or of the organisation, but in the end there is a tendency for them merely to take a percentage without particular benefit and sometimes they even obfuscate straightforward solutions.' (Moxley 1993)

This illustrates the potential existence of resistance to the project manager, if project management were seen to fall into this middlemen category, being paid without adding benefit, or worse, for making the construction process more difficult.

The traditional method used in the management of construction projects has usually involved the architect, or on occasion an engineer, being both principal designer and manager of the process. This role included coordinating and programming the work of co-consultants such as services engineers, structural engineers and quantity surveyors. Whichever discipline led the team, and normally it was the architect, there was a need to perform both a management role and design role simultaneously. Unfortunately, in practice, it is difficult to achieve this and generally the design aspect was given pre-eminence. That this situation existed is perhaps no surprise given the apparent lack of attention to management subjects in UK Schools of Architecture.

The traditional relationships that have existed between the client, consultants and contractors have increasingly become strained over the years. This has frequently resulted in the dissatisfaction of the client, a central concern highlighted in both the Latham and Egan reports. The problems are largely due to the increasing complexity of design and construction, the importance of completion on time or early completion if desirable, the need for acceptable levels of building performance, and the increasing concern for financial control in its entirety. It is apparent from these demands that there is an urgent need for a much greater understanding and interpretation of the client's requirements. In addition, there has been a necessity for improved communications, together with a closer coordination of the work of all those involved in the design and construction processes. This is in part due to the changed nature of the construction industry and the different procurement routes adopted. Project management adds value by fulfilling the management role within the context of a modern and increasingly complex construction industry and in recognition of client demand.

Terminology

Whatever name is given to the role of the project manager, and alternatives may include project controller, project administrator, or project coordinator, the general intent is usually the same. The idea is that one person or organisation should take overall control and responsibility for coordinating the activities of the various consultants, contractors, subcontractors, processes and procedures for the full duration of the project. The project duration in this context starts at inception and ends on the completion of the rectification period. The management process may also extend into the time when the building is in use and thus link with the facilities management role. This whole-life view will be beneficial in bringing the design and development function more closely together with that of occupation and use.

Whilst the general intent of the project management role is understood, irrespective of designation, the title given to this function may denote differing levels of service. For example, the term 'project co-ordinator' is likely to indicate the exclusion of the responsibility for the appointment of the other consultants. In any event, the terminology used will always require precise definition in terms of the service to be provided that is stated in the terms of engagement. Clarification is certainly required to ensure that there is no misunderstanding as to the level of professional indemnity cover required. Figure 17.1 indicates a comprehensive list of project management duties differentiating between the project co-ordinator and project manager designations in regard to both external and in-house situations (CIOB 2002).

When acting in the dedicated role of project manager, there should be no attempt to perform any of the functions normally undertaken by the design team, including the traditional duties of the quantity surveyor. These should always be separate to avoid having to make any compromised decisions that might otherwise occur.

The term project manager may be seen within several contexts and is commonly used throughout industry. Within construction, it is important to distinguish between a contractor's project manager, who will primarily manage the construction process, and the client's project manager to whom this chapter is dedicated.

A further development of the management function may be seen in management-based contracts, for instance, construction management. Such a procurement route necessitates the appointment of a construction manager whose role involves the management of the design, procurement and construction. No main contractor exists; instead, the construction manager manages the trade contractors, and, as reimbursement is by way of a fee, this role can be carried out in a more objective manner with due recognition of the client's and project's interests. Although the duties of the construction manager differ from those of the project manager, with the former having a detailed responsibility for the management of the construction process, there is some commonality in the roles.

Attributes of the project manager

It is difficult to be dogmatic about the attributes of the project manager because there are no real rules. However, there are certain qualities that are both desirable and helpful in staff who have to fulfil the particular processes of management. There is a need, for example, for integrity, and for clarity of expression when speaking or writing. Loyalty, fairness and resourcefulness are also necessary. The following are some of the more important attributes that a good project manager should possess.

Personal traits

There are those who will attempt to argue that good managers are born and not made and that particular inherent qualities must already be present in an individual. They will equally argue that no amount of education and training or even experience can produce good managers. However, improvement in performance can always be achieved by encouragement in the right direction, and hence further study.

Duties*	Client's requirements			
	In-house project management		Independent project management	
	Project management	Project co-ordination	Project management	Project co-ordination
Be party to the contract	●		○	
Assist in preparing the project brief	●		●	
Develop project manager's brief	●		●	
Advise on budget/funding arrangements	●		○	
Advise on site acquisition, grants and planning	●		○	
Arrange feasibility study and report	●	○	●	○
Develop project strategy	●	○	●	○
Prepare project handbook	●	○	●	○
Develop consultant's briefs	●	○	●	○
Devise project programme	●	○	●	○
Select project team members	●	○	○	○
Establish management structure	●	○	●	○
Co-ordinate design processes	●	○	●	○
Appoint consultants	●	●	●	○
Arrange insurance and warranties	●	●	●	○
Select procurement system	●	●	●	○
Arrange tender documentation	●	●	●	○
Organise contractor pre-qualification	●	●	●	○
Evaluate tenders	●	●	●	○
Participate in contractor selection	●	●	●	○
Participate in contractor appointment	●	●	●	○
Organise control systems	●	●	●	●
Monitor progress	●	●	●	●

(continued)

Duties*	Client's requirements			
	In-house project management		Independent project management	
	Project management	Project co-ordination	Project management	Project co-ordination
Arrange meetings	●	●	●	●
Authorise payments	●	●	●	○
Organise communication/ reporting systems	●	●	●	○
Provide total coordination	●	●	●	○
Issue safety/health procedures	●	●	●	○
Address environmental aspects	●	●	●	○
Coordinate statutory authorities	●	●	●	○
Monitor budget and variation orders	●	●	●	●
Develop final account	●	●	●	●
Arrange pre- commissioning/ commissioning	●	●	●	●
Organise handover/ occupation	●	●	●	●
Advise on marketing/ disposal	●	○	●	○
Organise maintenance manuals	●	●	●	○
Plan for maintenance period	●	●	●	○
Develop maintenance programme/staff training	●	●	●	○
Plan facilities management	●	●	●	○
Arrange for feedback monitoring	●	●	●	○

*Duties vary by project and relevant responsibility and authority

Symbols ● = suggested duties; ○ = possible additional duties

Fig. 17.1 A comprehensive list of project management duties (*Source:* CIOB 2002).

Self-motivation and the ability to motivate others are very important, as is the relationship with those with whom the manager has to work. Own personal goals and moral values, coupled with own attitude to work and that of others, also need to be considered. The response to the various aspects of the project will vary depending on the project manager's own professional allegiance.

Appropriate knowledge and skills

The project manager is required to have a good balance of technical and managerial skills. Figure 17.2 indicates some of the key areas of knowledge and their interrelationships that contribute to the project management function. Whilst it is not a realistic expectation that the project manager is proficient in each aspect of the construction process, a working knowledge through indirect experience should be requisite.

Technical knowledge

Technical knowledge is an important element of project management. There is something unique about the construction industry, and project management within it is a rather specialised form of management. The techniques derived from the manufacturing industry, for instance, often do not work on a construction site, as construction is concerned with one-off projects undertaken on the client's premises. Ideally therefore, the project manager will already be a member of one of the construction professions. An understanding of the process and the product of construction, and a working knowledge of the structure of the industry, will clearly be advantageous if not essential. The importation of managers with no knowledge of the industry or its workings has drawbacks, and their appointment should be approached with caution. Quantity surveyors have a thorough overview of the

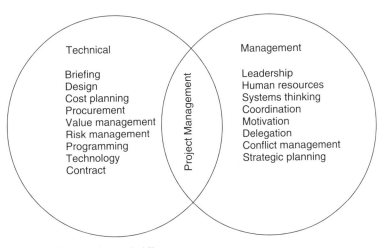

Fig. 17.2   Balance of knowledge and skills.

design and construction process and are technically well positioned for the role of project manager.

Management skills

Management is essentially a human matter, and this fact overshadows all other considerations. No one professional group therefore has a monopoly of the skills required, although the quantity surveyor is well suited to this role and perhaps better suited than many others. Various individuals have, however, chosen to specialise in management, having emerged from backgrounds in architecture, engineering, building and surveying.

In considering the performance of the function of project management, it is important to stress the importance of management skills. This is in the realisation that management is a job in its own right and is not something that other professions are able to do without certain qualities and additional skills. The manner in which managers are selected should also be borne in mind. Business organisations, including those within the construction sector, will generally appoint management personnel on the basis of the achievements of staff performing at a more operational level. Unfortunately, the range of skills and personal qualities necessary for such achievement often differs from those required in a management position. A successful project architect, quantity surveyor or engineer may be less successful in a management role. As previously emphasised, good management skills are an essential attribute of a good project manager.

Leadership qualities

Whenever a group of people work together in a team, the situation demands that one of the members becomes the leader. In many situations, leadership is an outcome of the inherent qualities of each member of the group; however, with the appointment of a project manager, the choice of leader is made. The project manager is the designated leader of the project team, whose duty it is to ensure that the whole work is carried out as efficiently as possible. The responsibility of combining the various human resources and obtaining the best from them is the project manager's, who must seek to complement the attributes of the various members of the team and keep conflict to a minimum.

It is important to recognise the difference between project administration and project management, which includes this essential function of leadership. The ability to lead and motivate others while commanding their respect is an essential characteristic, not required in administration but essential to effective management. Project management incorporates leadership and recognises the need to deal with people; project administration is a bureaucratic function, considerably less complex and often confused with true management.

In aspiring to project management, the responsibility for leadership should be accepted. However, conformity to a particular leadership style is not desired since there is no one perfect approach and various project situations may demand some flexibility. In the extreme, leaders may be autocratic or manage by some means of

---

Example: A project manager's need for flexibility in management style
Scenario: Development of an inner city commercial residential scheme, part new build, part refurbishment. An in-house project manager was involved from the outset to manage the project.

Objective A
The project team needed to resolve a range of issues at sketch design stage, including design issues, budget, procurement, etc. The management style used by the project manager was very consensus focused and successful in bonding the group. In addition, solutions were found to all of the key problems, including those relating to design, budget and procurement approach. There were positive contributions from all members of the team.

Objective B
During the course of design development, several problems were encountered with the project, largely due to the constraints imposed by the refurbishment element of the scheme. At this stage, programme difficulties were encountered and confrontation had arisen between key members of the design team. Several design team meetings were held in an attempt to resolve key issues and monitor/ encourage progress. As with objective A, a consensus approach was adopted; however, this failed to achieve clear direction. The end result was a significant and costly extension to the design period and a less than perfect working atmosphere, which continued throughout the project.

A key reason for this situation was the management style of the project manager at this crucial stage in the project. His democratic approach, successful in earlier stages of the scheme, was inappropriate in the latter phase when a more direct and firm approach is likely to have mitigated many of the difficulties.

The management approach should be a function of the situation, not the opposite. Flexibility in approach is preferred to dogmatism.

Fig. 17.3   Example of need for flexibility.

consensus. Clearly there are advantages and disadvantages to each. This is not to suggest that managers in practice execute a choice between these extremes since they will be governed by their natural tendencies. They should however recognise their own strengths and weaknesses and adjust their approach to meet the circumstances. Figure 17.3, which relates to the project management for a large commercial development, illustrates this need for flexibility.

With regard to construction project management, it should be accepted that the other team members will be experienced and competent, otherwise their appointment should not have occurred. It is therefore probable that in most situations a good project manager will tend toward the consensus approach, relying on the contribution of other team members and recognising their independence.

A good leader rarely needs to act ruthlessly, as this creates conditions of stress, strain and insecurity. The project manager's character and ability will set the tone from the beginning, and from this the loyalty of the other members of the team will be gained. The project manager must of course know when to praise and when to reprove, and must also have the courage to admit a mistake, to make changes or to proceed against opposition.

Clarity of thought

The ability to think clearly is also an important aspect of management. A confused mind creates confusion around it and may result in confused instructions to other members of the project team, which signifies that the manager is not in full control of the project situation. The inability in the first place to think clearly originates from an inadequate understanding of the objectives and priorities associated with the problem.

Effective delegation

The total amount of knowledge required in the management of a project is beyond the scope of a single individual. The manager must therefore be able to delegate certain tasks and duties to others involved in the project, and be able to rely on and receive advice from them. The inability to delegate is likely to result in overwork for the project manager, frustration on the part of others and a generally badly run project. Everyone must feel that they are able to make a valid contribution to the overall success of the project.

Decision-making

Decisions need to be made at all levels in the organisation of a construction project. Those made at the top will be concerned more with policy, client objectives and the framework for the project as a whole. At lower levels, they tend to relate more to the solving of particular problems. The aptitude for making decisions is an important quality that distinguishes the manager from the technician. The general level of complexity of construction projects and the number of consultants involved make the decision-making process for the project manager particularly difficult. Making a rapid decision requires a certain amount of courage. Sticking to a decision in the face of criticism, opposition or apparent failure requires a large amount of conviction. It is, of course, vitally important for the project manager to make the right decision, and this can only consistently occur through experience. The ability to sense a situation and exercise correct judgement will always improve with practice.

There has been much behavioural related research carried out over many years incorporating various aspects affecting the decision-making process. Whilst practitioners are unlikely to afford the time to participate in such activities themselves, they should be aware of the complexities surrounding the decision-making process and should understand the significance of this research in the context of their work.

Duties and responsibilities of the project manager

The project manager's terms of engagement, extent of authority and basis of fee reimbursement must be established prior to appointment. The experienced project manager will realise the importance of unambiguous conditions of appointment. Notwithstanding the cautionary comments outlined further below regarding the single system approach, the *Project Management Memorandum of Agreement and Conditions of Engagement* 1999 and associated guidance note, both issued by the RICS, will be beneficial in this regard.

The duties of a project manager in the construction industry will vary from project to project. Different countries around the world will also expect a different response to the situation, depending on the contractual systems that are in operation. The mistake, and perhaps the reason for the failure, of the traditional system in certain instances is that an attempt is made to use a single system to suit all circumstances. Any contractual arrangement, however good, must be adapted to suit the needs of the client and the project, and not vice versa. The project manager will need to employ a wide variety of skills and options for a whole range of different solutions. The key duties of the project manager identified in the RICS *Project Management Memorandum of Agreement and Conditions of Engagement* 1999 are to:

- Communicate to the consultants the requirements of the client's brief
- Monitor the progress of design work, and the achievement of function by reference to the client's brief
- Monitor and regulate programme and progress
- Monitor and use reasonable endeavours to coordinate the efforts of all consultants, advisors, contractors and suppliers directly connected with the project
- Monitor the cost and financial rewards of the project by reference to the client's brief.

Although not stated in the above outline of duties, it is implicit that the project manager will also lead the project, and therefore other expectations of the project manager, previously considered, should not be forgotten. The list above is relevant but is a bureaucratic response to the need to formalise the service provided.

The following sections provide more detailed consideration of some, but not necessarily all, of the duties of the project manager. In considering these duties, please note the significance of other sections of this book that are of major importance in project management, for instance, procurement, value management and risk management.

Client's objectives

The starting point of the project manager's commission is to establish the client's objectives in detail. The success of any construction project can be measured by the degree to which it achieves these objectives. The client's need for a building or engineering structure may have arisen for several reasons: to meet the needs of a manufacturing industry, as part of an investment function, or for social or political demands. In an attempt to provide satisfaction for the client, three major areas of concern will need to be considered. The weighting given to these factors will vary depending on the perception of the client's objectives:

- *Performance*
  The performance of the building or structure in use will be of paramount importance to the client. This priority covers the use of space, the correct choice of materials, adequate design and detailing, and the aesthetics of the structure. Attention will also need to be paid to future maintenance requirements once the building is in use. Clients are more likely to be concerned with the functional standards of the project, and to a lesser extent, the aesthetics.
- *Cost*
  All clients will have to consider the cost implications of the desired building's performance. The price that they are prepared to pay will temper, to some extent, the differences between their needs and wants. Clients today are also more likely to evaluate costs not solely in terms of initial capital expenditure, but rather on a basis of life cycle cost management. This is increasingly the case with central government and local government construction activity, as well as projects that are procured via an arrangement within the structure of the Private Finance Initiative.
- *Time*
  Once clients decide to build, they are generally in a hurry for their completed building. Although they may spend a great deal of time deliberating over a scheme, once a decision to build has been reached they often require the project to be completed as quickly as possible. In any event, in order to achieve some measure of satisfaction, and to prevent escalating costs, commissioning must be achieved by the due date.

The project manager's strategy for balancing the above three factors will depend on an interpretation of the client's objectives. It would appear, however, that there is some room for improvement in all three areas. The improvement of the design's completeness, particularly, should reduce the contract time and hence the constructor's costs. The correct application of project management should be able to realise benefits in these areas.

The client's objectives should be used as a goal for the broader issues involved in the design and construction of the project. The discernment of these objectives will assist the project manager to decide which alternative construction strategies to adopt. It is very important that an adequate amount of time is allowed for a proper evaluation of the client's needs and desires. Failure to identify these properly at the outset will make it difficult for the project to reach a successful conclusion upon completion.

Client's brief

This involves the evaluation of the user requirements in terms of space, design, function, performance, time and cost. The whole scheme is likely to be limited one way or another by cost, and this in turn will be affected by the availability of finance or the profits achieved upon some form of sale at completion. It is necessary therefore for the project manager to be able to offer sound professional advice on a large range of questions, or to be able to secure such information from one of the professional consultants who are likely to be involved with the scheme. This will include the coordination of all necessary legal advice required by the client. It is most important that the client's objectives are properly interpreted, as at this stage ideas, however vague, will begin to emerge, and these will often then determine the course of the project in terms of both design and cost. It is the project manager's responsibility to ensure that the client's brief is clearly transmitted to the various members of the design team, and also that they properly understand the client's aims and aspirations.

An increasingly used technique at this stage of a project is value management (see Chapter 8). This will assist in establishing clearly what the client's objectives are and is an excellent approach to the achievement of a consensus view of the brief, incorporating an evaluation of the needs and desires of all key stakeholders. Similarly risk management (see Chapter 9) is also an important consideration at this stage of the project.

Contractor involvement

The client will probably require some initial advice on the methods available for involving the contractor in the project. The necessity for such advice will depend on the familiarity of the client with capital works projects, although the growth and complexity of procurement options is such that some advice will almost certainly be of benefit. The correct evaluation of the client's objectives will enable the project manager to recommend a particular method of contractor selection. It may be desirable, for example, to have the contractor involved at the outset or to use some hybrid system of contractor involvement (see Chapter 10). The project manager will be able to exercise expert judgement in this respect by analysing the potential benefits and disadvantages of the project concerned. This decision will need to be made reasonably quickly, as it can influence the entire design process and the necessity of appointing the various consultants.

Design team selection

The project manager may be responsible for the selection of the design team. If this is the case, the task should be carried out in a professional manner, with the same amount of care as in the selection of the contractor. Although some situations will demand prompt negotiation with a proven team, if circumstances allow, proposals should be sought from three to six consultants. Ideally, the information submitted by the consultants should include matters relating to design and supervision methodology, and, possibly as a separate submission, a fee proposal. It will assist the decision-making process if matters relating to method and level of service provision are considered before and apart from the fee submission. The fee element should only be considered if these key aspects are acceptable or preferred.

If the client has been involved in capital works projects previously, they may already have designated consultants with whom the project manager will need to work. If this is the case, the project manager should make clear, at the outset, relevant concerns relating to any of the client appointed consultants, including past performance, location and resources. Where the consultants are appointed on a regular basis by the client, the project manager may experience problems due to existing relationships and lines of authority. This will be particularly difficult in situations where the client is using project management for the first time.

The project manager is likely to be responsible for agreeing fees and terms of appointment of all consultants on behalf of the client. In certain instances, the project manager may appoint the consultants direct as subconsultants. Under such circumstances, the only contractual link is between the client and the project manager. Where the contractor is to be appointed during the design stage, then the project manager will need to consider the means of selection. Whatever the circumstances, the project manager must control rather than be controlled by either the contractor or any of the consultants. The relationships between the contributions from each consultant must be clear at the outset to avoid any misunderstandings that may occur later.

Feasibility and viability reports

During the early stages of the design process, it will be necessary for the project manager to examine both the feasibility and the viability of the project. Sound professional advice is very important at this stage, as it will determine whether or not the project should proceed. A feasible solution is one that is capable of technical execution and may only be found after some site investigation and discussion with the designers. A feasible solution may, however, prove not to be viable in terms of cost or other financial consideration. Unless the project is viable in every respect, it will probably not proceed. The investigation work should be sufficiently thorough while taking note of the fees involved, particularly if the project should later be abandoned.

At this stage of the project it will be beneficial to carry out a detailed risk analysis of the proposed scheme or various options that are under consideration. Risk is an inherent part of every project and whilst a development may be determined as feasible, such a decision is reliant on assumptions and predictions relating to uncertainties that exist. Project management should incorporate a professional approach to the identification, assessment and management of risk. It should not rely solely on instinct, which may be subject to optimism, pessimism or other elements of bias (see Chapter 9), Opportunity for value management at feasibility stage also exists.

Planning and programming

Once the project has been given the go-ahead it will then become necessary to prepare a programme for the overall project, incorporating both design and construction. The programme should represent a realistic coordinated plan up to the commissioning of the scheme. The project manager must carefully monitor, control and revise where necessary. Several useful techniques exist for programming purposes, and since these can be computer-assisted, rapid updating can easily be achieved. The selection of the appropriate technique will allow the project to be properly controlled in terms of time.

The construction industry has a poor reputation with regard to achieving project completion dates. It is a difficult task to predict the completion date of a proposed development at inception. There are likely to be many unknown factors, not least those relating to site conditions, design solution and construction method. Notwithstanding the difficulties, project management will only be considered a success if project deadlines are met.

With regard to the programming of the construction phase, subject to the method of procurement, additional difficulties exist throughout the project's duration. This is due to the method of construction being largely unknown until tenders are received and, in any event, forecasts are reliant on the contractor's expertise and cooperation.

Design process management

Project information is often uncoordinated, and this leads to inefficiency, a breakdown in communications between the design team, frequent misunderstandings and an unhappy client. For example, the delayed involvement of service engineers often results in changes to the design of the structure both to accommodate the engineering work and also incorporate good engineering ideas. An important task therefore for the project manager is to ensure that the various consultants are appointed at appropriate times and that they easily and frequently liaise with each other while maintaining their own individual goals. There is not much room in the design team for those who wish to go it alone; teamwork is very much underrated, but it is vital for the success of the project. The project manager will therefore need to exercise both tact and firmness in ensuring that the client's objectives remain paramount.

The project manager, although not directly involved in the process of designing in its widest sense, must nevertheless have some understanding of design in order to appreciate the problems and complexities of the procedures involved. Responsibility for the integration and control of the work from various consultants rests with the project manager who, in the first instance, will be directly answerable to the client for all facets of the project. This will include ensuring quality control of all aspects of the design (and construction) process and carrying out regular technical audits on the developed design solutions.

Problems with the design process are more likely to occur when construction market conditions are very active. Although the project manager will endeavour to establish, prior to appointment, that each of the appointed consultants has the resources available to complete the commissioned work, circumstances may quickly change. This may be due to the acceptance by one of the consultants of additional work, perhaps for a larger client with the promise of additional work in the future. The cyclical nature of the construction industry and human nature may make it very difficult for some consultants to decline the opportunity of such commissions, despite concerns they may have about available resources. Although the terms of engagement should clarify time-scales, there may be problems with their enforcement, and delays in the design process can result when consultants become over committed. Project managers are likely to be in a better position to deal with this situation than a client, particularly those clients who are only involved in construction work occasionally. In this respect, it is important for the project manager to properly inform the client about the work of the consultants, something clients often feel is lacking.

The project manager must also be kept informed of the cost implications as the design develops, usually the function of an independent quantity surveyor. The design team must be informed of what can or cannot be spent and promptly advised when problems are envisaged. In this respect the control of the costs should be more effective than when relying on the efforts of the architect alone. The project manager must, of course, have a very clear understanding of the client's intentions and will also need to advise the client in those circumstances where the original requirements cannot be met in terms of design, cost or time. The project manager will always have an eye on the future state of the project and must keep at least one step ahead of the design team.

During this stage, unless the contractor has been appointed earlier, the project manager will need to consider a possible list of firms who are capable of carrying out the work, and to ensure that the proper timely action is taken to obtain all statutory approvals.

Supervision and control during construction

Subject to the selected method of procurement, the project manager should try to make sure that the design of the works is as near complete as possible prior to tendering. This is likely to result in fewer problems on site, a shorter contract period with a consequent reduction in costs, and commissioning at the earliest possible date. During the contract period the project manager will need to have regular meetings with the consultants and contractor and his subcontractors. Progress of the works must be monitored and controlled and any potential delays identified. The effect on the programme and the budget of any variations will also need to be monitored. The project manager must be satisfied that the project is finished to the client's original requirements; although one of the consultants may be responsible for the quality control, the project manager will need to be careful about accepting substandard or unfinished work. Some problems may need to be discussed with the client, but early decisions should be sought to bring the project to a successful conclusion. The project manager might have an ongoing role after the main construction contract to administer fitting-out work for occupiers and tenants.

Evaluation and feedback

This represents the final stage of the project manager's duties. It should be ascertained that all commissioning checks have been carried out satisfactorily, that the accounts have been properly agreed and that the necessary drawings and manuals have been supplied to the client. The project manager will need to advise on the current legislation affecting the running of the project, on grants, taxation changes and allowances. It may also be necessary to 'arbitrate' between consultants and contractors in order to safeguard the client's interests. The client should be issued with a 'close-out' report to identify that responsibilities of all parties have been satisfactorily discharged, and this will also assist in any future capital works that the client might undertake.

Quantity surveying skills and expertise

The skills of the quantity surveyor traditionally included measurement and valuation and to these were later added accounting and negotiation. As the profession evolved, these skills were extended to include forecasting, analysing, planning, controlling and evaluating, budgeting, problem solving and modelling. Knowledge has also been considerably developed both by a better understanding of the design and construction process and by having a broader base. The quantity surveyor in a rapidly changing work environment continues to increase the expertise base further and a survey of the contents of this book will reveal the extent to which the profession is growing, with the development of a wide range of skills and techniques to meet the varying demands of clients.

This provides the quantity surveyor with an excellent background, which is appropriate for project management. Indeed, a significant number of those already engaged in this work are members of the quantity surveying profession. The traditional role of the quantity surveyor, including that within the contracting organisation, is usually seen as advisory and reactive rather than managerial and proactive. In essence, the difference between the traditional role of the quantity surveyor, one that is diminishing, and that of the project manager, is one of attitude and method of approach. Consideration of the detailed list of duties of the project manager shown in Fig. 17.1 will serve to demonstrate the relative proximity of the two disciplines of quantity surveying and project management. This is certainly true from the perspective of larger firms of consultants that are able to provide a wide spectrum of specialist services.

Fees

In most countries, including the UK, the fees charged for professional services include the management of projects for clients. If the service being offered is enhanced by project management, then some extra charge will be deemed equitable. As the existing management organisation in construction is to some extent responsible for the poor quality, time delays and extra costs, then a process that attempts to rectify these problems should be worth paying for. The professional fee involved may in any case show a saving to the client overall. This, to a certain extent, will depend on the 'discount' given by other consultants for the omission of the management element from their service. However, it is clear that the type of management envisaged was never deemed to be included in the existing traditional fee structures.

The project manager will in the first instance need to negotiate a fee with the client. This fee will need to take into account the type of service being offered, the complexity of the project, its size and its duration. In periods of heavy competition, clients tend to require lump sum fees to be agreed for project management services. It is essential therefore that the basis of the fee in terms of scope and time-scale is clearly stated in the terms of appointment.

Education and training for the project manager

Quantity surveying education is constantly changing to meet the needs of the profession and industry of the future. Subjects that were once seen as being of paramount importance have been demoted as new subjects vie for space in the curriculum. This process is rarely without controversy and debate, partially due to the widening remit of the quantity surveying discipline.

A significant element of this change may be attributed to the development of project management skills within the profession. Quantity surveying undergraduate courses now include a range of subject matter that may once have been considered beyond the traditional role. This material, including topic areas such as risk management, human resource management, procurement management and value management, is provided with the view that many graduates may be practising in these areas in their early careers. This may be either in a project management role or in other management related areas of the construction industry.

Despite the acknowledgement given to the development of project management within undergraduate programmes, as with facilities management, there are many who regard the training and education as a postgraduate issue, supplementing a suitable first degree. There are several of these available, including master's courses, and in addition, for those surveyors who have not had any formal education in management, opportunity exists for attendance at short training courses and seminars.

The Association of Project Management (APM), which was established in 1972, is dedicated to the development of the discipline and provides control of the standards that must be attained to gain certification. The qualification gained from the APM is awarded on the basis of proven competence rather than academic achievement, and is therefore dependent on actual project management experience. The process for application involves submission of a report on a project for which the applicant had 'carried appropriate executive authority' and attendance at an interview. The association is broad in terms of subject matter; however, it has catered for particular specialisms by the establishment of special interest groups. Membership of the association is intended to bring with it advantages similar to those provided by other institutions, including the recognition of competence, opportunity for professional development, the marketing of services, a benchmark to employers and the control of standards of entry.

The RICS has also recognised the significance of project management by the establishment of a separate project management faculty in 2000. This is a significant development in that discipline may now be developed within a major international organisation and well directed resources and significant membership should enhance the abilities and status of all project managers within the institution.

Discussion topic

Consider the role of the project manager and discuss the potential knowledge and skills gap for the quantity surveyor aspiring to the discipline.

There is a well-recognised overlap in the knowledge and skills of the quantity surveyor and project manager. Indeed, one of the drivers behind the development of knowledge and skills in the QS profession has been its link with project management. Many QS practitioners have already elected to move into project management and, since this emerging discipline is recognised as a popular career path open to the QS, it is worthwhile giving consideration to the relative strengths and weaknesses of the QS in making this transition.

The activities of the project manager and the knowledge and skills base required to perform this role is discussed earlier in this text. From the perspective of the QS wishing to move into project management, the competence and confidence to perform the range of PM activities outlined will vary and be dependent upon the background and experience of the individual practitioner. To consider this in more detail, Fig. 17.4 is a schedule of the PM technical knowledge and skills base shown in Fig. 17.2 with a commentary on the perceived ability of the typical QS practitioner to relate to each function within a project management context.

Although there are apparent weaknesses in the traditional QS base of technical knowledge and associated skills of application, for example, with regard to design and briefing, there are also unique strengths such as cost planning and contract. The relevance of QS knowledge to the role of project management is further supported by consideration of the review carried out by the RICS in 1992 (*The Core Skills and Knowledge Base of the Quantity Surveyor*) (Fig. 17.5).

It is important to note that the above schedule, which was published in 1992, makes no specific reference to certain areas of practice, which although not yet considered as standard QS practice, are nonetheless well understood by most QS practitioners at present, for example, risk management and value management.

However, not to understate the importance of technical capability, as with project management relating to any industry, the factor most critical to the success of construction project management is the range of skills used in carrying out the role. Support for this opinion can be seen in a comparison of relevant research findings shown in Fig. 17.6. Although this indicates a difference in the relative ranking of the varying skills identified, the emphasis on human and organisational skills is clearly identified by all researchers (RICS 2003).

PM knowledge/ skill	QS level of knowledge and understanding	QS capability in terms of PM function (high/med/low)
Briefing	For most QS practitioners there is little involvement with the brief despite the impact this can have on all aspects of a project	Low
Design	Little appreciation of design and, more importantly, the design process. Design management can be a major issue in terms of deadlines	Low
Cost planning	Cost planning is a key aspect of QS activity, particularly those in private practice, and should be well understood across the profession	High
Procurement	Most practitioners have a broad knowledge of procurement, however, as with anyone involved in construction, the QS may suffer from bias and limited experience to alternative forms	High
Value management	Although this has been practised within the UK construction sector since the late 1980s, there are still many practitioners with little appreciation of its application	Medium
Risk management	In some respects similar to value management. However, the benefits of risk management seem to be widely acknowledged and most practitioners have exposure to basic risk management tools and processes	Medium
Programming	Within private practice, detailed programming is rarely encountered other than where claims may arise. Contractor programming activity is usually a separate role beyond the normal activity of the QS. Notwithstanding, the QS has some of the relevant skills	Medium
Technology	Typically, both contractors and consultants share the same academic base and develop their understanding through regular contact with the construction process	High
Contract	Both contractors and consultants share the same academic base and develop their understanding and skills through regular contract involvement	High

Fig. 17.4   Quantity surveying appreciation of project management knowledge.

QS knowledge (RICS 1992)	Relevance to project management
Construction technology	Knowledge of construction technology is of importance to the PM role. It would be challenging to lead a team of construction specialists otherwise. Furthermore, aspects of technology, for example, health and safety, are essential
Measurement rules and conventions	This is a special skill dedicated to the QS profession. However, knowledge of the measurement process, similar in this respect to knowledge of the design process, is useful to anyone involved in the PM of construction
Construction economics	Whilst the QS may prepare cost plans and give budget advice, the PM needs to understand the factors that influence cost (and revenue) and be able to assemble relevant cost data, and interpret relevant cost reports
Financial management	The project manager will be closely involved with the financial management of a project and ultimately report to the client on financial matters albeit at a more strategic level and dependent upon the QS for data and advice
Business administration	An appreciation of how businesses function, and the business objectives of clients is a pre-requisite to the provision of relevant and sound advice – a fundamental requirement for the project manager
Construction law	In the day to day management of projects a good knowledge of construction law is important. Where issues arise, the PM will be required to understand and relate to legal advisors

Fig. 17.5   Relevance of QS knowledge base to project management.

Rank	Young & Duff (1990)	Edum-Fotwe & McCaffer (2000)	Egbu (1999)
1	Supervision of others	Leadership	Leadership
2	Communication	Planning and scheduling	Communication (written/oral)
3	Motivation of others	Delegation	Motivation of others
4	Leadership	Chairing meetings	Health and safety
5	Organisation/site	Negotiation	Decision-making
6	Health and safety	Presentation	Forecasting and planning
7	Programme construction	Technical knowledge	Site organisation
8	Programme maintenance	Establishing budgets	Budgetary control
9	Management of quality	Drafting contracts	Supervision of others
10	Manpower planning	Decision-making	Team building

Fig. 17.6   The 10 most important skills in construction and development projects (*Source:* RICS 2003).

Therefore, the key question now remains: although the QS is able to show close proximity to the technical demands of the project management role, what about abilities with regard to the more important human and organisational skills?

The traditional *generic* skills considered to belong to the QS profession which were highlighted in 1992 (management, documentation, analysis, appraisal, quantification, synthesis and communication) do contain both human and organisational elements, for example, reflected in the reference to both management and communication. However, in the main, the listed items relate to the technical demands of the QS role, although relevant aspects of these are in part organisational and transferable to project management.

As with any professional moving to a new function within an organisation or starting new and different employment, there are inevitable gaps in technical knowledge which will need to be addressed. Fortunately, much of this can be achieved via academic study and peer support. However, despite this confidence in terms of the acquisition of the *hard* technical skills, there are many who believe that good managers are born and not made and, with this in mind, it may be that those QSs who do make the transition to PM successfully, do so because of inherent personal characteristics rather than professional knowledge and skills. Notwithstanding this viewpoint, there is also good opportunity to develop management attributes through academic pursuit and experience. The skills relating to the management function are discussed in detail in the main text.

Project management as a distinct discipline within the construction sector is relatively recent. To function well as a project manager it is necessary to have a wide range of knowledge and skills, some of which are shared by QS practitioners. As the project management discipline matures, so will its skills base and with the support of research, development and education this will continually grow. Competition in industry will generate this development through the need of organisations to gain market edge. Much of the project manager's attention is devoted to the delivery of projects on time, to a high standard of quality and to budget, clearly, all crucial in the eyes of most clients. However, in order to achieve these targets there is more to the PM role than establishing procedures and lines of communication, monitoring progress and addressing issues as they arise. There is a need to manage organisations, teams and individuals, making sure that they appreciate their role and the performance level required. In dealing with people, there is a need to acquire 'softer' management skills and this is perhaps the biggest gap that all aspiring to the project management role, from any discipline, need to address.

References and bibliography

Bennett J. *International Construction Management*. Butterworth-Heinemann. 1991.
Centre for Strategic Studies in Construction. *Trusting the Team*. University of Reading. 1995.
CIC. *Project Management Skills in the Construction Industry*. Construction Industry Council. 1996.
CIOB. *Code of Practice for Project Management for Construction and Development*. Blackwell Publishing/Chartered Institute of Building. 2002.

Edum-Fotwe F.T. and McCaffer R. Developing project management competency: perspective from the construction industry. *International Journal of Project Management* 18, 111–124. 2000.

Egbu C.O. Skills, knowledge and competencies for managing construction refurbishment works. *Construction Management and Economics* 17, 29–43. 1999.

Fryer B., Egbu C., Ellis R. and Gorse C. *The Practice of Construction Management.* Blackwell Publishing. 2004.

Langford D., Hancock M.R., Fellows R. and Gale A.W. *Human Resources Management in Construction.* Longman. 1995.

Lavender S. *Management for Building.* Longmans. 1996.

Moxkey R. *Building Management by Professionals.* Butterworth-Heinemann. 1993.

Powell C. *The Challenge of Change.* The Royal Institution of Chartered Surveyors. 1998.

RICS *The Core Skills and Knowledge Base of the Quantity Surveyor.* The Royal Institution of Chartered Surveyors. 1992.

RICS. *Project Management Agreement and Conditions of Engagement: Guidance Note.* RICS Books. 1999.

RICS. *Project Management Memorandum of Agreement and Conditions of Engagement.* RICS Books. 1999.

RICS. *Learning from Other Industries.* The Royal Institution of Chartered Surveyors. 2003.

Walker A. *Project Management in Construction.* Blackwell Science. 2002.

Woodward J.F. *Construction Project Management: getting it right first time.* Thomas Telford. 1997.

Young B.A. and Duff A.R. Construction management: skills and knowledge within a career structure. *Building Research and Practice* 18 (3), 183–192. 1990.

Website

The Association of Project Management. Certification available on line at http://www.apm.org.uk/ac/cert.htm.

18 Facilities Management

Introduction

The emergence of facilities management is in response to the growing realisation that a company's property assets are of vital importance to business success. This may be considered in two ways:

- In most companies, the costs associated with real estate are a major overhead, often representing the largest cost after wages and salaries. The control of this overhead, via cost management and use optimisation, is therefore likely to make a significant contribution to profitability.
- In addition, the impact that real estate may have upon income generation may be significant, indeed, more significant than the direct costs of provision. For instance, a car factory with an ageing and inefficient layout may be unable to compete with a modern facility that is able to accommodate new techniques, for example, robotics.

Recognition of the importance of facilities management is a vital stage in the development of this relatively new profession. Not that the function of facilities management is new; every company has needed to manage the facilities requirements of their business in some way. For example, consider the facilities management occurring in a typical small office situation, described in Fig. 18.1.

In this situation neither the appointment of a dedicated in-house facilities manager nor the outsourcing (see later in this chapter) of the facilities management (FM) role (much though the senior partner of the practice would probably like it) is practical due to the size and resources of the firm. Nevertheless, the FM function is occurring, and the cost of this to the practice can be evaluated in terms of the time of the senior partner, which, if considered in terms of opportunity cost, may be very significant.

Until recent years, the majority of facilities management, including that relating to larger concerns, would be handled from within a company's own organisation. This continues to be a common situation; however, there is a growing realisation that the outsourcing of this management function offers several benefits to the client including:

- It allows a company to concentrate on its core business activities
- It allows expertise to be developed through specialism

A small firm of accountants operating from a leased, single office location comprising a late Victorian terraced house located on the fringe of a provincial city centre. The decisions relating to the office facility are made by the senior partner and/or the office manager and are likely to include matters relating to:
- The property lease, renewal, rent and rates
- Repairs and maintenance to the building within the terms of the lease
- General furniture and fittings
- Office equipment including IT, copiers, telephones, etc.
- Insurances
- Health and safety aspects
- Security
- Ancillary services including kitchen provision, WC provision
- Car parking arrangements for staff
- Operational expenses relating to utilities, etc.

Fig. 18.1   Facilities management in a typical small office situation.

- It allows smaller organisations, unable to justify the direct full-time employment of specialist FM staff, to access experienced FM personnel
- It may provide economies of scale.

Facilities managers have evolved from the existing professions and those practising include personnel from various surveying disciplines, other construction professionals, contractors, and professions beyond the construction domain, e.g. accountants – such is the size and attractiveness of the market opportunity. Despite the scale and variety of this competition, it is accepted that the quantity surveying profession is well placed to provide a facilities management service. This may be done by adapting and expanding the existing quantity surveying skills and knowledge base, a large part of which is central to the facilities management process.

In a book of this type it is by no means possible to adequately cover the entire subject matter that falls under the umbrella of facilities management. As shown below, the range of activities is potentially diverse and the knowledge and skills necessary for the fulfilment of each requires specialist attention. The purpose of this chapter therefore, is to introduce the reader to the world of facilities management, examine some of the general practice and procedures relating to it and highlight the opportunity it provides to the quantity surveyor.

The work of the facilities manager

There is some confusion as to exactly what facilities management is and there is an array of published definitions, the variety of which displays a range of differing interpretations. These are likely to be influenced by the particular interests of their originating body, reflecting professional strengths and institutional perspectives. One of the simplest and most direct definitions of facilities management is that given by the Royal Institution of Chartered Surveyors, which states that 'FM involves the total management of all services that support the core business of an organisation'. This emphasises the management role and indicates the sphere of activity as that relating to services 'supporting' the core business.

In practice, the scope of facilities management service may range from the provision of a single service at operational level, for example maintenance management, to that in which a comprehensive range of services is provided including those at a strategic level. Although it is likely that quantity surveyor involvement in facilities management is at present largely restricted to operational matters, typically maintenance management, it is important to recognise the discipline as much more. Failure to do so demeans the role of facilities management and deprives both the client body and consultant of value adding opportunity. The remainder of this section therefore, considers the work of the facilities manager in the wider context and identifies the opportunity that this relatively new discipline provides to the quantity surveying profession.

Space planning

The pace of change in the business environment in recent years has been great and is generally seen to be accelerating. This places a continual demand upon the workplace to adapt. Companies must respond to change; otherwise they will deteriorate in their importance and in the worst situations disappear entirely.

In terms of property, it is essential that this changing demand be reflected in the space provided for the business to operate effectively, which in terms of survival also means competitively. Facilities managers need to research and analyse the space requirements of an organisation, which may include a complete range of working environments, for example, manufacturing, retailing, storage, administration, health and education.

Workspace is an expensive commodity and therefore its provision must be fully justified in terms of need and output. The process of space planning should incorporate the following stages.

Identifying and evaluating demand

As a company grows and develops, its demands on property will change. The nature of property is such that there is a need to plan for future needs since there is invariably a time lag between need and delivery. Unfortunately, it is a common feature within organisations that space provision becomes inadequate and inefficient before action is properly considered. In forecasting future needs, the factors to be considered are wide and include anticipated company growth, changes in working practices (e.g. hot-desking), the impact of technology, changes in function and expected duration of need. In some businesses, much of this changing need can be seen in the concept of the 'virtual office' that to many people is already becoming a reality.

Since the cost of space is a high proportion of a company's outlay, there is likely to be some emphasis on the reduction of space costs to an optimum; perhaps this may be particularly so if managed from a traditional quantity surveying perspective. Whilst this approach may lead to a reduced cost, the effect that such a policy may have on revenue should also be considered. It has already been established (Chapter 8, Value Management) that value is increased by the elimination of unnecessary costs. Therefore, within the context of space planning, it is only unnecessary space costs that should be eliminated. The elimination or downgrading of space necessary to support creative work, marketing, image, and other aspects that may affect productivity or sales is unlikely to add value. The impact that the correct space configuration may have on efficiency may be considered by comparison of the relative attributes and constraints of cellular and open plan office layouts. For example, the level of privacy achieved in a cellular layout is likely to enhance concentration, however at the same time deprive the workforce of the benefits derived from the close proximity of colleagues, such as support, learning and team working.

In order to identify and evaluate demand, an objective research process should be carried out which will allow the determination of the optimum space allocation and layout for the various working activities within a business organisation. In assessing the needs of staff, there are many factors that will need to be taken into consideration including the need for privacy; access and use of equipment; status; communications; anticipated churn; meeting requirements; storage requirements; functional needs; staff interaction; patterns of working; power supplies; and security provision. The methods used in acquiring the required information will include staff interviews and/or questionnaires, observation of space use, e.g. to assess room or workspace vacancy, and the application of benchmarked data.

Determining supply

Once the accommodation needs of a business are established, the availability of premises may be considered via the consideration of existing properties owned or leased by the company, or by the procurement of additional space. To allow the optimum use of space, an assessment of the attributes and constraints of the available premises is necessary. The considerations will include those of efficiency, quality of environment and aspects relating to aesthetics and image. In addition, factors at a relatively micro level will also need to be considered. For example, in attempting to provide a satisfactory space solution for an office premises, the following factors may impact on the eventual layout:

- Positions of windows limiting the position of partitions in a cellular layout
- Depth of office space
- Points of access and emergency egress
- Lift provision and location
- Position of electricity mains distribution
- Existing serviced amenities
- Building structure
- Position of cables and ducts.

Frequently, the provision of new space will come from existing building stock and therefore involve the refurbishment of existing premises. The decision to refurbish rather than procure new premises is unlikely to be simply that of a straightforward choice depending on the relative merits of the space provided; existing stock must be used or disposed of and capital costs, existing lease arrangements and time factors are likely to have a significant impact. In deciding to refurbish existing premises, the conflicting aspects of continued occupation, construction speed and safety must be considered. Where a building will continue in use during refurbishment works, aspects of comfort and safety should be to the fore and the impact that such provisions will have on speed of construction understood and fully accommodated within the business programme. The problems associated with working in existing buildings are considered in more detail in the next section.

In addition to providing new accommodation, either from the refurbishment of existing stock or the procurement of new, the need to deal with excess space due to obsolescence may also occur.

Maintenance management

One area of facilities management for which the traditional quantity surveyor skills can be successfully applied is that of maintenance management. This is a very large market opportunity with a total expenditure estimated by Building Maintenance Information in 1997 to be approximately £40bn (Building Maintenance Panel of the RICS 2000). In addition to the size of the market, the workload activity is less vulnerable to market fluctuations in that maintenance is required to existing building stock and is not reliant on investment in new buildings.

An illustration of the areas and distribution of maintenance expenditure relating to a modern air-conditioned building is shown in Fig. 18.2.

There are two broad categories of building maintenance: preventative maintenance and corrective maintenance. In terms of any piece of equipment, there are some items that will be attended to following 'failure'; however, there are distinct advantages in adhering to a policy of preventative maintenance. For example, consider the example of a service to a car. Figure 18.3 highlights typical items of work that may be carried out and shows the category of service provided with reference to preventative and corrective maintenance. The benefits of carrying out the preventative maintenance work are clear. For instance, the timing chain, which may cost say £200 to replace, could lead to a repair bill in excess of £1 000 if its failure occurred during road use. Likewise, the oil change, air filter and oil filter are relatively cheap measures that will protect the engine and extend its life. In addition to the long-term benefits, preventative maintenance provides peace of mind and improves reliability.

This philosophy works equally well in buildings; however, the complex nature of buildings and the business organisations that use them may result in a less well-organised approach.

The maintenance of buildings is an important element of facilities management in that it is essential to the efficient use and costs of operation. The production of a maintenance policy, similar in concept to the list of checks at various service intervals during the life of a car, will provide the direction necessary to achieve good

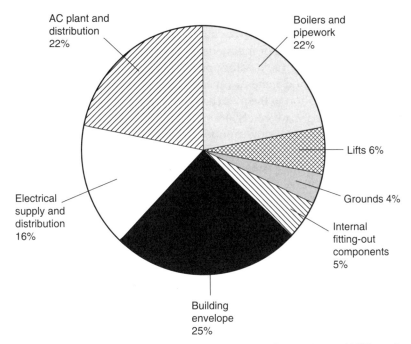

AC plant and distribution 22%

Boilers and pipework 22%

Lifts 6%

Electrical supply and distribution 16%

Grounds 4%

Internal fitting-out components 5%

Building envelope 25%

Fig. 18.2 Maintenance expenditure relating to air-conditioning (*Source*: Bernard Williams Associates, 1994).

Preventive maintenance	Corrective maintenance
Replace timing chain	Replace indicator bulb
Oil change	Renew illegal tyre
Replace air filter	Touch up paintwork
Replace oil filter	Replenish screen wash
Replace brake pads	Adjust handbrake

Fig. 18.3 80 000-mile service on a car.

standards of maintenance. The policy should consider the various categories of maintenance work, the standards to be attained, health and safety requirements, security factors and building access. In addition, an action plan with regard to each category of work, including response times and lines of communication, needs to be established, as well as methods of budgeting and payment.

Problems in existing buildings

There are unique problems to be overcome when working in existing buildings. The difficulties may be extended to the design, planning, costing and execution of the work and incorporate the following considerations:

- Working conditions: Work in existing buildings is likely to be hindered by continued occupation. Whilst this may not impact on some types of work, the execution of any major repair or maintenance will need to accommodate the needs of existing users. Health and safety provisions (see later in this chapter) must be strictly enforced and general nuisance factors such as noise and dust can cause serious disruption to the main activities of a business. Similarly, the image of a company may be adversely affected due to building operations. To compound the problems, working around occupants can be very time consuming and thus costly. Careful planning of work in existing buildings is essential if disruption and costs are to be minimised.
- Abnormal hours: Some aspects of a business cannot be interrupted during normal business hours; for example, any work to an existing IT installation may necessitate the closure of a networked system preventing internal and external communications, business transactions and the like as well as the general use by all members of staff of the IT facility. This is major disruption that could result in a substantial loss to a business. Therefore, work of this type should be carried out beyond normal working hours where possible. Similarly, power supplies are vital to any organisation and must be maintained to allow continued production. Consider the cost implications of a loss in power in a large manufacturing organisation resulting in standing time of one hour.
- Layout constraints: The installation of new floor layouts, services and specialist installations may be dramatically constrained by the existing floor heights, internal walls and structures. Many older buildings are unsuited to adaptation and this factor may ultimately result in their obsolescence and disposal.
- Legal constraints: Planning restrictions and building regulations may restrict refurbishment potential. In particular, listed buildings, whilst possibly providing a unique atmosphere and enhancing company image, pose particular problems by preventing alteration and generating additional expense for simple repairs and maintenance.
- Unique building structures: Depending on the age and status of a building, repairs and maintenance may be problematic due to external controls, the availability of suitable materials and the inability to accurately assess and plan repairs. Some typical problems encountered are considered in the example in Fig. 18.4.

Due to the uncertainties that may exist, procurement of work to existing structures needs careful consideration. Without the opportunity to prepare adequate detailed designs, which as outlined above is not always achievable, it is likely that the client will be required to accept more risk than with new works.

Procurement

The approach to the execution of maintenance work will be influenced by the size and type of client. Large client organisations may have a comprehensive in-house team fulfilling the professional, technical and construction roles of building maintenance, whilst smaller businesses will be unable to sustain such an internalised

Premises: Large landmark Victorian building being used as main administration head-quarters

Site problems:

- Repairs to existing façade required following storm damage; availability of matching imperial bricks caused delay and expense
- Outbreak of dry rot on upper floor detected during the works. Extensive access required to check for dry rot throughout the building, causing disruption to the occupiers due to remedial work and adherence to health and safety provisions
- Installation of secondary glazing throughout the west elevation of the building to exclude external noise. Irregularity of the window sizes and types involved much customisation and delay
- Large ornate window requires replacement. Due to listed building status, the replacement window must comply with the requirements of the planning authority and be an exact replica – a very expensive purpose-made unit taking several weeks to manufacture
- All external stonework needs to be cleaned and repaired. The organisation and budgeting of this work is difficult due to the inability to accurately assess the scope of the works. Full details will not be known until scaffolding is erected and cleaning occurs.

Fig. 18.4   Problem of unique building structures: example.

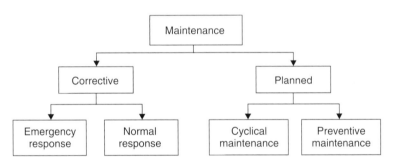

Fig. 18.5   Types of maintenance (adapted from Spedding 1994).

operation and need to outsource all of the maintenance requirements. The issues surrounding the merits and concerns relating to outsourcing are discussed later in this chapter.

With regard to maintenance work, the employer may choose to contract with several contractors to carry out individual projects or maintenance contracts, or appoint a single management contractor responsible for the entire maintenance requirement.

The types of maintenance work required by the employer will vary, and as stated above, broadly include work of a preventative nature or corrective nature. Although there is varying terminology in use, including that defined in BS 3811, the outline shown in Fig. 18.5 provides a clear indication of the range of maintenance work encountered.

The procurement of maintenance work will depend on the type of work required, as follows.

- *Corrective maintenance*
  Emergency response   Where sudden breakdown or accident occurs, prompt action may be necessary, for example in order to maintain a service or make a building safe, secure or watertight. This may extend to a 24 hour/365 day service. *Normal response*   It would be inefficient and costly to categorise all corrective maintenance as 'emergency' work. Some items of work may be deferred until a regular visit occurs, for example the replacement of a light bulb or ceiling tile, or attendance to a dripping tap.

  The organisation for the execution of corrective maintenance work will require the pre-definition of items of emergency work, financial limits, response times and clear lines of action. There may be a desire for local client facilities representatives to pursue emergency action in non-emergency situations. Surveyors involved in the maintenance process may be required to give appropriate direction in accordance with corporate policy and arrangements when a possible emergency situation is referred.

  The most appropriate contractual response for corrective maintenance work is likely to be some sort of term arrangement (see later in this chapter) with a general building contractor and/or specialist contractors.

- *Planned maintenance*
  Cyclical   It is usual to carry out some areas of maintenance work at regular intervals, for example, painting and decorating, irrespective of the actual condition of the building fabric. This type of work may also be influenced by cash-flow factors, e.g. leaving the repainting work until the following summer when a better financial position is anticipated. To some extent, this approach may defeat the purpose of the planned interval of work, which is the long-term protection of the premises. The choice of procurement for this type of work will be situation dependent. Term contracting may be appropriate but individual contracts are likely to be more suitable.
  Preventive   This work is essential to the ongoing safety and reliability of key components, for example a boiler or air conditioning plant. The suppliers of the equipment will usually recommend a programme of maintenance and also offer the maintenance service itself. Some care should be taken to ensure that neither under nor over-maintenance occurs. In some instances, statutory requirements must also be adhered to, for example with lift installations. The most appropriate method of procurement is likely to be via a term contract, probably with the supplier of the equipment or other suitable specialist.

Tendering and contractual arrangements

Methods of selecting the contractor and tendering procedures have been discussed in Chapter 10, Procurement, the contents of which are equally applicable to maintenance work.

With maintenance work, particularly that relating to emergency situations, the location of the contracting organisation, relative to the properties to be included within the maintenance contract, should be carefully considered to ensure that logistics are consistent with an adequate response.

The choice of contract will relate to the nature of the work outlined above. The employer may appoint a contractor on a single project basis or on a term basis whereby the agreement will stand for a given period of time.

Project contracts

Individual project work may be let via one of the many arrangements that are discussed elsewhere in this book, depending on the procurement objectives relating to the specific project. These will include both measurement and cost reimbursement based contracts.

Measured term contracts

A measured term contract may be awarded to cover a number of different buildings. It will usually apply for a specific period of time, say 2 to 3 years, although this may be extended depending on the necessity of maintenance standards and the acceptability of the contractor's performance. The contractor will at the outset be offered the maintenance work for various trades. The work when completed will then be paid for using rates from an agreed schedule. This schedule may have been prepared specifically for the project concerned, or it may be based on a standard document such as the PSA schedule of rates or the NSR (National Schedule of Rates) or BMI (Building Maintenance Information) Price Book. This incorporates labour constants and current prices for major trades in the maintenance field from demolitions and alterations to external works and drainage. Such rates may be updated at monthly intervals, for instance by use of the BMI indices.

Where the client supplies the rates for the work, the contractor is given the opportunity of quoting a percentage addition to or deduction from these rates. The contractor offering the client the most advantageous percentage will usually be awarded the contract. An indication of the amount of work involved over a defined period would therefore seem appropriate for the contractor's assessment of the prices quoted. The JCT publishes a Measured Term Contract (2005) suitable for use with any schedule of rates.

The nature of maintenance work is such that it is likely to be difficult and expensive to cost manage. The surveyor is unlikely to find it practical, or the client to find it cost efficient, to verify every item of maintenance work carried out. A pragmatic approach often used is to check, say, 10% to 20% of the total work, relying on the contractor for the remainder of the account (Building Maintenance Panel of the RICS 2000). This approach is a valid compromise since the measurement sampling process permits an acceptable audit that would reveal any errors requiring an alternative form of cost management.

Managed contracts

A single management contractor may be appointed to manage the entire maintenance requirements of a client. Some clients may prefer this strategy since it reduces
their input significantly and allows the benefits of outsourcing to be more fully
achieved. With this approach, individual sections of maintenance work will be suitably packaged, procured and managed on behalf of the client by the management
contractor.

Facilities management contract

In 2001, the CIOB updated its Standard Form of Facilities Management contract
published in partnership with Cameron McKenna. This contract is intended to
focus on facilities management issues and may be used for a range of private
and public sector facilities management work including maintenance, cleaning,
security, etc.

The CIOB contract is compliant with all current legislation and deals with
employment issues, the Construction Act and fair payment provisions. The contract allows the parties to agree the services to be provided and the specification
to be met. (Copies of the contract may be obtained from Construction Books
Direct, CIOB.)

Budget and cost control

As mentioned earlier, the definition of facilities management from the RICS states
that 'FM involves the total management of all services that support the core business of an organisation'. Since cost management is a key element of 'total management', the facilities manager must therefore be concerned with this aspect in
delivering a satisfactory service to clients. Indeed, to most clients the single most
important justification of facilities management is one of economics whereby value
is improved through the desire to provide an optimum service at the minimum
cost. This requires sound cost management skills.

This is a particular strength of the quantity surveying profession incorporating
traditional cost management skills, including those of whole life costing, extended
to accommodate the needs of facilities management. Cost control is considered in
detail in Chapter 6 of this book; however, it is important to consider its application
within the context of the facilities manager.

The costs relating to property and associated services are significant and, to many
business organisations, the second largest cost after staffing. Cost management
requires the forecast of costs and the control of such costs to achieve budget targets.
In facilities management, it is vital that this will incorporate whole life costing considerations. The process of cost management is outlined as follows.

Setting budgets

As with any form of cost prediction, budget setting in facilities management is
a difficult task and is accompanied by attendant risks. Many budgets will be

established by reference to historical data, i.e. 'How much did it cost last year or last time?'. Provided these estimates are adjusted to incorporate changes in circumstance and policy and also account for inflation, this is a sensible approach for many parts of the total FM budget. However, it will be of little value where budgets are to include items of major expenditure, for example, purchase of new property or equipment.

Budgeting for a large business organisation will require sound organisation and the identification of cost centres, heads of expense and a clear understanding of the channels of the collection and communication of necessary data. The main headings of cost will be case specific but will generally include the following.

Budget heads likely to relate to historical data subject to adjustment for changes due to inflation, tax, condition surveys, new rental agreements, maintenance policy, etc.

- Maintenance costs: incorporating buildings' mechanical and electrical components and external works
- Repair costs: although clearly random in occurrence, larger organisations will be able to forecast overall budget allowances from previous data
- Running costs: energy; water; rent and rates; cleaning and security; waste disposal
- Service charges; costs associated with rented property, for example, the upkeep of common areas in shared office accommodation
- Facilities management costs.

When researching costs from previous data, more accuracy will be achieved by considering costs over a period, say the preceding 4 to 5 years, than merely examining last year's accounts. A one-year view may conceal exceptional circumstances which are likely to distort future budgets if wrongly applied.

Budget heads likely to require detailed estimates/quotations:

- Improvements/alterations: these will relate to the requests of departments and other relevant subdivisions within a business organisation and may be budgeted for accordingly
- Anticipated churn: the costs associated with reorganisation and the relocation of staff within an organisation
- Provision of new space requirements
- Provision for additional equipment.

Risk should be accommodated within the overall budget provision via appropriate contingencies, determined in response to planned activities and levels of certainty relating to them. In addition to cost risks, there is a need to consider both schedule risks and risks to the quality of the facility maintained, which may interfere with the output of the core business of the client. The principles and practice of risk management that are discussed in Chapter 9 should be considered in the context of the facilities management service.

Whole life costing

Whole life costing is discussed in Chapter 7 and is of key importance in facilities management. Although there is less opportunity to fully apply whole life costing principles to existing premises stock, there are many situations where it is a valuable and important technique, for example when the choice between repair and replacement of a major piece of equipment occurs. The scenario in Fig. 18.6 illustrates the significance of whole life costing in this context.

The application of life cycle costing is problematic in practice. For example, predicting the life span of a component may be confused by conflicting data: a manufacturer of equipment may provide a guarantee of 2 years, stating that it will last 8 years, however, if well maintained, perhaps much longer, say 12 to 15 years. Determining an appropriate life cycle in such circumstances thus involves a certain amount of faith rather than certainty.

Cost control

Costs relating to each of the heads referred to above should be monitored at regular intervals, or with single projects as and when the work is being carried out. Actual expenditure needs to be related to forecast, and, where necessary, adjustments in planned expenditure should be made to avoid over-spend or problems, which often occur due to end of year budget surfeit. Once a budget has been prepared, a programme of planned works is required. This will be necessary if a cashflow forecast is required, which can be used as a monitoring tool.

Scenario: The refrigeration plant in an existing food processing factory is under-performing and requires attention. Following the visit of the service engineer, the following two options become clear:

- *Option A* Repair the existing plant, now 10 years old, at a cost of £20 000. The engineer is able to guarantee the repair for 5 years; however, the expected life of the remainder of the plant is less certain. 15 years is a normal life expectancy.
- *Option B* Replace the entire plant at a total cost of £150 000. The new equipment, also with an expected life of 15 years, will provide several operating advantages and relative to the existing plant will reduce energy costs by approximately 15%.

The cost considerations to be made are clearly more than capital cost considerations. With reference to Chapter 7, the costs must be discounted to present day values, including all future maintenance and running costs, to allow a fair comparison of the options. In making the decision, advantages and disadvantages which cannot be costed also need to be taken into consideration, for example, possible disruption, reliability, improvement in operation and image.

Fig. 18.6 Example of significance of whole life costing.

One of the problems in applying sound cost control techniques is the lack of adequate information relating to an organisation's property portfolio. It is suggested that data should be provided in accordance with the following (Spedding 1994):

- Asset registers: a register of key information relating to property assets. The scale of this register will depend on the requirements of the organisation and will typically include information relating to location, access, areas of accommodation and occupancy, as well as details relating to construction. The register should only include information that will be used since the inclusion of unwanted information will add unnecessarily to the expense of establishing and maintaining the database.
- Condition registers: as implied, these are intended to record the condition of properties. This information may be used for several purposes, for instance, in valuing assets at the time of property transactions or to assist in the scheduling of repairs and maintenance.
- Costs relating to previous maintenance and repair, improvements and alterations
- Operating costs including energy costs, rent and rates, waste disposal, cleaning, insurances and sundry service charges.

Taxation considerations

Since investment in facilities is likely to have a major impact on the tax liability of a business organisation, it is important to highlight the need to understand the potential taxation implications within this context (see also Chapter 7).

For example, with regard to capital allowances, issues that may be a concern to the facilities manager include:

- The distinction between capital expenditure and revenue expenditure: revenue expenditure, for example expenditure on salaries, rent, cleaning costs, repairs, etc. may be deducted from revenue in the assessment of taxable profit. Alternatively, capital expenditure, for example the construction of new buildings or the purchase of new equipment, is subject to a different set of rules established within the prevailing capital allowances arrangements. In terms of tax allowances these are generally much less advantageous.
  The distinction between these two types of expenditure may sometimes be difficult. For instance, repairs to an existing stonework façade will be considered as revenue expenditure, while the cost of its total renewal may become a capital item. The point at which a repair becomes replacement may, in practice, often be unclear.
- The capital allowances system: this shows recognition of depreciation by prescribing, by statute, a range of varied allowances against expenditure on fixed assets. For example, items of 'plant and machinery' receive a preferential tax allowance and, as such, must be identified and valued separately, incorporating allowances for associated builders' work and fees.

The distinction, for taxation purposes, between what may or may not be allowed as plant and machinery is thus very significant; however in practice, it is often less than clear.

The facilities manager should also be aware of the implications with regard to both capital gains tax (CGT) and value added tax (VAT).

Benchmarking to improve value

It may be important and/or necessary to demonstrate that the facilities strategy and organisation is achieving good value. Benchmarking techniques may be used in this respect to recognise areas of poor service and assist in establishing measures required to improve performance. An important element of benchmarking is the identification and agreement of relevant and measurable key performance indicators. With reference to maintenance work, examples may include:

- Total cost of maintenance work per m^2
- Cost of maintenance work per unit of occupancy (e.g. bed spaces)
- Percentage of emergency responses within the stated response time.

This approach to monitoring value is particularly helpful within large organisations that have sufficiently large portfolios to enable internal benchmarking comparisons to be made. Benchmarking can be applicable to all areas of facilities management.

Health and safety

The need for any business organisation to correctly attend to health and safety matters could hardly be overstated. The extent of legislation, including that derived from the European Commission, and the increasing prevalence of associated litigation are such that most organisations are rightly zealous in their pursuit of high standards of health and safety in the workplace. In addition, the impact that certain health and safety matters may have on the efficiency of the workforce are well recognised, for instance in connection with sick building syndrome or problems relating to VDUs. This latter point may be unheeded in some organisations; the psychological effects of lighting or the impact of incorrect temperature or humidity levels on physical activity may be overlooked whilst the slavish adherence to legislation, necessary though it is, is seen as paramount.

Although the overall management for health and safety matters is likely to lie with a senior member of staff or, in larger organisations, a contingent of staff dedicated to this function, the facilities manager may have a significant responsibility in many related areas. For example, consider the design, or management of the design, of new accommodation. Under the terms of the Construction (Design and Management) Regulations 1994, the client, design team and constructors are obliged to consider health and safety in the use of the completed building as well as during construction. Therefore, the facilities manager, if acting on behalf of the client, will need to ensure that the regulations are fully complied with.

The surveyor acting in the role of facilities manager must be aware of client expectations and legal responsibilities with regard to health and safety matters. A sound understanding of the issues and an appreciation of the ownership of the various risks involved are of great importance. Whilst specialist knowledge may lie with others, the development of the facilities management discipline and its attendant expert status brings with it close involvement with health and safety affairs, particularly when problems may occur.

Sustainability

There is a rapidly growing realisation and corresponding expectation from various interested bodies to ensure that facilities are provided and managed in a much more sustainable way. In the years that have passed since the publication of the eleventh edition of this book, the amount of pressure upon organisations to react to environmental concerns has grown dramatically, both in the form of government intervention and society demands. Concerns about energy now seem to be at the top of the corporate agenda for a variety of reasons, including: short-term costs with substantial price increases predicted to continue; declining energy reserves, e.g. the UK will soon return to being a net oil importer; exposure to international events beyond our control, e.g. Middle East conflicts; the upsurge in demand from emerging economies, e.g. China; the obligation under Kyoto to hit energy reduction targets and the adherence to new building regulations, e.g. Part L April 2006 and the Energy Performance in Buildings Directive, January 2006. Clearly, the facilities manager has a significant part to play in energy reduction. It is also important to appreciate that irrespective of engagement in the provision of a dedicated service such as facilities management, the issues surrounding sustainability and the built environment are of importance to all quantity surveyors. An extensive supply of literature exists on this topic.

Outsourcing

In recent years, there has been an increasing trend to contract out or outsource certain support functions, usually regarded as non-core to the main raison d'etre of the business, rather than provide such services via the employment of in-house personnel. This is consistent with the downsizing philosophy that many companies follow and is said to offer an efficient and cost-effective alternative to large, internally staffed empires.

The policy of outsourcing can be applied to many support services including catering, security, fleet management, IT management, maintenance and repairs to buildings and services, waste management, landscape management, travel, recruitment . . . the list is extensive. Although outsourcing may be applicable to a wide range of functions, it should not be seen as synonymous with facilities management, merely one approach to be considered in obtaining necessary services. There may be strong reasons for a service to remain under direct staff control, for example, where commercial security is vital and would be endangered by the outsourcing

of office cleaning contracts. Some of the issues surrounding the outsourcing philosophy include:

- *Competition* In-house provision is prone to become less efficient due to the
  absence of competition. Outsourcing will promote economies of scale, right-
  sizing, and is generally seen to result in a reduction in costs. Where services are
  provided internally, benchmarking may assist in identifying inefficiencies and
  provide a means of comparison when competition is absent.
- *Specialisation* In contracting out services, companies will have access to experts
  they would otherwise be unable to adequately justify as full-time personnel on
  their payroll. For example, a small company will be unable to sustain the
  employment of a quantity surveyor. In practice, the range of facilities management related skills required are such that some outsourcing is inevitable.
- *Limited experience* In-house personnel are generally restricted in exposure to
  one set of systems and are likely to work in isolation of other organisations and
  professional associates, thus limiting technical and managerial development.
  Alternatively, the in-house team are immersed in the business operations of their
  company which may be advantageous compared to the external organisation
  that is poorly acquainted with the client organisation.
- *Quality* An external contractor is motivated toward profit maximisation, albeit
  via the provision of an adequate service, and can achieve this by providing the
  minimum acceptable service at the lowest possible cost. It is reasonable to assert
  that in-house teams are motivated simply to provide a service to the employer
  and that the quality objective cannot be compromised by the profit motive.
  Alternatively, in-house staff may be in sheltered employment positions and as
  a consequence may be poorly motivated generally.
- *Confidentiality/security* Some aspects of an organisation may be vulnerable to
  poor security/confidentiality. These would be better served by the provision of
  in-house services.
- *Community culture* Personnel employed on a contract basis may be very tran-
  sient and be uninvolved with anything other than the service they are contracted
  to provide. Rapport with other staff and extended knowledge of the organisation will be absent. As a consequence, social benefits enhancing the service provided may be absent.
- *Flexibility* In-house staff will be more widely involved within an organisation
  which may result in more flexibility of service, for example, cleaners may be
  requested to move furniture.
- *Employment law* In deciding to outsource a particular service, attention should
  be given to relevant employment law.

Once the decision to outsource particular services has been taken, consideration as
to how to organise and procure each particular service needs to be made. The organisation of contracted-out services may be dealt with in a variety of ways:

- *Individual contracts* Each service required to be outsourced may be done on
  an individual contract basis, for example, to supplement an otherwise

comprehensive in-house facilities management provision. This may be managed either centrally or by several appropriate management or cost centres within an organisation. For instance, photocopying costs, furniture costs and cleaning costs may be handled by a sectional office manager whilst IT provision is managed centrally. Centralising the outsourcing of service contracts promotes more efficient administration and performance monitoring.

- *Bundling* Bundling involves the collection of particular facilities management services that may be grouped on the basis of efficiency or convenience and placed with a single supplier, for instance, cleaning, security and catering. This approach reduces the points of responsibility and may lead to economies of scale.

- *Total outsourcing* With this approach, which is an extension to 'bundling', all aspects of facilities management are outsourced to a single facilities management company, including those that are operational as well as managerial. This approach allows a business organisation to benefit from a single point of contact, although the facilities management company appointed will be unable to provide all of the services required by the client and will in turn subcontract some of these out. There are further advantages to the appointment of single contractors. Administration should be reduced, economies of scale are likely to be achieved and, due to the increased commercial significance to the contractor, a larger commitment and better service should result.

The quantity surveyor has considerable knowledge and skills that may be applied to the procurement of services and is well placed to manage the outsourcing requirements of a business. An understanding of the principles surrounding effective procurement, including selection procedures and the preparation of relevant documentation, are important aspects and are within the traditional remit of the profession. Matters relating to the specification of particular services are likely to need the support of in-house advisors and other professional consultants.

Facilities management opportunities for the quantity surveyor

Many quantity surveyors are now employed in the sector of facilities management, which relative to the traditional quantity surveyor market offers great opportunity in that it provides the prospect of being involved beyond the initial procurement of a building. The facilities management market has spread from the USA throughout Europe and Japan and therefore on an international scale is immense.

The quantity surveyor has many of the skills required in the provision of a facilities management service, although when considering the discipline in the wider context, there are clear limitations in existing quantity surveying expertise. This position is also true of other construction professionals, as a considered review of the work of the facilities manager may reveal. There are some aspects of the role that will be better served by designers or contractors, electrical engineers and building surveyors. Likewise, management consultants and accountants are well placed with regard to strategic management consultancy and are beginning to enter the FM market. It seems that the facilities management function may naturally divide into operational and managerial affairs. Without investment in extensive additional human resources, which is likely to be unavailable to all but the larger practices, the majority of opportunities for the quantity surveyor will remain at the operational level. Previous research has identified that considerable barriers exist to surveying firms becoming strategic consultants. Presently, a capability for a more operational consultancy has been identified (Hinks *et al.* 1999).

When considering the service of facilities management, emphasis should be placed on the term management. There is no assertion that the FM role demands the knowledge and ability to apply all of the required skills, any more than a project manager is able to carry out all of the design aspects of a proposed project. However, knowledge of the range of FM functions is important to the adequate fulfilment of the management role.

Consideration of the management and specialist skills required in facilities management as listed below (Bernard Williams Associates 1994) will demonstrate the impossibility of entire FM knowledge:

- Business management
- Man management
- Building design
- Interior design
- Space management
- House management
- Office services management
- Energy management
- Catering management
- Purchasing management
- Motor fleet management
- IT management
- Legal advice
- Facilities audit
- Project management
- Financial management
- Tax management
- Risk management
- Documentation
- Construction technology
- Building maintenance management
- Services maintenance management
- Property management
- Grounds management
- Security management
- Health and safety management
- Cost planning and management
- Knowledge of procurement options and processes
- Knowledge of tendering
- Value management
- Contract formation and administration
- Life cycle costing

It is important to note that whilst the quantity surveying profession may possess some of the skills necessary to provide the facilities management service, the discipline of facilities management should be seen apart. The skills relating to cost management can be outsourced in the same way as design or engineering requirements may be. In appointing a facilities manager, clients are seeking managers not technicians.

Education and training for the facilities manager

Despite the acknowledgement given to the development of facilities management within undergraduate programmes, as with project management there are many who regard the training and education as a postgraduate issue, supplementing a suitable first degree. There are several of these available, including master's courses, and in addition for those surveyors who have not had any formal education in management, opportunity exists for attendance at short training courses and seminars.

There are several institutions with an interest in facilities management, which not only promote its professional standing but also provide vehicles for formal training and qualification:

- The British Institute of Facilities Management (BIFM) is dedicated to the development of the discipline and provides control of the standards that must be attained to obtain a recognised qualification. The route to qualification with BIFM is quite flexible with opportunity for entry via direct examination, accredited undergraduate or postgraduate higher education programmes of practical experience. Irrespective of entry route, all applicants are expected to complete their training by submission of a portfolio demonstrating practical skills and experience. The qualification offered is nationally recognised.
- The RICS has also recognised the significance of facilities management by the establishment of a separate facilities management faculty in 2000. This is a significant development in that the discipline may now be developed within a major international organisation, and directed resources and significant membership should enhance the abilities and status of all facilities managers within the institution.
- Within the CIOB, a FM Society has been established which recognises and promotes the role of the facilities manager. The Society also provides opportunity for CPD courses. The CIOB also examines the subject of facilities management via a Facilities Management Professional Option. This is included within the CIOB education framework, leading to a qualification with chartered designation that is widely recognised in the FM industry.

Discussion topic

Facilities managers should pay full attention to sustainability issues in carrying out their role. Discuss the rationale for this statement and outline how the FM can support the environmental agenda.

It is generally accepted that not only can we not continue to increase our greenhouse gas omissions as we have done over many years, but also, we must reduce them significantly to avoid the severe consequences of future climate change. Even to those who doubt the scientific evidence, the potential impact of climate change is so great that any risk whatsoever must surely be eliminated.

The construction and use of buildings accounts for a very large proportion of our natural resources. Recognition of this has given rise to the green building movement, which is now influencing how we design and construct our buildings, and is challenging the industry to become more efficient in the management of the built environment.

The position of facilities managers, in that they are central to the management of the majority of energy use in buildings, is of great importance in both developing and implementing sustainable policy. Facilities managers are therefore obliged to embrace the issues relating to sustainability in carrying out their role. The separation of economic management and environmental management is no longer a long-term sustainable option.

In addition to the contribution to environmental policy, the pursuit of sustainability can bring a range of other benefits, some of which support the more immediate business objectives of organisations. These include:

- A reduction in operating costs, for example through the introduction of low energy design. This is also likely to result in an increase in rental incomes as tenants recognise the benefits of energy efficient buildings.
- An improvement in the image of an organisation via adherence to green policies. Depending upon the nature of the core activity of the organisation concerned, green policy is seen to give a positive message to consumers and can be used to improve market position.
- An increase in the quality of the working environment, resulting in an improvement in occupant well-being and improved performance. For example, there are some well known buildings where the use of natural ventilation and lighting has not only reduced the costs associated with heating, ventilation and cooling systems, but is considered to have resulted in an improvement in productivity.

Although there are no valid arguments against the pursuit of green principles, the facilities manager is likely to be faced with the dilemma of balancing the traditional metrics of business success i.e. cost and revenue, with the need or desire to achieve a high standard of environmental quality. Since the FM is not in an independent position, it is important to recognise the significance of the commitment of top management in implementing environmentally friendly policy.

In practice, the facilities manager will need to oversee the development of a formal corporate environmental strategy or, where one exists, inform and review its content. Subject to the principles established in this strategy document, the FM may then be confident in implementing policy at a range of different levels by key actions including:

- Identifying the key aspects which may influence environmental performance. These may seem obvious to most organisations, however, it is important to take a wide perspective in determining a strategy that will be most effective. For example, switching the production of manufactured goods to a region of low labour costs may improve short-term profitability but increase the associated transport costs substantially with detrimental effects on the environment. This underlines the dilemma referred to above.

- Establishing and communicating environmental goals. At a macro level, this may be the reduction in associated transport costs referred to in the example above. At the micro level, it may relate to employee use of the private car. For example, one worthwhile goal for an organisation may be the reduction in car journeys in commuting to the workplace. These may then be translated into actions by both positive and negative communication, for example, by simultaneously promoting the benefit of car sharing schemes whilst at the same time, increasing car parking charges. At a more general level, it is critical that individuals adopt an appropriate sustainable lifestyle in the workplace and thus need to be educated in the use of environment friendly principles.

- Embracing the philosophy of environment friendly design and construction. In terms of building design, it is important to remember that what we are planning today is likely to be with us in 40–60 years time and, given the need for conservation, perhaps much beyond. With this in mind, design needs to reflect upon predicted temperature rises, rises in sea level, increased rainfall and more frequent severe weather occurrences. In designing buildings, clients and designers need to look at how long-term energy demand can be reduced, for example, through more efficient climate control and power management. Likewise, the use of buildings and urban planning is a major factor in reducing energy waste through achieving reduced transportation costs and emissions. The environmental problems caused by construction are also significant: manufacturing processes, for example, use of plastics, cause toxic waste; energy expended in transportation is often unnecessary; waste materials find their way through landfill into rivers; quarrying and mining lead to a degradation of the landscape; the need for timber leads to deforestation and loss of habitat.

- Ensuring that whole life costs are an intrinsic element of all value judgements. As discussed elsewhere, there is a need to consider whole life costs in designing buildings, in the knowledge that in the life of a building these will be much greater than the capital cost of construction. This is increasingly the case; prior to the 1960s, energy costs were relatively low, as were labour costs. However, since the first energy crisis in the 1970s, this relative position has changed dramatically and the cost of both energy and human resources in operating buildings has risen dramatically.

- Eliminating waste from an organisation's processes. Examples of excessive waste within any organisation abound and are well acknowledged but normally unresolved: lighting 20-person rooms occupied by 2; printing documents with less than a 1-day life; wearing jackets to compensate for the chill of an air-conditioning (AC) system. In terms of the FM, there is much that can be done to reduce and even eliminate such waste through reviewing relevant systems throughout an organisation and building incentives into the management structure, e.g. benchmarking inter-departmental energy use.

- Complying with environmental legislation. The scope of environmental law is wide and includes all regulations which are designed to control the use of natural resources by humans. Adhering to environmental legislation is, of course, not an option and the FM has a clear responsibility to ensure that relevant law is not transgressed. Whilst such legislation may be driven and derived globally, it is more visible in action locally, for example, via policy implemented by planning authorities with regard to land use, or in compliance with building regulations relating to double glazing.

- Benchmarking environmental performance through auditing environmental impact. An example of this, which is specific to building performance, is BREEAM (the Building Research Establishment Environmental Assessment Method for buildings) an approach which is growing in use in the UK. BREEAM takes into account: overall management policy; operational energy and carbon dioxide (CO_2) issues; indoor and external issues affecting health and well-being; air and water pollution issues; transport-related CO_2 and location-related factors; the use of green-field and brown-field sites; ecological values; conservation and enhancement of the site; the environmental implications of building materials, including life-cycle impacts; water consumption and efficiency. The system motivates the client and design team towards environmentally friendly design which, where successful, is recognised by a high BREEAM rating and the award of a certificate which can be used to improve the image of the organisation. With reference to the BREEAM website (http://www.breeam.org), the benefits of this approach to environmental auditing are extensive:
  - *Clients, planners and development agencies and developers* are using BREEAM to specify the sustainability performance of their buildings in a way that is quick, comprehensive and visible in the marketplace.
  - *Property agents* are using it to promote the environmental credentials and benefits of a building to potential purchasers and tenants.
  - *Design teams* are using it as a tool to improve the performance of their buildings and their own experience and knowledge of environmental aspects of sustainability.
  - *Managers* are using it to measure the performance of buildings and develop action plans, monitor and report performance at both the local and portfolio level.

There is a compelling case for the facilities manager to be involved in environmental management. The activities of the organisations they represent need to be managed to ensure that environmental objectives are achieved. As a consequence, successful policy and implementation will contribute to an improvement in the well-being of a sustainable built environment.

References and bibliography

Alexander K. (ed.) *Facilities Management: Theory and Practice*. Taylor and Francis. 2000.
Ashworth A. *Cost Studies of Buildings*. Longmans. 2004.
Ashworth A. and Hogg K.I. *Added Value in Design and Construction*. Pearson Education. 2000.
Atkin B. and Brooks A. *Total Facilities Management*. Blackwell Publishing. 2005.
Barrett P. (ed.) *Facilities Management: Towards Best Practice*. Blackwell Publishing. 2003.
Bernard Williams Associates. *Facilities Economics*. Building Economics Bureau Ltd. 1994.

Building Maintenance Panel of the RICS. *Building Maintenance: Strategy, Planning and Procurement guidance note*. RICS Business Services Ltd. 2000.

Grigg J. and Jordan A. *Are you managing facilities – getting the best out of buildings*. Allied Dunbar Financial Services Limited; Nicholas Brealy Publishing Ltd. 1993.

Hinks J. *et al. Facilities Management and the Chartered Surveyor: an investigation of chartered surveyors' perceptions*. RICS. 1999.

Langston C. and Ding G.K.C. (eds) *Sustainable Practices in the Built Environment*. Butterworth Heinemann. 2001.

McGregor W. and Then Shiem-Shin D. *Facilities Management and the Business of Space*. Arnold. 1999.

Park A. *Facilities Management: An Explanation*. Macmillan. 1998.

RICS. Research findings number 33. RICS. 1999.

Samuels R. and Prasad D.K. (eds) *Global Warming and the Built Environment*. E & FN Spon. 1994.

Spedding A. (ed.) *CIOB Handbook of Facilities Management*. Longmans. 1994.

Website

http://www.breeam.org

Index

Academic research, 106
Accountability, 233
Accounts, 68
Accreditation, 28, 185
Accuracy, 130, 160, 288
Addendum bill, 278
Adding value, 376
Adjudication, 358
Advance payment, 295
Alternative dispute resolution, 358, 371
APC, 31
Approximate estimate, 130
Approximate quantities, 125
Arbitration, 356, 361
Articles of Agreement, 246
Assets, 70
Audit, 328

Balance sheet, 69
Bankruptcy, 349
Barlow, 4, 113
Benchmarking, 12, 65, 73, 412
Best value, 132
Bills of quantities, 220, 248, 256, 265, 268, 275
Briefing, 170
Budget, 408
Budgetary control, 138
Budgeting, 71
Building, 6
Building engineering services, 6
Building life, 152, 158
Building renewal, 151
Built environment professions, 45, 47
Business, 2

Capital, 71
Capital allowance, 135

Cash flow, 139, 305
Cash forecasting, 71
Certificate, 287, 299
Change, 2, 17, 34, 102, 112, 113, 159, 369
Changing industry, 3
CIC, 33
Civil engineering, 7
Claims, 295, 363, 367
Clarity of thought, 384
Clerk of works, 316
Client, 10, 11, 39, 78, 85, 232, 233, 354, 369, 385, 386
Code of Practice, 222
Collateral warranties, 85
Company failures, 338
Competition, 216, 224
Component life, 153
Conditions of employment, 37, 40
Conduct, 51, 91
Construction management, 231
Construction sector, 6
Consultant, 216, 354
Contract bills, 248, 265–6
Contract documents, 244
Contract drawings, 247, 271, 313
Contract strategy, 232
Contracting surveying, 40
Contractor, 216, 221, 354, 366, 387
Contractor insolvency, 340
Contractor selection, 221
Contractor's cash flow, 140
Contractor's cost control, 139
Contracts of employment, 88
Contractual arrangement, 406
Contractual claims, 363, 367
Contractual dispute, 353
Control, 390
Coordinated project information, 245

Copyright, 268
Corporate image, 60
Corporation tax, 134
Correction of errors, 273, 277
Cost, 233
Cost advice, 122
Cost control, 121, 302, 408, 410
Cost limits, 127
Cost management, 192, 285
Cost modelling, 127
Cost planning, 126
Cost records, 56
Cost reduction, 169
Cost reporting, 302, 304
CPD, 32, 185
Creditor, 340
CRISP, 100
Customer, 2

Dayworks, 317
Decision making, 384
Defined work, 322
Depreciation, 154
Design and build, 42, 227
Design and manage, 231
Design process management, 389
Design team selection, 387
Determination of contract, 340
Disability Discrimination Act 2004, 90
Disciplinary Board, 92
Discount rate, 153
Dispute, 369, 370
Dispute resolution boards, 360

Education, 24, 66, 392, 417
Effective delegation, 384
Egan, 4
Elemental bills, 264–5
Employer, 57, 354
Employer insolvency, 346
Employer's liability insurance, 59
Employer's loss, 345
EMV, 200
Estimating accuracy, 130
E-tendering, 280
Ethics, 50

European Union, 258
Examination of priced bill, 275
Ex-gratia payment, 364
Expert witness, 361
Expertise, 391
Extension of time, 364

Facilities management, 398, 399
FAST, 177
Feasibility, 388
Fees, 82, 330, 391
Final account, 310, 315, 330
Finance, 68
Financial assistance, 136
Financial report, 139
Fire insurance, 58
Fluctuations, 323
Forecasting, 157
Form of tender, 269
Forms of contract, 38, 218, 245, 341, 366
Formula adjustment, 325
Functional analysis, 178
Future role, 1
Fuzzy set theory, 163

Government, 101, 147
Graduates, 13

Health and safety, 58, 412
Heavy and industrial engineering, 8

ICT, 3, 20, 73, 107, 110, 112, 264, 281
Industry characteristics, 4
Information release schedule, 254
Innovation, 95, 114
Insolvency, 302, 335
Interim payment, 329
Internet, 62
Invitation to tender, 269

Knowledge, 15, 381
Knowledge management, 21
Kondratiev, 115

Law, 78
Lay advocacy, 362

Leadership qualities, 382
Lean construction, 21, 141
Liabilities, 71
Limited liability, 35
Limited liability partnership, 36
Liquidated damages, 298, 330
Liquidation, 349
Liquidator, 339
Litigation, 355, 361
Loss and expense, 363

Maintenance management, 402
Management, 19
Management contract, 229
Management skills, 382
Manufacturer, 355
Market, 234
Market conditions, 234
Marketing, 59, 61
Master programme, 254
Materials, 293, 325
Measured term, 406
Measured works, 250, 317
Measurement, 217, 218, 254, 314, 320
Mediation, 360
Methods of measurement, 251
Mini-trial, 359
Modernising Construction, 4, 113
Monte Carlo, 163
Mutual fact finding, 360

National Building Specification, 253
Negligence, 83
Negotiation, 216, 222
Non-cognate student, 30
NVQ, 30

Obsolescence, 154
Office organisation, 55
Open competition, 222
Outsourcing, 413
Overtime working, 320

PAQS, 47
Partnering, 76, 234
Partnership, 35

Payments, 287
Performance bond, 87, 348
PFI, 5, 75, 237
Planning, 388
Post-contract methods, 138
Practice brochure, 62
Practice note, 218
Practice size, 14, 53
Preambles, 249, 266
Precontract methods, 123
Preliminaries, 249, 266, 291
Preparing the contract, 279
Presentations, 62
Price adjustment, 296
Private sector, 34
Probability, 163, 196
Procurement, 21, 100, 215, 238, 404
Procurement options, 226
Professional indemnity, 59, 87, 88
Professions, 44
Programming, 388
Project management, 375
Project management duties, 385
Project manager, 392, 393
Project manager traits, 378
Projects, 2
Proof reading, 268
Provisional measurement, 320
Provisional sum, 321
Public relations, 61
Public sector, 34, 39

QS 2000, 1, 9, 16, 95
QS and law, 78
QS death, 84
QS education, 24, 25
QS trends, 104
Quality, 233
Quality management, 62
Quantity Surveying Institute, 46
Quantity surveying, new techniques, 16
Quantity surveying, profession, 45, 281
Quantity surveyor, 18, 191, 237, 337,
    348, 364, 369, 391, 415
Quantity surveyors in education, 27
Query sheet, 261

Receipt of drawings, 259
Receipt of tender, 274
Reference books, 56
Reimbursement, 217, 218, 220
Reporting on tenders, 274
Research, 10, 95, 98
Research dissemination, 107
Resource allocation, 65
Retention, 296
Rethinking Construction, 4, 99, 113
RICS, 16, 17, 19, 20, 22, 48, 92, 96
RICS Partnership and Accreditation, 28
Risk management, 9, 132, 190, 194, 203, 208
Risk register, 197
Role of quantity surveyor, 8

Schedule, 220, 248, 253, 254, 261, 266–7
Schedule of rates, 220, 278
Schedules of work, 253, 266
Selective competition, 222
Sensitivity analysis, 162
Serial tendering, 223
Skills, 3, 15, 17, 381, 391, 393
Space planning, 400
Specification, 219, 252
Staffing, 53
Stage payment, 298
Standard, 29
Statement of affairs, 336
Statutory adjudication, 358
Strategy, 102

Subcontractor, 42, 251, 292, 293, 306, 307, 354
Superficial method, 124
Supervision, 390
Supplier, 355
Sustainability, 413, 417

Taxation, 133, 153, 411
Technological change, 159
Tender, 329
Tendering, 406
Time, 233
TQM, 117
Training, 24, 66, 392, 417
Tribunal, 361
Two-stage tendering, 223

Uncertainty, 161
Undefined work, 323
Understanding, 15
Unit method, 123

Valuation, 286, 302, 307
Value added tax, 134
Value management, 9, 132, 141, 166, 208
Variations, 295, 311, 316, 329, 331
Viability, 388

Whole life cost forum, 160
Whole life costing, 9, 131, 146, 149, 151, 410, 419
Whole life value, 148
Work of QS, 1

UNIVERSITY OF WOLVERHAMPTON
LEARNING & INFORMATION SERVICES

Willis's
Practice and Procedure
for the Quantity Surveyor

WITHDRAWN